REDEFINING DIVERSITY AND DYNAMICS OF NATURAL RESOURCES MANAGEMENT IN ASIA

VOLUME 2

Upland Natural Resources and Social Ecological Systems in Northern Vietnam,
Mai Van Thanh, Tran Duc Vien, Stephen J. Leisz and Ganesh P. Shivakoti; editors

Volume 2 is dedicated to Nobel Laureate Elinor Ostrom who is the source of inspiration in drafting these volumes and all chapter authors of these volumes have benefited from her theoretical framework.

Book Title: Re-defining Diversity and Dynamism of Natural Resource Management in Asia

Book Editors: Ganesh P. Shivakoti, Shubhechchha Sharma and Raza Ullah

Other volumes published:

1) Sustainable Natural Resources Management in Dynamic Asia Volume 1
   Ganesh P. Shivakoti, Ujjwal Pradhan and Helmi; editors
2) Natural Resource Dynamics and Social Ecological System In Central Vietnam: Development, Resource Changes and Conservation Issues Volume 3
   Tran N. Thang, Ngo T. Dung, David Hulse, Shubhechchha Sharma and Ganesh P. Shivakoti; editors
3) Reciprocal Relationship between Governance of Natural Resources and Socio-Ecological Systems Dynamics in West Sumatra Indonesia Volume 4

Rudi Febriamansyah, Yonariza, Raza Ullah and Ganesh P. Shivakoti; editors

REDEFINING DIVERSITY AND DYNAMICS OF NATURAL RESOURCES MANAGEMENT IN ASIA

Upland Natural Resources and Social Ecological Systems in Northern Vietnam

VOLUME 2

EDITED BY

MAI VAN THANH, TRAN DUC VIEN, STEPHEN J. LEISZ AND GANESH P. SHIVAKOTI

ELSEVIER

AMSTERDAM • BOSTON • HEIDELBERG • LONDON • NEW YORK • OXFORD
PARIS • SAN DIEGO • SAN FRANCISCO • SINGAPORE • SYDNEY • TOKYO

Elsevier
Radarweg 29, PO Box 211, 1000 AE Amsterdam, Netherlands
The Boulevard, Langford Lane, Kidlington, Oxford OX5 1GB, United Kingdom
50 Hampshire Street, 5th Floor, Cambridge, MA 02139, United States

Library of Congress Cataloging-in-Publication Data
A catalog record for this book is available from the Library of Congress

British Library Cataloguing-in-Publication Data
A catalogue record for this book is available from the British Library

ISBN: 978-0-12-805453-6

For information on all Elsevier publications visit our website at https://www.elsevier.com/

Publisher: Candice G. Janco
Acquisition Editor: Laura S Kelleher
Editorial Project Manager: Emily Thomson
Production Project Manager: Mohanapriyan Rajendran
Cover Designer: Matthew Limbert

Typeset by SPi Global, India

Contents

Contributors

F. Affholder CIRAD UR SCA, Montpellier, France

A. Aijaz Indiana University Bloomington, Bloomington, IN, United States

E. Boere Wageningen University, Wageningen, The Netherlands

P.T. Dung Ministry of Agriculture and Rural Development, Hanoi, Vietnam

N.T.T. Dung Office of Natural Resource and Environment, Gia Lam, Hanoi, Vietnam

H.H. Duong Vietnam Academy for Water Resources, Hanoi, Vietnam

N.H. Duong Center for Agricultural Research and Ecological Studies, Hanoi, Vietnam

C. Jacobson University of the Sunshine Coast, Sippy Downs, QLD, Australia

D. Jourdain Asian Institute of Technology, Bangkok, Thailand; CIRAD UR SCA, Montpellier, France

M.A. Kamran Nuclear Institute for Agriculture and Biology, Faisalabad, Pakistan

N.T. Lam Vietnam National University of Agriculture, Hanoi, Vietnam

S.J. Leisz Colorado State University, Fort Collins, CO, United States

A.T.T. Nguyen The University of Queensland, Brisbane, QLD, Australia; Vietnam National University of Forestry, Ha Noi, Vietnam

D.V. Nha Vietnam National University of Agriculture, Hanoi, Vietnam

N.H. Nhuan The University of Queensland, Brisbane, QLD, Australia; Vietnam National University of Agriculture, Hanoi, Vietnam

O. Nicetic The University of Queensland, Brisbane, QLD, Australia

D.D. Quang Northern Mountainous Agriculture and Forestry Science Institute (NOMAFSI), Phu To, Vietnam

L. Ribbe Cologne University, Cologne, Germany

H. Ross The University of Queensland, Brisbane, QLD, Australia

S. Sharma WWF-Nepal, Kathmandu, Nepal

G. Shivakoti The University of Tokyo, Tokyo, Japan; Asian Institute of Technology, Bangkok, Thailand

C.T. Son Vietnam National University of Agriculture, Hanoi, Vietnam

M.V. Thanh The University of Queensland, Brisbane, QLD, Australia; Center for Agricultural Research and Ecological Studies, Hanoi, Vietnam

C.P. Thanh Thainguyen University of Economics and Business Administration, Thainguyen (TUEBA), Vietnam

T. Thuc Vietnam Institute of Meteorology, Hydrology and Climate Change, Hanoi, Vietnam

V.X. Tinh Institute of Anthropology, Hanoi, Vietnam

E. van de Fliert The University of Queensland, Brisbane, QLD, Australia

M. van den Berg Wageningen University, Wageningen, The Netherlands

T.D. Vien Vietnam National University of Agriculture, Hanoi, Vietnam

Words From Book Editors

CONTEXT

Elinor Ostrom received the Nobel Prize in Economics for showing how the "commons" is vital to the livelihoods of many throughout the world. Her work examined the rhetoric of the "tragedy of the commons," which has been used as the underlying foundation in privatizing property and centralizing its management as a way to protect finite resources from depletion. She worked, along with others, to overturn the "conventional wisdom" of the tragedy of the commons by validating the means and ways that local resources can be effectively managed through common property regimes instead of through the central government or privatization. Ostrom identified eight design principles relating to how common pool resources can be governed sustainably and equitably in a community. Similarly, the Institutional Analysis and Development (IAD) framework summarizes the ways in which institutions function and adjust over time. The framework is a "multilevel conceptual map," which describes a specific hierarchical section of interactions made in a system. The framework seeks to identify and explain interactions between actors and action situations.

As a political scientist, Ostrom has been a source of inspiration for many researchers and social scientists, including this four volumes book. Her theories and approach serve as the foundation for many of the chapters within these volumes. Following in her footsteps, the books is based on information collected during fieldwork that utilized quantitative as well as qualitative data, and on comparative case studies, which were then analyzed to gain an understanding of the situation, rather than starting from a formulated assumption of reality. The case studies in these volumes highlight the issues linked to the management of the environment and natural resources, and seek to bring about an understanding of the mechanisms used in managing the natural resource base in the regions, and how different stakeholders interact with each other in managing these natural resources. The details of the books are as follows:

Volume title		Editors
"Re-defining Diversity and Dynamism of Natural Resources Management in Asia"		Ganesh P. Shivakoti, Shubhechchha Sharma, and Raza Ullah
Volume I	Sustainable Natural Resources Management in Dynamic Asia	Ganesh P. Shivakoti, Ujjwal Pradhan, and Helmi
Volume II	Upland Natural Resources and Social Ecological Systems in Northern Vietnam	Mai Van Thanh, Tran Duc Vien, Stephen J. Leisz, and Ganesh P. Shivakoti
Volume III	Natural Resource Dynamics and Social Ecological Systems in Central Vietnam: Development, Resource Changes and Conservation Issue	Tran Nam Thang, Ngo Tri Dung, David Hulse, Shubhechchha Sharma, and Ganesh P. Shivakoti

Continued

Volume title		Editors
Volume IV	Reciprocal Relationship between Governance of Natural Resources and Socio-Ecological Systems Dynamics in West Sumatra Indonesia	Rudi Febriamansyah, Yonariza, Raza Ullah, and Ganesh P. Shivakoti

These volumes are made possible through the collaboration of diverse stakeholders. The intellectual support provided by Elinor Ostrom and other colleagues through the Ostrom Workshop in Political Theory and Policy Analysis at the Indiana University over the last two and half decades has provided a solid foundation for drafting the book. The colleagues at the Asian Institute of Technology (AIT) have been actively collaborating with the Workshop since the creation of the Nepal Irrigation, Institutions and Systems (NIIS) database; and the later Asian Irrigation, Institutions and Systems (AIIS) database (Ostrom, Benjamin and Shivakoti, 1992; Shivakoti and Ostrom, 2002; Shivakoti et al., 2005; Ostrom, Lam, Pradhan and Shivakoti, 2011). The International Forest Resources and Institutions (IFRI) network carried out research to support policy makers and practitioners in designing evidence based natural resource polices based on the IAD framework at Indiana University, which was further mainstreamed by the University of Michigan. In order to support this, the Ford Foundation (Vietnam, India, and Indonesia) provided grants for capacity building and concerted knowledge sharing mechanisms in integrated natural resources management (INRM) at Indonesia's Andalas University in West Sumatra, Vietnam's National University of Agriculture (VNUA) in Hanoi, and the Hue University of Agriculture and Forestry (HUAF) in Hue, as well as at the AIT for collaboration in curriculum development and in building capacity through mutual learning in the form of masters and PhD fellowships (Webb and Shivakoti, 2008). Earlier, the MacArthur Foundation explored ways to support natural resource dependent communities through the long term monitoring of biodiversity, the domestication of valuable plant species, and by embarking on long-term training programs to aid communities in managing natural resources.

VOLUME 1

This volume raises issues related to the dependence of local communities on natural resources for their livelihood; their rights, access, and control over natural resources; the current practices being adopted in managing natural resources and socio-ecological systems; and new forms of natural resource governance, including the implementation methodology of REDD+ in three countries in Asia. This volume also links regional issues with those at the local level, and contributes to the process of application of various multimethod and modeling techniques and approaches, which is identified in the current volume in order to build problem solving mechanisms for the management of natural resources at the local level. Earlier, the Ford Foundation Delhi office supported a workshop on Asian Irrigation in Transition, and its subsequent publication (Shivakoti et al., 2005) was followed by Ford Foundation Jakarta office's long term support for expanding the knowledge on integrated natural resources management, as mediated by institutions in the dynamic social ecological systems.

VOLUME 2

From the early 1990s to the present, the Center for Agricultural Research and

Ecological Studies (CARES) of VNUA and the School of Environment, Resources and Development (SERD) of AIT have collaborated in studying and understanding the participatory process that has occurred during the transition from traditional swidden farming to other farming systems promoted as ecologically sustainable, livelihood adaptations by local communities in the northern Vietnamese terrain, with a special note made to the newly emerging context of climate change. This collaborative effort, which is aimed at reconciling the standard concepts of development with conservation, has focused on the small microwatersheds within the larger Red River delta basin. Support for this effort has been provided by the Ford Foundation and the MacArthur Foundation, in close coordination with CARES and VNUA, with the guidance from the Ministry of Agriculture and Rural Development (MARD) and the Ministry of Natural Resources and Environment (MONRE) at the national, regional, and community level. Notable research documentation in this volume includes issues such as local-level land cover and land use transitions, conservation and development related agro-forestry policy outcomes at the local level, and alternative livelihood adaptation and management strategies in the context of climate change. A majority of these studies have examined the outcomes of conservation and development policies on rural communities, which have participated in their implementation through collaborative governance and participatory management in partnership with participatory community institutions. The editors and authors feel that the findings of these rich field-based studies will not only be of interest and use to national policymakers and practitioners and the faculty and students of academic institutions, but can also be equally applicable to guiding conservation and development issues for those scholars interested in understanding a developing country's social ecological systems, and its context-specific adaptation strategies.

VOLUME 3

From the early 2000 to the present, Hue University of Agriculture and Forestry (HUAF) and the School of Environment, Resources and Development (SERD) of AIT supported by MacArthur Foundation and Ford Foundation Jakarta office have collaborated in studying and understanding the participatory process of Social Ecological Systems Dynamics that has occurred during the opening up of Central Highland for infrastructure development. This collaborative effort, which is aimed at reconciling the standard concepts of development with conservation, has focused on the balance between conservation and development in the buffer zone areas as mediated by public resource management institutions such as Ministry of Agriculture and Rural Development (MARD), Ministry of Natural Resources and Environment (MONRE) including National Parks located in the region. Notable research documentation in this volume includes on issues such as local level conservation and development related policy outcomes at the local level, alternative livelihood adaptation and management strategies in the context of climate change. A majority of these studies have examined the outcomes of conservation and development policies on the rural communities which have participated in their implementation through collaborative governance and participatory management in partnership with participatory community institutions.

VOLUME 4

The issues discussed above are pronounced more in Indonesia among the Asian countries and the Western Sumatra is such typical example mainly due to earlier logging concessionaries, recent expansion of State and private plantation of para-rubber and oil palm plantation. These new frontiers have created confrontations among the local community deriving their livelihoods based on inland and coastal natural resources and the outsiders starting mega projects based on local resources be it the plantations or the massive coastal aqua cultural development. To document these dynamic processes Ford Foundation Country Office in Jakarta funded collaborative project between Andalas University and Asian Institute of Technology (AIT) on Capacity building in Integrated Natural Resources Management. The main objective of the project was Andalas faculty participate in understanding theories and diverse policy arenas for understanding and managing common pool resources (CPRs) which have collective action problem and dilemma through masters and doctoral field research on a collaborative mode (AIT, Indiana University and Andalas). This laid foundation for joint graduate program in Integrated Natural Resources Management (INRM). Major activities of the Ford Foundation initiatives involved the faculty from Andalas not only complete their degrees at AIT but also participated in several collaborative training.

1 BACKGROUND

Throughout Asia, degradation of natural resources is happening at a higher rate, and is a primary environmental concern. Recent tragedies associated with climate change have left a clear footprint on them, from deforestation, land degradation, and changing hydrological and precipitation patterns. A significant proportion of land use conversion is undertaken through rural activities, where resource degradation and deforestation is often the result of overexploitation by users who make resource-use decisions based on a complex matrix of options, and potential outcomes.

South and Southeast Asia are among the most dynamic regions in the world. The fundamental political and socioeconomic setting has been altered following decades of political, financial, and economic turmoil in the region. The economic growth, infrastructure development, and industrialization are having concurrent impacts on natural resources in the form of resource degradation, and the result is often social turmoil at different scales. The natural resource base is being degraded at the cost of producing economic output. Some of these impacts have been offset by enhancing natural resource use efficiency, and through appropriate technology extension. However, the net end results are prominent in terms of increasing resource depletion and social unrest. Furthermore, climate change impacts call for further adaptation and mitigation measures in order to address the consequences of erratic precipitation and temperature fluctuations, salt intrusions, and sea level increases which ultimately affect the livelihood of natural resource dependent communities.

Governments, Non-governmental organizations (NGOs), and academics have been searching for appropriate policy recommendations that will mitigate the trend of natural resource degradation. By promoting effective policy and building the capacity of key stakeholders, it is envisioned that sustainable development can be promoted from both the top-down and bottom-up perspectives. Capacity building in the field of natural resource management, and poverty alleviation is, then, an urgent need; and several policy alternatives have been suggested (Inoue and Shivakoti, 2015; Inoue and Isozaki, 2003; Webb and Shivakoti, 2008).

The importance of informed policy guidance in sustainable governance and the management of common pool resources (CPRs), in general, have been recognized due to the conflicting and competing demand for use of these resources in the changing economic context in Asia (Balooni and Inoue, 2007; Nath, Inoue and Chakma, 2005; Pulhin, Inoue and Enters, 2007; Shivakoti and Ostrom, 2008; Viswanathan and Shivakoti, 2008). This is because these resources are unique in respect to their context. The management of these resources are by the public, often by local people, in a partnership between the state and the local community; but on a day-to-day basis, the benefits are at the individual and private level. In the larger environmental context, however, the benefits and costs have global implications. There are several modes of governance and management arrangement possible for these resources in a private-public partnership. Several issues related to governance and management need to be addressed, which can directly feed into the ongoing policy efforts of decentralization and poverty reduction measures in South and South East Asia.

While there has been a large number of studies, and many management prescriptions made, for the management of natural resources, either from the national development point-of-view or from the local-level community perspectives, there are few studies which point toward the interrelationship among other resources and CPRs, as mediated by institutional arrangement, and that have implications for the management of CPRs in an integrated manner, vis-a-vis poverty reduction. In our previous research, we have identified several anomalies and tried to explain these in terms of better management regimes for the CPRs of several Asian countries (Dorji, Webb and Shivakoti, 2006; Gautam, Shivakoti and Webb, 2004; Kitjewachakul, Shivakoti and Webb, 2004; Mahdi, Shivakoti and Schmidt-Vogt, 2009; Shivakoti et al., 1997; Dung and Webb, 2008; Yonariza and Shivakoti, 2008). However, there are still several issues, such as the failure to comprehend and conceptualize social and ecological systems as coupled systems that adapt, self-organize, and are coevolutionary. The information obtained through these studies tends to be fragmented and scattered, leading to incomplete decision making, as they do not reflect the entire scenario. The shared vision of the diverse complexities, that are the reality of natural resource management, needs to be fed into the governance and management arrangements in order to create appropriate management guidelines for the integrated management of natural resources, and CPR as a whole.

Specifically, the following issues are of interest:

a. How can economic growth be encouraged while holding natural resources intact?

b. How has the decentralization of natural management rights affected the resource conditions, and how has it addressed concerns of the necessity to incorporate gender concerns and social inclusion in the process?

c. How can the sustainability efforts to improve the productive capacity of CPR systems be assessed in the context of the current debate on the effects of climate change, and the implementation of new programs such as Payment for Ecosystem Services (PES) and REDD+?

d. How can multiple methods of information gathering and analysis (eg use of various qualitative and quantitative social science methods in conjunction with methods from the biological sciences, and time series remote sensing data collection methods) on CPRs be integrated into national natural resource policy guidelines, and the results be used by local managers and users of CPRs, government agencies, and scholars?

e. What are the effective polycentric policy approaches for governance and management of CPRs, which are environmentally sustainable and gender balanced?

2 OBJECTIVES OF THESE VOLUMES

At each level of society, there are stakeholders, both at the public and private level, who are primarily concerned with efforts of management enhancement and policy arrangements. Current theoretical research indicates that this is the case whether it is deforestation, resource degradation, the conservation of biodiversity hotspots, or climate change adaptation. The real struggles of these local-level actors directly affect the management of CPR, as well as the hundreds of people who are dependent upon them for a living. This book is about those decisions as the managers of natural resources. Basically, the authors of these chapters explore outcomes after decentralization and economic reforms, respectively. The volumes of this Book scrutinize the variations of management practices with, and between, communities, local administration, and the CPR. Economic growth is every country's desire, but in the context of South and South East Asia, much of the economic growth is enabled by the over use of the natural resource base. The conundrum is that these countries need economic growth to advance, but the models of economic growth that are advanced, negatively affect the environment, which the country, depends upon. Examples of this are seen in such varied contexts as the construction of highways through protected areas, the construction of massive hydropower dams, and the conversion of traditional agricultural fields into rubber and oil palm plantations.

The research also shows that the different levels of communities, administration, and people are sometimes highly interactive and overlapping, for that reason, it is necessary to undertake coordinated activities that lead to information capture and capacity building at the national, district, and local levels. Thus the impacts of earlier intervention efforts (various policies in general and decentralization in particular) for effective outcomes have been limited, due to the unwillingness of higher administrative officials to give up their authority, the lack of trust and confidence of officials in the ability of local communities in managing CPR, local elites capturing the benefits of decentralization in their favor, and high occurrences of conflicts among multiple stakeholders at the local level (IGES, 2007).

In the areas of natural resource management particular to wildlife ecology monitoring and climate change adaptation, the merging of traditional knowledge with science is likely to result in better management results. Within many societies, daily practices and ways of life are constantly changing and adapting to new situations and realities. Information passed through these societies, while not precise and usually of a qualitative nature, is valued for the reason that it is derived from experience over time. Scientific studies can backstop local knowledge, and augment it through the application of rigorous scientific method derived knowledge, examining the best practices in various natural resource management systems over spatial and temporal scales. The amalgamation of scientific studies and local knowledge, which is trusted by locals, may lead to powerful new policies directed toward nature conservation and livelihood improvements.

Ethnic minorities, living in the vicinity to giant infrastructure projects, have unequal access, and control over, resources compared to other more powerful groups. Subsistence agriculture, fishery, swiddening, and a few off-farm options are the livelihood activities for these individuals. But unfortunately, these livelihood options are in areas that will be hit the most by changing climatic scenarios, and these people are the least equipped to cope; a situation that further aggravates the possibility of diversifying their livelihood options. Increasing tree coverage can help to mitigate climate change through the sequestering of carbon in trees. Sustainably planting trees requires technical, social, and political dimensions that are mainly possible through the decentralization of power to local communities to prevent issues of deforestation and degradation. The role of traditional institutions hence becomes crucial to reviving social learning, risk sharing, diversifying options, formulating adaptive plans and their effective implementation, fostering stress tolerance, and capacity building against climate change effects.

Though, the role of institutions in managing common pool resources has been explained in literature, it is also worth noting that institutions play significant roles in climate change adaptation. A study conducted by Gabunda and Barker (1995) and Nyangena (2004) observed that household affiliations in social networks were highly correlated with embracing soil erosion retaining technologies. Likewise, Jagger and Pender (2006) assumed that individuals involved in natural resource management focused programs were likely to implement land management expertise, regardless of their direct involvement in particular organizations. Friis-Hansen (2005) partially verifies that there is a positive relationship among participation in a farmer's institution and the adoption of smart agriculture technology. Dorward et al. (2009) correspondingly notes that institutions are vital in shaping the capability of local agrarians to respond to challenges and opportunities. This study has also shown that institutions are the primary attribute in fostering individuals and households to diversify livelihoods in order to adapt to a changing climate. In the context of REDD+, a system is required that can transcend national boundaries, interconnect different governance levels, and allow both traditional and modern policy actors to cooperate. Such a system emphasizes the integration of both formal and informal rule making mechanisms and actor linkages in every governance stage, which steer toward adapting to and mitigating the effects of local and global environmental change (Corbera and Schroeder, 2010).

Based on the above noted discussions, the volumes in this book bring these issues forward for a global audience and policy makers. Though earlier studies show that the relationship between scientific study and outcomes in decision making are usually complex; we hope that the studies examined and discussed here can have some degree of impact on academics, practitioners, and managers.

G. Shivakoti, S. Sharma, and R. Ullah

References

Balooni, K.B., Inoue, M., 2007. Decentralized forest management in South and Southeast Asia. J. Forest. 2007, 414–420.

Corbera, E., Schroeder, H., 2010. Governing and Implementing REDD+. Environ. Sci. Pol. http://dx.doi.org/10.1016/j.envsci.2010.11.002.

Dorji, L., Webb, E., Shivakoti, G.P., 2006. Forest property rights under nationalized forest management in Bhutan. Environ. Conservat. 33 (2), 141–147.

Dorward, A., Kirsten, J., Omamo, S., Poulton, C., Vink, N., 2009. Institutions and the agricultural development challenge in Africa. In: Kirsten, J.F., Dorward, A.R., Poulton, C., Vink, N. (Eds.), Institutional Economics Perspectives on African Agricultural Development. IFPRI, Washington DC.

Dung, N.T., Webb, E., 2008. Incentives of the forest land allocation process: Implications for forest management in Nam Dong District, Central Vietnam. In: Webb, E., Shivakoti, G.P. (Eds.), Decentralization, Forests and Rural Communities: Policy outcome in South and South East Asia. SAGE Publications, New Delhi, pp. 269–291.

Friis-Hansen, E., 2005. Agricultural development among poor farmers in Soroti district, Uganda: Impact Assessment of agricultural technology, farmer empowerment and changes in opportunity structures. Paper presented at Impact Assessment Workshop at CYMMYT, Mexico, 19–21. October. http://citeseerx.ist.psu.edu/viewdoc/download?doi=10.1.1.464.8651&rep=rep1&type=pdf.

Gautam, A., Shivakoti, G.P., Webb, E.L., 2004. A review of forest policies, institutions, and the resource condition in Nepal. Int. Forest. Rev. 6 (2), 136–148.

Gabunda, F., Barker, R., 1995. Adoption of hedgerow technology in Matalom, Leyte Philipines. Mimeo. In: Bluffstone, R., Khlin, G. (Eds.), 2011. Agricultural Investment and Productivity: Building Sustainability in East Africa. RFF Press, Washington, DC/London.

IGES, 2007. Decentralization and State-sponsored Community Forestry in Asia. Institute for Global Environmental Studies, Kanagawa.

Inoue, M., Isozaki, H., 2003. People and Forest-policy and Local Reality in Southeast Asia, the Russian Far East and Japan. Kluwer Academic Publishers, Netherlands.

Inoue, M., Shivakoti, G.P. (Eds.), 2015. Multi-level Forest Governance in Asia: Concepts, Challenges and the Way Forward. Sage Publications, New Delhi/California/London/Singapore.

Jagger, P., Pender, J., 2006. Impacts of Programs and Organizations on the Adoption of Sustainable Land Management Technologies in Uganda. IFPRI, Washington, DC.

Kijtewachakul, N., Shivakoti, G.P., Webb, E., 2004. Forest health, collective behaviors and management. Environ. Manage. 33 (5), 620–636.

Mahdi, Shivakoti, G.P., Schmidt-Vogt, D., 2009. Livelihood change and livelihood sustainability in the uplands of Lembang Subwatershed, West Sumatra, Indonesia, in a changing natural resource management context. Environ. Manage. 43, 84–99.

Nath, T.K., Inoue, M., Chakma, S., 2005. Prevailing shifting cultivation in the Chittagong Hill Tracts, Bangladesh: some thoughts on rural livelihood and policy issues. Int. For. Rev. 7 (5), 327–328.

Nyangena, W., 2004. The effect of social capital on technology adoption: empirical evidence from Kenya. Paper presented at 13th Annual Conference of the European Association of Environmental and Resource Economics, Budapest.

Ostrom, E., Benjamin, P., Shivakoti, G.P., 1992. Institutions, Incentives, and Irrigation in Nepal: June 1992. (Monograph) Workshop in Political Theory and Policy Analysis, Indiana University, Bloomington, Indiana, USA.

Ostrom, E., Lam, W.F., Pradhan, P., Shivakoti, G.P., 2011. Improving Irrigation Performance in Asia: Innovative Intervention in Nepal. Edward Elgar Publishers, Cheltenham, UK.

Pulhin, J.M., Inoue, M., Enters, T., 2007. Three decades of community-based forest management in the Philippines: emerging lessons for sustainable and equitable forest management. Int. For. Rev. 9 (4), 865–883.

Shivakoti, G., Ostrom, E., 2008. Facilitating decentralized policies for sustainable governance and management of forest resources in Asia. In: Webb, E., Shivakoti, G.P. (Eds.), Decentralization, Forests and Rural Communities: Policy Outcomes in South and Southeast Asia. Sage Publications, New Delhi/Thousand Oaks/London/Singapore, pp. 292–310.

Shivakoti, G.P., Ostrom, E. (Eds.), 2002. Improving Irrigation Governance and Management in Nepal. Institute of Contemporary Studies (ICS) Press, California, Oakland.

Shivakoti, G.P., Vermillion, D., Lam, W.F., Ostrom, E., Pradhan, U., Yoder, R., 2005. Asian Irrigation in Transition-Responding to Challenges. Sage Publications, New Delhi/Thousand Oaks/London.

Shivakoti, G., Varughese, G., Ostrom, E., Shukla, A., Thapa, G., 1997. People and participation in sustainable development: understanding the dynamics of natural resource system. In: Proceedings of an International Conference held at Institute of Agriculture and Animal Science, Rampur, Chitwan, Nepal. 17–21 March, 1996. Bloomington, Indiana and Rampur, Chitwan.

Viswanathan, P.K., Shivakoti, G.P., 2008. Adoption of rubber integrated farm livelihood systems: contrasting empirical evidences from Indian context. J. For. Res. 13 (1), 1–14.

Webb, E., Shivakoti, G.P. (Eds.), 2008. Decentralization, Forests and Rural Communities: Policy Outcomes in South and Southeast Asia. Sage Publications, New Delhi/Thousand Oaks/London/Singapore.

Yonariza, Shivakoti, G.P., 2008. Decentralization and co-management of protected areas in Indonesia. J. Legal Plur. 57, 141–165.

Foreword

It was during the 1980s that the debate on how the global resource system should be managed was initiated, coming to the attention of the conservation and scientific communities, as well as the broader public. Over the 25 years that followed, society's efforts to halt the environmental tragedies of deforestation, unsustainable land use, and excessive carbon dumping in the atmosphere have sparked numerous researchers to work together, and to agree on the problems of analyzing complex systems. The dominant challenge is that many of the natural interactions that impact ecological functions and society occur at multiple hierarchies. A series of studies have also revealed that this complexity is likely to grow simultaneously increasing analytical worries, because social and natural systems interact together. The mosaic of communities managing forests, agriculture, land, fisheries, and other related common pool resources (CPRs) across a broad range of landscapes offers the perfect illustration of this complication. Understanding these complexities is vital, given the human induced impacts that are being detected on the environment and the significant effort that communities could put into managing these resources across diverse socioeconomic settings.

CPR or simply known as Ostrom's "Commons," have, in some particular niches, maintained their productivity and become more robust than they were some decades ago. This may be due, in part, to the vast ecological differences found from one landscape to the next. The range of these "commons" areas is governed by a variety of private, state, and community institutional arrangements, making any one idealized solution, or management solution, more complex and fleeting. While substantial commons are jointly managed or comanaged by governments and communities, the provisional arrangements regarding the particular arrangements for state, private, and community property may either prosper or be unsuccessful in supporting resources and delivering better economic yields, depending upon a variety of conditions within each particular setting. While some might assume that open access resources would be over exploited, as foreseen by Garrett Hardin, examples from the natural world show that this assumption of unavoidable tragedy is too comprehensive.

Land degradation and land use changes can be considered as being CPR problems, they are also linked to the ability of the environment to function, or not, in its role as a pollutant sink for greenhouse gases (GHG). While the global atmosphere has the ability to absorb considerable flows of GHG emissions, fulfilling this role also results in global temperature rises.

Mitigating and adapting to climate change is one of the major challenges of our times. Especially in Southeast Asia, it is vital to develop a scenario for collective action that can be achievable, because it is linked to information provisions, policy actions, and interaction among the variety of stakeholders, while concurrently nurturing trust and reciprocity. Information plays an important role at all levels of government by facilitating localized adaptation plans and formulating localized policies.

Latin America, Africa, South Asia, and Southeast Asia are on similar platforms when it comes to developing systematic methodologies for studying climate change and its impacts on agricultural production and development. This research needs to include vulnerability assessments in the context of multiple stressors, vulnerability mapping, and local-level case studies across the regions. In Northern Vietnam, which is home to diverse ethnic communities, there are similarities with traditional communities found in Latin America and Africa. Responses forged by these communities include home gardens that incorporate local cultivars that have been adapted to individual microniches, offering crucial nutrition and sustainable livelihoods. The value of this local biodiversity must be measured by the services that it provides for these vulnerable traditional communities.

Management of transboundary resources, water, and wildlife, in particular, are gaining popularity due to the upscaling impacts of climate change. A primary issue arising in the management of these transboundary resources is the mismatch between the scales at which the problems are experienced (watershed, ecosystem) and the scales at which decisions are made (political boundaries, for instance). Differences in institutional capacity, a lack of political cooperation, and brittleness arouse difficulties in promoting larger-scale joint management for a broader ecosystem scale.

Economic and demographic changes are altering the quality and availability of land. This is especially true in Vietnam, as land use changes have been profound after the introduction of the *Doi Moi* policies in the mid-1980s. This has impacted the existence and sustainability of land as a CPR. Landscape level modification is prominent in the deltas of the midlands and uplands of Northern Vietnam, and urban centers encroach into areas that were previously agricultural communes. The literature on land use in Vietnam mainly documents that the land use and land cover changes have led to less carbon sequestration, and lesser biodiversity than was the case prior to *Doi Moi*. These changes can have devastating consequences on the grasslands, as well as for pastoralists and agriculturists, as their production is mainly dependent on CPRs. This conversion of land use has affected vast tracts of forests, wetlands, and pastures, becoming permanent croplands that are managed by farmers with new longer-term land use rights. Under the pressures of rapid population growth, erosion of customary laws, conflicting land uses, uncertain land tenure, poor land market development, deforestation, erosion of farming systems, and migration of people and livestock, effective land use planning becomes more difficult.

Ecosystem services also function under the features of CPRs, although the difference between ecosystem services and CPRs is subtle. CPR delivers benefits to humans, whereas ecosystem services are the "procedure of ecosystems that deliver benefits" (Fisher et al., 2010). All these services could flow from one system to another, and it could be assumed that CPR management and the delivery of ecosystem services are closely related. For instance, "water regulation" is not managed, but the process that arranges the regulation of water is managed. A tool that connects the two, and successfully intervenes in Vietnam, is payment for ecosystem services (PES). PES connects conservation with market incentives. Past experience has shown that direct payment is more cost effective than integrating conservation and development related projects. PES programs have been successful in terms of "double dividend benefits of poverty reduction and biodiversity conservation" (van Wilgen et al., 1998).

Vietnam's shift in CPR governance has devolved natural resource management into local communities and user groups. The sustainability and endurance of self-governing institutions and user groups are connected to the satisfaction of the communities with the responsibility for CPR management and sustainability. Research on CPR governance in Vietnam has often been limited to local users that do not consider the higher governmental and institutional levels. Provincial and regional hierarchies and power asymmetries between CPR users and those making the policies are often ignored. An analysis about collective and individual actions on institutional performance in CPR management and governance is vital. The agriculture and forestry sectors in Vietnam have been decentralized through the onset of Forest Land Allocation (FLA) policies, however the differences between the practice and implementation remains significant. FLA policies are often designed on a nationwide basis that does not fully consider diverse population groups, farming cultures, and biophysical conditions, using an orientation toward market oriented communities rather than traditional hamlets in remote areas.

These are all issues that are important and timely in understanding the use, governance, and protection of landscapes and CPR. I am, therefore, delighted to read this volume, which addresses these issues through the lens of Ostrom's Social Ecological Systems Framework. It arrives at major policy implications for polycentric governance and management of natural resources through the better understanding of multilevel and reciprocal relationships among resources and resource-dependent communities. Northern Vietnam's landscape and its ethnic diversity is a small world in itself that should be a motivating factor for researchers, development practitioners, policy makers, and academics from Vietnam and beyond to read and learn from this volume.

David L. Hulse
Former Representative for Vietnam and Thailand
Ford Foundation

References

Fisher, B., Kulindwa, K., Mwanyoka, I., Turner, R.K., Burgess, N.D., 2010. Common pool resource management and PES: Lessons and constraints for water PES in Tanzania. Ecol. Econ. 69 (2010), 1253–1261. Available online http://www.fao.org/fileadmin/user_upload/kagera/resource/Tanzania-PES.pdf.

Van Wilgen, B.W., Le Maitre, D.C., Cowling, R.M., 1998. Ecosystem services, efficiency, sustainability and equity: South Africa's Working for Water Programme. Trends Ecol. Evol. 13 (9), 378.

Preface

Human and nature essentials respond consecutively and dynamically as a sequential phenomenon, a key feature of socioecological systems (SES). The trajectory changes in the global environment are hard to foresee, as the problem lay in exposure, sensitivity, and vulnerability of complex SES, where a sole incident concurrently cuts a swath through multiple components and results in nonlinear responses. Policy makers and natural resource managers are therefore under increasing pressure to make the right management decisions. Disregarding these features falsifies the policies and attempts to be counterproductive and less effective. This sheds light on the understanding of the SESs, especially when South East Asia is facing the challenge of balancing conservation and development.

Natural resources in South East Asia are continuously changing and so are the management approaches. A multidisciplinary approach to embarking on these dynamisms is necessary because natural resource degradation is happening at higher rates particularly in Northern Vietnam, which is visible through deforestation, land use and land cover changes, livelihood transition, reduced agricultural productivity, and biodiversity losses. Climate changes have also posed challenges in conservation in Northern Vietnam, where natural resources are huge but technical and financial resources are inadequate, which is further heightened by the ambiguity in the projected consequence, especially for people who depend on natural resources. Conversely, policies are least effective in achieving the dual goal of sustainable development and effective adaptations.

In the past, through economic reforms and political settings, attempts have been made to generate knowledge and formulate sustainable resource governance initiatives that extensively recognize the connections between nature and social systems. Though conservation, infrastructure development, and land management plans are formulated, most of the time they remain unimplemented, because there is less thought on the social courses that impact decisions. At times, though, SES robustness is believed to be enhanced through involving users into the economic bargaining process, by means of payment for ecosystem services, sustainability is always an issue. These issues remained less considered in the policy formulation process, as a comprehensive document that considers multidisciplinary approaches to natural resource manage is least available.

We are glad to share that this book has put forth an effort to provide a broader understanding on the dynamics and complexity of natural resources management for those interested in interdisciplinary work. This book has provided insights on effective measures that are inexpensive and prompt acting, and increases the resilience for both ecosystems and social systems through multiple case studies. Modern day readers interested in natural resources management, looking for comprehensive linkages among ecological and social factors, are likely to benefit from this book. This book has delivered broader knowledge in order to equally support such inadequacies at the local, national, and regional level. With growing concerns about natural resource management in Africa, Australia, and Central America, this book predominantly addresses those issues concerning global scales. The contents have proficiently explained the difference between problem solving tools and sustainability, which is vital for incorporating both scientific and collective encounters, but "operationalization has been elusive" (Leslie et al., 2015).

This book is a world within itself. The authors and editors have demonstrated facts from a dynamic world, together with the management of transboundary resources, and have summarized switches in resource use. They have successfully intermingled best practices also for land use and land cover changes, which in long run have also helped diverse natural resources management through out the world where natural resources are under massive pressure. These areas are also struggling with proper tenure arrangement and devolution rights which have hindered local people's participation. For strengthening local participation strengthening should be done.

During this course of change, a number of collaborative research studies were intended to integrate a balance between development and conservation. The basic inspiration comes from Elinor Ostrom's work on resource "commons" as vital to livelihoods. Her theory inspired academic collaboration between the Asian Institute of Technology (AIT) and the Vietnam National University of Agriculture (VNUA). Through their partnership, with additional support from Ford Foundation and the MacArthur Foundation, they have developed a methodology for exchanging ideas that has benefited both the livelihoods and ecosystems in Northern Vietnam. Prominent through the chapters enlisted in this volume, this academic research collaboration has developed a modern lens on polycentric approaches for effective governance and the management of sustainable and inclusive natural resources.

The School of Environment, Resources, and Development (SERD) at AIT has responded to regional and global research on issues related to natural resource management through strengthening technical capacities for the environment, and through social development partnerships. This publication has yet again added a milestone to AIT and SERD's research list. Both of us acknowledge the researchers, authors, and stakeholders of AIT, VNUA, and the Ford Foundation and MacArthur Foundation for their continued support, with the greatest appreciation to Elinor Ostrom for her theory, on which most of the research on natural resource management is based on. Also, our special gratitude to the editors and authors for their hard work, our duties have not finished yet. We still need to reach as many natural resource dependent communities as possible, and continue to look for ideas improving both livelihoods and natural resources.

Worsak Kanok-Nukulchai
President
Asian Institute of Technology
Bangkok, Thailand

Tran Duc Vien
Rector
Vietnam National University of
Agriculture
Hanoi, Vietnam

References

Leslie, M.H., et al. 2015. Operationalizing the social-ecological systems framework to assess sustainability. Proceedings of the National Academy of Sciences of the United States of America 112 (19), 5979–5984. Available online: http://dukespace.lib.duke.edu/dspace/bitstream/handle/10161/11470/PNAS%20Leslie,%20Basurto%20et%20al%202015%20%2B%20SI.pdf?sequence=2.

SECTION I

INTRODUCTION

CHAPTER

1

Toward Transforming the Approach to Natural Resource Management in Northern Vietnam

S.J. Leisz, M.V. Thanh*[†,‡], *T.D. Vien*[§]

[*]Colorado State University, Fort Collins, CO, United States [†]International Centre for Applied Climate Sciences, University of Southern Queensland, Toowoomba, QLD, Australia [‡]Center for Agricultural Research and Ecological Studies, Hanoi, Vietnam [§]Vietnam National University of Agriculture, Hanoi, Vietnam

1.1 INTRODUCTION

From the early 1990s through to the present the Center for Agricultural Research and Ecological Studies (CARES) at the Vietnam National University of Agriculture (VNUA) and the School of Environment, Resources, and Development (SERD) of the Asian Institute of Technology (AIT) have collaborated in both studying and understanding the participatory process that has occurred, and in some cases not occurred, as farming systems and natural resources management (NRM) changed or transitioned in rural northern Vietnam. Some of these changes and transitions have been to socially acceptable and ecologically sustainable livelihood systems by local communities in the northern mountains in the emerging context of climate change, while other changes have not been so benign. This collaborative effort, which is aimed at reconciling the standard concepts of development with conservation, has focused on the small microwatersheds within the larger Red River Delta basin. Support for this effort has been provided by the Ford Foundation and the MacArthur Foundation, in close coordination with CARES and VNUA, and guidance from Vietnam's Ministry of Agriculture and Rural Development and Ministry of Natural Resources and Environment at national, regional, and community levels. Notable research documentation in this volume includes issues such as local level land-cover and land-use transitions, conservation- and development-related agroforestry policy outcomes at the local level, and alternative livelihood adaptation and management strategies in the context of climate change. A majority of these studies have examined the outcomes of conservation and development policies on the rural communities that have participated in their implementation through collaborative governance and participatory

Redefining Diversity and Dynamics of Natural Resources Management in Asia
Volume 2
http://dx.doi.org/10.1016/B978-0-12-805453-6.00001-2

management in partnership with participatory community institutions. The editors and authors feel that the findings of these rich field-based studies will not only be of interest and use to national policy makers and practitioners and the faculty and students of academic institutions, they can also be equally of use to those scholars interested in understanding developing country socioecological systems and their context-specific adaptation strategies.

Besides this introduction, which is the first section of the book, this volume is categorized into six other sections including fourteen additional chapters. Thus, the second section discusses climate change vulnerability, risk reduction and adaptation strategies that are taking place in Vietnam. The third section focuses on projects that have piloted efforts to pay for ecosystem-related services in northern Vietnam. The fourth looks at land-use planning in the region. The fifth explores adaptive livelihood strategies that have either been studied or introduced, while the sixth explores aspects of decentralization relevant to the region. The seventh section discusses systems modeling as a new way of thinking through management of the complex natural resource systems of the area. The underlying theme that ties almost all of the chapters in the volume together is the importance of local involvement in all approaches toward NRM, whether it is planning for or reacting to climate changes, payment for environmental services, managing forestland, alleviating food shortages, living with biodiversity, or assessing the development projects and policies that are being implemented. Without the involvement of local communities, households, and ultimately individual people, the needed action will not be effectively taken. In this sense the chapters are also arguing for the need to do away with "one-size-fits-all" policies, and in some cases "models" for NRM, that have previously been promoted by the government of Vietnam as the way to manage all of the country's natural resources.

1.2 CLIMATE CHANGE

Following the introductory chapter, Chapter 2, *Responding to Climate Change in the Agriculture and Rural Development Sector in Vietnam*, points out that Vietnam is one of the countries that is likely to be the most seriously affected by climate change, and the agriculture and rural sectors are the most vulnerable sectors to climate change in the country. Recognizing this reality, the country has developed a comprehensive framework of policy and institutional systems for climate change adaptation and mitigation. The chapter reviews this framework and its implementation so far and details the national-level laws and regulations that are in place, rightly noting that much work at the administrative level has been completed. In reviewing these policies, the author illustrates that the majority of this system is oriented from a top-down direction and points out that there are two significant weaknesses to the policy formulations. These involve a lack of addressing the private sector in the policy formulation and in recognizing and addressing the need for local involvement in responding to climate change, noting "there are no concrete policies to encourage or create incentive for local people and private sectors to participate in climate change adaptation and mitigation." Vietnam's private sector is fast growing and dynamic and has a lot to offer in helping the country, and specifically the agriculture and rural development sectors respond to climate change. The private sector also can help develop mitigation mechanisms, as it is the private sector in other parts of the world that is leading the way in response to the pressures of climate change. Thus, it is well noted by the author that this weakness needs to be addressed.

Chapter 3, *Assessing and Calculating a Climate Change Vulnerability Index for Agriculture Production in the Red River Delta, Vietnam*, initiates a discussion on one strategy for integrating local people into the workings of the national framework on climate change mitigation and adaptation. This chapter looks at ways to assess and calculate climate change vulnerability from a local perspective up through the district levels. The chapter discusses both the context of climate change vulnerability in Vietnam and how much of the vulnerability assessments to date have only taken place at higher levels and not looked at the local/community level, and the development of software that can be used as a decision support system tool to facilitate assessment of climate vulnerability at the local/community level. The initial trials of this methodology and software produced analyses that detailed the levels of vulnerability to climate change variables recognized by the Intergovernmental Panel on Climate Change (IPCC) at commune levels in the provinces studied, providing an example of how the local/community level can be reached in initiating climate change adaptation and mitigation work in the rural areas of Vietnam.

1.3 PAYMENT FOR ECOSYSTEM SERVICES

Recognizing that resources are needed for local communities to effectively take part in managing and ultimately protecting resources, Chapter 4, *Cash-Based versus Water-Based Payment for Environmental Services in the Uplands of Northern Vietnam: Potential Farmers' Participation Using Farm Modeling*, investigates how and what kind of payments for protecting environment services might work in the context of Vietnam's uplands. Payment for Environmental Services (PES) is seen as a potentially effective approach for improving NRM and is increasingly being promoted as part of the process by which NRM is devolved to local community levels. The authors look at the possible ways that PES could work in the uplands of Vietnam to protect forests through a study of who would participate and a modeling exercise devoted to projecting how a specific PES would impact on different farm types and translate into protected forestland and household revenues. The authors conclude that rather than paying smallholder and poor farmers directly with cash for protecting forests, if the PES (for protecting the forest) results in increasing access to irrigated terraces as a way to compensate for forestland protection and expansion, then participation of smallholder farmers and the poorest farmers in protecting forests will increase. They conclude that this type of payment for environmental services could ultimately be beneficial to both forest protection and poverty alleviation in Vietnam's uplands. Finally, the chapter shows the validity of modeling in helping natural resources managers evaluate possible policy proposals.

Chapter 5, *A Voluntary Model of Payment for Environmental Services: Lessons from Ba Be District, Bac Kan Province of Vietnam*, also explores a project that initiated PES. The project encouraged villagers, home-based entrepreneurs, and boatmen to participate in the project. The villagers received money from the environmental service users, through the project, to protect the forest so as to preserve the natural landscape and the watershed. Initially it appeared that the project worked; however, after the project had run its course the PES system broke down and the villagers stopped protecting the watershed and landscape. The study documented in this chapter found that the model that was implemented in the project failed as it identified inappropriate environmental service users, resulting in too little fees being collected, so the moderators who were supposed to pass the payments from the users to the suppliers stopped performing this service after the project stopped, and ultimately the villagers didn't receive payments and stopped protecting the resources that the users needed. Unlike in Chapter 4, the PES in Ba Be District ultimately did not work as a sustainable system was not put in place by the project. While ultimately failing, the documentation of the project does highlight lessons learned that may be valuable for future efforts that aim to effectively pay local people for environmental services that originate from their lands.

1.4 LAND-USE PLANNING

Since 1986 and the initiation of the *doi moi* reforms in Vietnam, there have been enormous changes in the way the government of Vietnam manages its natural resources base. Two chapters in this volume overview these policies and present a practical example of a land-use planning case that was carried out. Setting the stage for this discussion is Chapter 6, *Land-Cover and Land-Use Transitions in Northern Vietnam from the Early 1990s to 2012*, which overviews the changes in both land use and land cover, which have taken place in the northern part of the country through 2012. This chapter tracks the changes in land use and land cover and the policy drivers of these changes in lowland, midland, and upland, and in both urban and rural areas. The chapter notes that official policies have been a driver of some of these changes and have also acted to keep change from happening, especially in periurban areas where land is still devoted to agriculture that otherwise may have been converted to other uses if regulations restricting this conversion were lifted. It also notes that market demand for agriculture and forestry products and local reactions to these demands have played a part in how land use and land cover, especially in the uplands, have changed. The chapter concludes by discussing implications of the changes to land-use systems and the associated land cover in the context of climate change and future development.

Chapter 7, *The Role of Land-Use Planning On Socioeconomic Development in Mai Chau District of Northern Vietnam*, takes a case study approach to review the implications at a more local level of specific land-use planning that takes place. The author overviews the participatory process followed in the land-use plan that was created for Mai Chau District in 2000. A review of the reality of land use in 2010 illustrates that the land-use plan was generally followed over the time period with some straying from it in the areas of agricultural land use, which covered a larger area than planned, and forestland, which covered a slightly smaller area. The development in the district is then compared to the land-use plan and correlations are found between changes in the districts land use and socioeconomic development, providing evidence for the author to suggest that land-use planning, in this case, has aided socioeconomic development.

1.5 ADAPTIVE LIVELIHOODS IN RESPONSE TO CHANGE

A number of chapters in this volume reflect targeted studies of issues at a very local level. Chapter 8, *Coping Mechanisms of the Ethnic Minorities in Vietnam's Uplands as Responses to Food Shortages*, applies the sustainable livelihood approach (SLA) to identify similarities and differences in coping mechanisms that different ethnic minority groups (Thai and Kho Mu) have used to cope with food shortage situations. The chapter also examines factors that bring about the differences or similarities within and between ethnic groups. At the household level, the study looks into five types of assets (natural, social, human, financial, and physical) that are both indirectly and directly affected by household food shortages and can conversely also affect how the food shortage affects the household. The author also looks at the capital resources of the abovementioned ethnic communities in assessing the causes of food insecurity. The two villages are similar in three of these capitals, but vary in the areas of human capital and financial capital, with the Thai village studied having more of both. It is this difference in capital that the author concludes is the reason for more severe food shortages affecting the Kho Mu village than the Thai village.

Chapter 9, *Home Gardens in the Composite Swiddening Farming System of the Da Bac Tay Ethnic Minority in Vietnam's Northern Mountain Region*, examines the role that this very local land-use system plays in the livelihood system of individual households. The authors look at the structure and function of the home garden and how it is changing over time. They also note that the way the garden is managed has changed, and is changing, in response to outside pressures. They show that the management of the garden is done in combination with the larger management of each household's composite swiddening system. They conclude that the composition of the gardens change as the local environment changes and as the household becomes integrated into the wider regional market system. A place is seen for the changing gardens as the uplands change. That place can be to increasingly add a level of security for the household as the garden provides resources and products either for household use or for the market.

Chapter 10, *How Agricultural Research for Development Can Make a Change: Assessing Livelihood Impacts in the Northwest Highlands Of Vietnam*, addresses the issue of how agricultural research can help address local level issues. The authors argue that a holistic approach toward the assessment of agricultural research for development is needed to improve the impact that these types of research projects have on farmers in northern Vietnam's uplands. The holistic research approach that they promote is needed to increase the use of the participatory approach in agricultural research for development. This approach is in contrast to the short-term and quantitative impact indicators that are commonly used in mainstream assessment approaches by most project implementers. The approach the authors promote focuses on social, human, and other development indicators, utilizing the sustainable livelihoods framework developed by the United Kingdom's Department of International Development as part of the assessment of agriculture for development projects. The results of the research assessment are then fed back into the development process to quickly magnify the impact of the research by promoting aspects that are successful. To test this approach three recently completed agriculture for development projects were investigated using the holistic impact assessment framework. The findings of the research indicate that even though all three of the projects professed to be successful according to their own indicators, when the holistic approach was used to assess them, the project that utilized participatory communication strategies actually had better social, human, economic, and environmental outcomes and impacts for the local community than the other two projects. This successful project also did a better job of spreading the techniques that worked in the research project beyond the group of farmers involved in the research. The result shows that the holistic approach to assessing agriculture for development projects yields better insights into what works in culturally diverse rural areas than more traditional assessment tools.

Chapter 11, *Changes in the Nature of the Cat Ba Forest Social Ecological Systems*, looks at the way biodiversity and humans interact at a local level through carrying out a socioecological systems' analysis of the historical development of community interactions with biodiversity on Cat Ba Island. Historically, people on the island interacted with the biodiversity as a means of sustaining their livelihood. However, over time these interactions led to a degrading of the natural resource base. With the creation of Cat Ba National Park, the relationship was forced to change and limitations were placed on how the local communities could use the resource base. The rapid development of tourism on the island is another important factor that further influenced the ways the communities interact with their natural resource base. Through the social-ecological analysis of this system, the authors conclude that, "The dynamic linkages between livelihoods, forests and governance have been, and will continue to be, major elements of the island's state. The integrity of native ecosystems and the services that they provide will affect local livelihood activities such as agriculture and tourism. The future conservation of the unique biodiversity and ecosystems of Cat Ba Island will largely depend on local residents, who must ultimately use their knowledge and understanding in an ecologically responsible way to maintain positive economic and social practices."

1.6 DECENTRALIZATION

Decentralization of management of natural resources is often promoted as a means to encourage better and more responsive NRM. Two chapters in this volume examine this strategy. One of them focuses on a study of the decentralization process in Vietnam's uplands, while the other discusses how application of Elinor Ostrom's ideas on decentralization could be useful in resolving cross-boundary water disputes. Chapter 12, *Decentralization in Forest Management in Vietnam's Uplands: Case Studies of the Kho Mu and Thai Ethnic Community*, describes the process of forest devolution policy implementation by means of the forestland allocation (FLA) policy, which has been implemented in the country's uplands since the early 1990s. The chapter also examines the impacts of FLA policy on forest resources and local livelihoods. The author uses case studies of two different ethnic groups (Thai and Kho Mu) residing in the same ecological condition (Nghe An Province), and the same ethnic group (Thai) residing in different ecological conditions (Nghe An and Son La Province) to further examine how ethnicity and ecological conditions contribute to diverse local responses and environmental management results. The chapter points out some differences and similarities in how decentralization was done in both provinces. The chapter also highlights that the devolution policy places a strong emphasis on individual rights to land and forest resources and the allocation and contracting of these rights often come with very strong management obligations. This severely restricts the rights of villagers and can create opportunities for local elites to capture most of the benefits from the forest. The success of these decentralization efforts, with respect to protecting forest area and/or expanding it, was not uniform however. In the Son La case, the rotational swidden area decreased and forest area increased, while in the two Nghe An cases, one saw its forest area hold steady, while the other saw a decrease in forest area. These mixed results are attributable to a number of issues that the author also discusses; however, a few issues pertaining to the way that decentralization was, and wasn't, carried out are emphasized. In the end, the author notes that if decentralization is to work, the local government at the community level needs not only implementing power, but also decision-making power and resources to carry out the task of managing the resources, and participation by the whole community needs to be encouraged, not just by some sectors of the community.

Chapter 13 has a different geographic focus, but has been added to this volume to address the issue of transboundary NRM and how decentralization may help deal with this contentious border issue. This is an issue that Vietnam is dealing with in its relations with its northern neighbor, China, and its western neighbor, Laos. However, studies have not been done on this sensitive subject. Thus Chapter 13, *Institutions for Governance of Transboundary Water Commons: The Case of the Indus Basin*, looks at the issue of transboundary water governance between Pakistan and India in south Asia, with the hope that some of the lessons suggested in this case may one day be useful in the case of Vietnam and its neighbors. The chapter overviews the history of water as a conflictual resource in south Asia and specifically the issues that are raised between Pakistan and India with regards to access to water. The issue of above ground water and the role that dams have played as sources of conflict are reviewed and it is pointed out that increasingly groundwater, which was originally included in transboundary agreements, is becoming a contentious point because of its "hidden nature, lack of monitoring mechanism and data." To assess these issues the authors use Ostrom's design principles to investigate key points of how the transboundary resource is currently managed. They discuss the issues of clearly defined boundaries, congruence to local conditions, collective choice arrangements by the resource users, monitoring, use (or lack of) graduated sanctions, conflict resolution mechanisms, minimal recognition of rights to organize, and the nested enterprises. They conclude that major conflicts between India and Pakistan can be resolved if lessons are drawn from these design principles and a revised treaty for the resource use is devised based on the lessons. It is hoped that this analysis may someday be useful to Vietnam in dealing with its neighbors and the contentious issue of cross-border water resources.

1.7 NEW WAYS OF THINKING TO MANAGING COMPLEX NATURAL RESOURCES SYSTEMS

Given the dynamically complex, multilayered and multistakeholder problems intrinsic to the nature of NRM, which conventional reductioni\st ways of thinking are totally ineffective in dealing with, Chapter 14, *A System Dynamics Approach for Integrated Natural Resources Management*, points out a clear need to adopt systems-based thinking and an integrated approach to effectively manage natural resources. Through a real-life example of applications of the system dynamics approach to the study of natural resources-based tourism in Cat Ba Island of Vietnam, the author develops a conceptual model and its associated simulation model of tourism on the island through the interacting mental models of stakeholders. The conceptual model presents the "bigger picture" of the tourism system on the island and visualizes how the factors affecting the system are not isolated and independent, but are dynamically linked. The model is used as a platform for developing a common understanding of the issues influencing tourism development, and for improving stakeholders' engagement, communication, and collaboration in natural resources-based tourism development planning. The conceptual model also serves as a solid foundation to enable the identification of key leverage points in the system. In addition to this, the simulation model allows policy makers and tourism managers to test alternative tourism development strategies and scenarios; it can be used as a tool to enhance collective learning and improve the design of plans for the management of natural resources.

Finally, the concluding Chapter 15, *Navigating Complexities and Management Prospects of Natural Resources in Northern Vietnam*, summarizes and discusses the main findings of all chapters presented in the volume. The chapter includes some implications that can be drawn from these chapters toward better NRM in Vietnam's uplands in general and particularly in the northern mountains.

In conclusion, with this brief overview of all the chapters presented in this volume, we hope that the reader will delve deeper into each of them. The lessons from these studies are valuable as Vietnam progresses into the 21st century. The country will depend more on its natural resources base as it develops and based on the experience of the past four decades these chapters offer insights into what is needed for successful NRM in northern Vietnam.

SECTION II

CLIMATE CHANGE

CHAPTER

2

Responding to Climate Change in the Agriculture and Rural Development Sector in Vietnam

*P.T. Dung**, *S. Sharma*†

*Ministry of Agriculture and Rural Development, Hanoi, Vietnam †WWF-Nepal, Kathmandu, Nepal

2.1 INTRODUCTION

2.1.1 Background

Climate change has been taking place more and more severely on a global scale, whereby Vietnam is one of the most vulnerable countries. Viet Nam, with a long coastline and two large low-lying deltas, is among the most affected countries in terms of climate change impacts and sea level rise; a major concern for scientists and policy makers. Over the years, diverse stakeholders have come up with multiple approaches to aid communities to adapt themselves to climate change. This chapter explores the extent of policy options to support adaptation to climate change, especially in the agriculture and rural development sector as a major portion of the Vietnamese population is dependent on the agricultural system, which is highly affected by climate dynamism.

Traditionally, policy options have been taken as prime challenges to be addressed in climate change. However, modern theoretical and empirical studies focus on policy issues as effective adaptation strategy. This chapter explores diverse policies that mainstream climate change, especially into agriculture and rural development policies with the view to decrease impacts on people. The major issue is to formulate a policy that increases the adaptive capacity and stress-tolerating ability among people. With such endeavors, polices should be imperative in the context of global warming in which climatic and environment features and patterns are relatively changing. Adaptation is required in the view with long-term changes and extreme events observed in the Cuu Long and Red River Deltas. According to Dasgupta et al. (2007), if the sea level rises by 1meter (m), about 11% of the population will be directly

http://dx.doi.org/10.1016/B978-0-12-805453-6.00002-4

affected and the loss of GDP is estimated at 10%. If the sea level rises by 3m, about 25% of the population will be affected and GDP loss is estimated at 25%. The latest climate change and sea level rise scenarios developed by the Ministry of Natural Resources and Environment of Vietnam (MONRE, 2012) also shows the severe impacts of sea level rise to population, of which about 35% of the population in the Cuu Long River Delta and more than 9% of the population in the Red River Delta, 9% of the population in the coastal central region and 7% of the population in Ho Chi Minh City will be directly affected if the sea level rises by 1 meter. Some studies have predicted that by the year 2100, the Cuu Long River Delta that produces the highest amount of rice in Vietnam may lose 7.6 million ton of rice production per year, the equivalent to 40.52% of total rice yield of the whole region due to the effects of climate change (Tran Van The et al., 2011; Department of Crop Production, 2013a,b).

There is a growing concern that extreme climate events will threaten natural resources-dependent communities through vulnerabilities and risk exposures, especially in the agriculture and rural development sectors (IPCC, 2012). Countries like Vietnam where crops, livestock production, forestry, and fisheries form the livelihood options for 70% of the population are affected by climate change impacts. Among these, crop production is likely to be severely affected; rice production, in particular, could be reduced by 10%. With higher sea level rise, salt intrusion is likely to destroy more than 2.4 million ha (hectare) of cultivated land (Department of Crop Production, 2013a,b).

Some other regions, especially the mountainous areas in the northwest and the Central Highlands, are likely to suffer from severe drought and terrible water shortages. All these will ultimately threaten agriculture production, food security, and the lives of millions of people residing there. Recognizing these severe impacts and the critical need for response activities to climate change, the government of Vietnam has set up a quite comprehensive framework of policy and institutional systems for climate change adaptation and mitigation. However, the belief that polices will reduce climate change becomes problematic as a large number of policy statements do not get the attention they deserve. This chapter first explores involved policies to address issues of climate change in the broader agriculture and rural development context, followed by detailed analysis, and finally suggest some measures to further strengthen those policies in Vietnam.

2.1.2 Climate Change Impacts and Institutional Responses

A report by the Ministry of Agriculture and Rural Development of Vietnam shows that Vietnam may lose 2 million out of total 4 million ha of cultivated rice, seriously threatening the food security of millions of people (MARD, 2009). A study conducted by the Institute of Agricultural Environment in 2011 showed that the average damage caused by climate change in agriculture and rural areas to be 800 billion Vietnamese Dong (VND). In 2030, the yield of rice, maize, and soybeans in Vietnam may be reduced by 8.37%; 18.71%, and 3.51%, respectively (Tran Van The et al., 2011) due to high sea level rise accounting for flooding and soil salinization in coastal and low plain arable lands. This is likely to be further aggravated by disease and pest infestations. Higher sea level rise also decreases existing mangrove areas, which in turn seriously affects the forest planted in sulfate soil in the southern provinces of Vietnam. Flora and Fauna will be at higher risk of extinction due to forest fires ignited through higher temperatures (Vu Tan Phuong et al., 2011). The process of temperature fluctuations and extreme events results in flooding consequences that includes reduced natural fish yield that ultimately results in food insecurity and socioeconomic challenges in the coastal areas of Vietnam (Nguyen Viet Nam et al., 2011). Approximately 58% of the population lives in the coastal area that are likely also to be affected by reduced water supply and environmental sanitation issues.

2.1.2.1 National Arrangement for Climate Change Response

There have been a number of arrangements undertaken at different times to ensure better response to climate change. These are the National Committee on Climate Change, the National Target Program to Respond to Climate Change, the National Strategy on Climate Change, and lately, the National Action Plan to Respond to Climate Change during 2012–20.

The National Committee on Climate Change was established in accordance with Decision No. 43/QĐ-TTg, dated January 9, 2012, and signed by the prime minister. The committee has the prime minister as its chairman , the deputy prime minister as its standing vice-chairman, the minister of natural resources and environment as vice-chairman, and some members, including the minister of agriculture and rural development. The role of the committee is to consult and assist government and the prime minister to study, propose, direct, coordinate, and speed-up inter- and subsectoral strategies and national programs on climate change. They also have the responsibility of organizing and implementing international dialog and cooperation on climate change. The Office of National Committee on Climate Change under the Department of Meteorology, Hydrology, and Climate Change and MONRE are the supporting and standing agencies for the committee. Likewise, the National Target Program to Respond to Climate Change (NTP-RCC) was also approved for promulgation by the prime minister (in accordance with Decision No. 158/QĐ-TTg, dated February 11, 2008). The NTP-RCC is one of the initial legal documents of Vietnam showing the resolution of the Vietnamese government in response to climate change. The overall goal of NTP-RCC is to assess the magnitude of climate change on the sectors, fields, and localities during different time periods and to develop a feasible action plan to effectively respond to climate change for the short-term, medium-term, and long-term time frames to ensure economic development adopting a low carbon approach and to participate in an effort to mitigate climate change together with the international community.

Similarly, the National Strategy on Climate Change was approved by the prime minister as Decision No. 2139/QĐ-TTg, dated December 5, 2011. The specific objectives of the strategy are identified as (1) ensuring food, energy, and water security, poverty reduction, gender equity, social welfare, livelihood improvement, and protection of natural resources in the context of climate change; (2) promoting a low carbon economy and green growth for sustainable development; (3) enhancing awareness, mutual responsibilities, and capacity to respond to climate change while developing potentials for quality technology and natural resources through better institution and policies through capturing opportunities through climate change for socioeconomic development; and (4) actively contributing to the international community for climate change response while simultaneously implementation of greenhouse gas (GHG) emissions mitigation and adaptation activities to effectively respond to climate change.

In the same way, a national action plan to respond to climate change during 2012–20 was approved by the prime minister as Decision No. 1474/QĐ-TTg, dated October 5, 2012. The plan has 10 objectives and a list of 65 projects to be completed with participation through different ministries and departments at both central and local levels to incorporate components in response to climate change. The action plan also envisions developing scientific technologies and knowledge, as well as conducting impact assessments to identify climate change mitigation and adaptation measures both for national and international cooperation and to improve mobilizing financial and technological resources to respond to climate change. Nevertheless, Resolution No. 24-NQ/TW, dated June 3, 2013. stressed that, "By 2020 climate change and disasters will be proactively adapted, natural resources will be rationally utilized, and GHG emissions will be effectively reduced while environmental pollution will be relatively lowered to develop a green and environmentally friendly economy." Similarly, the same document also states that, "By 2050 proactive responses on climate change would be put on to ensure environmental criteria equivalent to the current state of industrial developed countries in the region." Similarly, the Steering Committee for Climate Change Mitigation and Adaptation of Agriculture and Rural Development Sector was established in 2007 in accordance with Decision No. 3665/QĐ-BNN-KHCN, dated November 21, 2007, by the minister of agriculture and rural development. The committee has been regularly adjusted to adapt to actual situations, and is now chaired by the minister of agriculture and rural development. The steering committee has the task of directing the organization to implement action plans to mitigate and adapt to climate change in agriculture with a mandate of a national target program to respond to climate change.

2.1.2.2 *Climate Change Response Targeted to Agriculture and Rural Development*

Agriculture is one of the most important sectors of the Vietnamese economy. Crop production, livestock, and fisheries, all under the agricultural system, are dependent on climatic conditions and often it is difficult to understand the impacts on individual units of the system. However, climate change impacts agriculture in many ways through temperature and rainfall fluctuations, rises in pests and disease infestations, and changes in the nutritional quantity and quality of some foods. Nevertheless, agriculture is also responsible for global GHG emissions and requires innovative practices to facilitate adaptation and mitigation in the agricultural sector. Formulating the necessary policies at multiple scales is important to promote local climate responses with global impact interlinkages. Hence, there are a number of legal and institutional arrangements set up by the government of Vietnam to promote climate responses in agriculture and rural development.

Action plan framework for adaptation to climate change of the ministry of agriculture and rural development during 2008–20

Upon realizing the explicit impacts of climate change on food security, agriculture, and rural development, and to improve adaptive capacity and mitigation potential, the minister of the Ministry of Agriculture and Rural Development (MARD) signed Decision No. 2730/QĐ-BNN-KHCN, dated September 5, 2008, to issue "Action Plan Framework for Adaptation to Climate Change of agriculture and rural development sector for 2008–2020." The plan focuses on ensuring stability and safety of residential areas in cities and regions, especially in the Cuu Long Delta, the Northern Delta, the Central Highlands, and mountainous regions, which ensures stable agriculture and food security with the production of 3.8 million ha of rice. This also focuses on ensuring safety of dyke systems, farmer serving, and technical and economic infrastructure to prevent, avoid, and mitigate natural disaster impacts.

Action plan to respond to climate change of ministry of agriculture and rural development during 2011–2015 and vision to 2050

Based on the Action Plan Framework for Adaptation to Climate Change during 2008–20, national strategy on climate change with support from the United Nations Development Program (UNDP), the action plan responding to climate change in agriculture and rural development was developed during 2011–15. MARD is the prime stakeholder to implement this action plan with seven objectives and key tasks.

This plan looks forward to enhancing the capacity of studying and forecasting climate change impacts on agriculture, irrigation, forestry, and fisheries to develop polices and measures that are adaptive to climate change. The plan also considers developing a policy system to integrate climate change into sectorial plans and involve stakeholders in an action plan. An opportunity for proposing solutions and mainstreaming disadvantageous regions into sustainable development is also promoted by this plan, which in fact develops human resources in the sectoral activities of climate change adaptation and mitigation. The major highlight of the plan is that it ensures organizations, individuals, and communities will be entitled to equitable benefits from climate change responses.

Integrating climate change into the development and implementation of strategies, planning, plans, programs, projects, and development proposals of the agriculture and rural development sectors during 2011–15

MARD issued Directive No. 809/CT-BNN-KHCN, dated March 28, 2011, for integrating climate change issues into the development, with approval and implementation of organizational strategies, plans, programs, projects of agriculture, forestry, salt production, aquaculture, irrigation, and rural infrastructure development in the entire country to make sure they are suitable with the national strategy, target program, and action plan to respond to climate change, focusing on GHG gas emissions mitigation and climate change adaptation.

Proposal on greenhouse gas emissions reduction in agriculture and rural development sector up to 2020

Although Vietnam is not on the list of countries obliged to reduce GHG emissions in accordance with Annex 1 of the Framework Convention on Climate Change, it has ratified the Kyoto Protocol and actively participated in climate change platforms. By so doing Vietnam has promoted GHG emissions reduction activities into the agricultural sector in particular and the entire economy in general as a manifestation of sharing responsibility with the world in the fight against climate change. In effect, many measures that bring in both economic development and climate change adaptation, results in GHG emissions reduction. They have been successfully transferred into productive activities such as effective irrigation technologies, less input-intensive mixed cultivation patters, a system of rice intensification (SRI), Integrated Pest Management (IPM), integrated crop management, contour cropping, and livestock waste management through biogas and compost. However, the implementation of the abovementioned actions mainly focuses on the economic attributes rather than on the possibilities of reducing GHG emissions. Therefore, the development and transfer of GHG emissions reduction measures in agriculture play an important part in sustainable development of agriculture together with climate change adaptation and mitigation.

By taking the initiative in implementing GHG emissions reduction activities that are suitable with national conditions and sustainable in terms of low carbon economic development objectives, the program on GHG emissions reduction in agriculture and rural development has three major combined objectives: stimulating green agriculture production by adopting safety orientations that produces less GHG emissions and is sustainable in terms of ensuring food security, contributing to poverty reduction, and effectively responding to climate change. It is assumed that by 2020, 20% of GHG emissions in agriculture and rural development will be reduced, which is the equivalent of 18.87 million tons of CO2e. At the same time, the sectoral growth and poverty reduction objectives are ensured in accordance with the sectoral development strategy that was approved through Decision No. 3119/QĐ-BNN-KHCN, dated December 26, 2012. The main task of the program is to promote green agriculture production with less GHG emissions through ensuring food security, poverty reduction, and effectively responding to climate change.

The plan of restructuring agriculture and rural development sector

The plan of restructuring agriculture and rural development sector toward greater added value and sustainable development was approved by the prime minister in Decision No. 899/QĐ-TTg, dated June 10, 2013, and has one among three objectives that is "to strengthen natural resources management, reduce GHG emissions and other negative impacts to environment, and enhance disaster risk management capacities." Regarding restructuring of the specific crop production subsector, one of the measures is to promote the application of science and technology, especially advanced technologies to improve crop productivity, quality, and reduce production cost and better adapt to climate change.

2.1.3 Actual Implementation of These Policies in Agriculture and Rural Development Sector

2.1.3.1 Adaptation to Climate Change

There have been numerous activities undertaken to respond to climate change and the key tasks undertaken for the period of 2008–15 have already shown approximated achievements. At a national scale, climate change impacts on agriculture have been evaluated to date, whereby MARD has proposed solutions in the form of providing options for cropping practices as per diverse ecological regions. Prior to that, a training framework for sectoral staffs and communities was organized to capture their learning, experience, and build capacity in responding to climate change in agriculture, fisheries, salt production, forestry, and rural infrastructure building through developing pilot models. Four studies and impact evaluations on agriculture have been conducted and published for wider circulation.

To ensure sustained crop production with low water inputs through SRI technologies, activities were undertaken to disseminate at a larger scale. Simultaneously, a production model for large-scale paddy fields has been stimulated to envision agricultural sector restructuring. A conference during 2012 revealed that northern provinces of Vietnam had applied this model for more than 12,575 ha of land, while 76,559 ha of southern provinces had also applied the model by late 2013. This model has also been replicated to upland vegetables and sugar cane. With technological advancement in a rice-shrimp model in the Cuu Long Delta, more than

120,000 ha of land have benefited as a rice-shrimp model has been asserted to have high adaptability and sustainability in the context of salt intrusions; products of these areas are ensured in terms of safety and quality due to limited use of plant protection chemicals (Department of Crop Production, 2013a,b; Ministry of Agriculture and Rural Development, 2013).

An attempt has been made to implement preferential credit programs for farmers for constructing livestock-biogas facilities under the Quality and Safety Enhancement of Agricultural Products-Biogas Program (QSEAP-BP) project. With this facility, credit farmers could be loaned as much as VND10 million to tens of VND millions for the expansion of livestock scale with waste treatment facilities. Likewise, procedural efforts for large-scale production of catfish, tiger shrimps, and water shrimps have been developed along with integrating factors for water monitoring, training on breeds, feeds, and enhanced cultivation practices. Boats and associated technologies have been reviewed, assessed, and classified accordingly to conduct modernization plans for tuna raising and investing in fishermen's equipment by 2020.

Forest growth and productivity is directly affected by climate change mainly due to fluctuations in temperatures, rainfall, and atmospheric carbon dioxide through multiple interactions occurring within forest ecosystems. Because climate change also affects the occurrence of many forest disturbances such as forest fires and pest outbreaks, there have been multiple projects and programs initiated for strengthening watershed and forest quality and capacity to adapt to these changes, especially the mangroves in the coastal areas. Simultaneously, plans to reduce deforestation and degradation through plantation and forest enrichment have been focused. Most of them have been implemented through Decree No. 99 in relation to payment of a forest environment fee. To date, 35 provinces have established provincial steering committees while 23 provinces have established provincial forest protection and development funds. By 2012, a total of VND1130.8 billion was collected throughout the country as an environment fee. This has been supported by developing programs on the effective use of barren hills and land to create employment opportunities through abolishing shifting cultivation and livelihood diversification for poverty reduction. To help further adaptation to climate change, special use forest systems are studied and strategies adjusted accordingly. This was followed by drafting a circular for regulating criteria and management of buffer zones of special use forests and rings of marine conservation zones. The government of Vietnam has continued to direct localities with the reinforcement of organizational mechanisms, operation, and planning of special use forests in accordance with Decree No. 117/2010/NĐ-CP, dated December 24, 2012. With a view to realign adaptation in agriculture with global mitigation approaches, the MARD have integrated, coordinated, and organized a GHG emissions reduction program through preventing forest losses and degradation and integrating emission reduction programs, plans, and planning on climate change into strategies and programs of MARD. Often it is stated that climate change adaptation and mitigation stake conjoint aims with disaster risk reduction as both are targeted to reducing vulnerabilities. Therefore, programs are launched on resident reallocation of different regions that are prone to natural disasters, especially difficult in terms of vulnerability of a bordering area or island through free mitigation and taking benefits through special use forests during 2013–15 (Decision No. 1776/QĐ-TTg, dated November 21, 2012).

Water resources are vital to ecosystems, including humans, for drinking, agriculture, manufacturing, and energy production. Much of these potential uses are likely to decline due to climate change. This is because climate change is expected to result in an upsurge in water demand, along with a reduction in water supplies. This fluctuating equilibrium rquires water managers to balance the needs of communities, ecosystems, and energy producers. The case is no different in Vietnam and a number of policy making attempts have been made considering climate change effects on water availability. A master plan for the Cuu Long Delta, Red River Delta, central region, and coastal provinces had been formulated in 2012. Simultaneously, an irrigation planning and sea dyke upgrading program from Quang Ninh to Quang Nam and from Quang Ngai to Kien Giang was strengthened due to climate change and sea level rise. Together with this, procedures for the integration of Muong chuoi and Thu Bo sluices in an antiflooding project for Ho Chi Minh City was added to the list of priorities of the Support Program to Respond to Climate Change (SP-RCC) in accordance with Decision No. 1719/QĐ-TTg, dated October 4, 2011, have been completed. Together, antiflooding planning for big cities such as Ho Chi Minh City, Ha Noi, Can Tho, Ca Mau, and Hai Phong have been developed by reviewing and developing new criteria and specifications for the construction of hydraulic works in adaptation to climate change. These structures are less effective until the respective community is involved in the process. So, community awareness-raising and community-based natural disaster management have been focused by the enhancement of awareness-raising programs on natural disaster prevention and control and climate change adaptation.

2.1.3.2 Mitigation to Climate Change

Vietnam is one of the most vulnerable countries to climate change with an agricultural sector that has been highly affected in terms of lower agricultural production and productivity. The agricultural sector at the same time also emits GHG, which in turn adds to the climate fluctuation process. The GHG emissions inventory of MONRE in 2010 shows that GHG emissions from agriculture make up the highest portion (43.1%) of total emissions in Vietnam. Within the agricultural sector, paddy is the largest source of GHG emissions (57.5%) while other agriculture land use and livestock husbandry simultaneously emit 12.15% and 21.8%, respectively. Apart from directly responding to climate change to ensure sustainable agricultural production and rural development, attempts have been made to reduce emissions by promoting green agriculture and infrastructure development. Within the emission reduction program, forest carbon deposits have been evaluated while a baseline for a GHG emissions curve for Reducing Emissions from Deforestation and Forest Degradation (REDD+) implementation has been developed. Likewise, integrated technical solutions for mangrove sustainability in coastal areas have been focused on developing a rice cultivation model through water saving irrigation techniques.

In terms of crop production, the SRI has already been extended to more than 812,345 ha of land, while the ineffective rice production areas been converted to short-term cash trees that relatively emits less. They have been further supported with reduced straw burning and forest fire interventions (Department of Crop Production, 2013a,b). Similarly, livestock wastes are used for anaerobic compost technology to reduce GHG emissions from livestock sectors. Approximately 500,000 biogas facilities have been installed with 170,000 directly funded and subsidized by MARD (Department of Livestock Production, 2013). To facilitate carbon absorption and reduce GHG emissions in forestry, forestation, and regeneration have been implemented in accordance with the forest development strategy for 2010–20 (Directorate of Forestry, 2013).

2.2 DISCUSSIONS AND IMPLICATIONS

2.2.1 Policy and Institutional Framework for Climate Change Response in Vietnam

So far, Vietnam in general, and the agriculture and rural development sectors in particular, have a quite comprehensive policy system for climate change response, of which both mitigation and adaptation are given priority attention. As the most vulnerable sectors to climate change, the MARD is a bit ahead in developing a policy framework for climate change adaptation. In recent years, the ministry has also paid great attention to the reduction of GHG emissions in the sectors. It has a twofold objective: (1) to develop a green and sustainable agriculture in accordance with the National Green Growth Strategy and (2) to reveal its strong commitment together with international communities in mitigating climate change, then to mobilize external support in terms of technologies, finance, and capacity building for climate change mitigation.

However, the current policy system still does not properly address private sector and local community involvement in the climate change response. There are no concrete policies to encourage or create incentives for local people and the private sector to participate in climate change adaptation and mitigation. Besides, there are no clear policies and mechanisms for financial investment (both internal and external funding) in climate change response; there is no concrete guidance to integrate climate change in current socioeconomic development plans of the sector; and the climate change action plan of the sector for the period 2011–15 lacks clear identification of priority activities in the context of financial shortage.

The abovementioned polices and implemented activities look forward to providing environmental adaptations and promoting stress tolerance capacity among the people. However, they are not inline with other issues surrounding the same policy arena. The policy fails to acknowledge the fact that climate change poses diverse challenges to different communities, groups, and societies; particularly, male and female issues as mentioned by the Intergovernmental Panel on Climate Change (IPCC)'s fourth assessment report. The psychosocial, economic, and physical impacts experienced are different and so are the responses and recovery procedures. Furthermore, women in a resource-deprived society tend to be the most vulnerable and have a reduced capacity to prepare themselves, and they are the least recognized group by the policy.

In addition, agriculture-dependent rural communities are the most vulnerable to climate change with reduced coping ability as they live in hazardous risk areas. The policy framework further needs to work on a policy specifying the need for information and knowledge targeted for these communities. Studies carried out at a national level are beyond the reach and understanding of these communities and need to be simplified and made accessible. This is necessary because vulnerable communities own fewer assets and have lower income with less access to community services. As such they are reliant on rain-fed agriculture that is prone to hazards. With easy information, these farmers may plan their agricultural practices accordingly while also adding their voices to those of their communities on their needs and priorities in terms of climate change adaptations. The second assessment report of the IPCC also focuses on the need for education and communication to encourage behavior change as part of adaptation. The policies should allow people to recognize and understand the fluctuating and unclear climatic surroundings and involve them efficiently in the overall process of setting adaptation priorities.

Adaptation is a continuous process and farming communities are constantly undertaking several local coping strategies, explicitly changing cropping pattern, rain water harvesting, enterprise diversification, and soil control mechanisms. Within a given policy framework, an attempt should be made to mainstream and prioritize such activities undertaken at the local level. For this purpose, a framework needs to be devised in such a way that climate adaptation is included in national plans that are inclusive, responsive, participatory, and that promote bottom-up planning. The same solution may not apply in all geographical area and at all scales, as the impact of climate change differs. Through the participatory approach of identifying existing vulnerabilities and the coping strategies undertaken, these can be prioritized and integrated into local development plans to create a climate resilient development. With this model of bottom-up planning and the formation of a national level adaptation framework, needs and resources of vulnerable communities and their skills could be mainstreamed into national polices for effective adaptation planning. The bottom-up approach also effectively identifies specific issues and challenges. Learning can also be extracted through other climate vulnerable countries like Nepal (GoN, 2011).

2.2.2 Science and Technology Development for Climate Change Response

There are quite a large number of projects, programs, and practices addressing climate change adaptation and mitigation in the agriculture and rural development sectors. However, so far, there are no official science and technology programs on climate change in the sectors. Current studies still stand at assessing climate change impacts and propose response measures with some pilot models of climate adaptation and mitigation. Production practices, most of them associated with other programs, not climate change, still lack a long-term vision with proper consideration of climate change adaptation and mitigation. There is no assessment of those agriculture practices in terms of climate change adaptation, mitigation effectiveness, or benefit sharing among relevant stakeholders, especially farmers.

Climate adaptation requires a contingency arrangement whereby agrobiodiversity components need to be incorporated that can provide for both long-term and short-term action. Provisions for insurance and irrigation schemes are some good interventions forwarded by the policy, but also important is capacity building at diverse levels. Specific consideration needs to be paid to strengthening traditional practices and knowledge in terms of agrobiodiversity management, which is missing in the available agricultural policies of Vietnam. Participatory research should be considered to identify and document stress-tolerant varieties that can withstand climate extremities. This is important because growing corporate seeds limits the resource poor's attempt to limit climate change impacts as they do not benefit when corporate seed companies use local seeds to discover high-yielding varieties through obtaining intellectual property rights as new seeds. Participatory research equitably benefits both local farmers and breeders in exchange for knowledge and seeds. Access to seed is the basic need for agricultural-dependent communities and strengthening the process of on-farm conservation through home garden diversification, promoting farmer's seed networks, and developing community seed banks needs to be incorporated within the agricultural strategies. Also missing is the research on ruminant animals regarding their biometabolism and manure capture technologies.

Most of the climate change responses consider adaptations and tend to focus less on mitigation potentials. Mitigation strategies in agriculture should not harm yields and be cost-effective. According to Dickie et al. (2014), "approximately 1.8 Gt CO2e annually GHG mitigation is possible at 2030 through effective manure management and reduced methane emissions from paddy plantations and improving fertilizer productions based on the activities possible today." Research on developing antimethane medication and vaccination for livestock and improving traditional nitrogen use efficiency crops substantially reduces emissions. Emission reductions are required to be feasible in theory and in their reach to resource poor farmers supported with strong forest conservation strategies.

The transfer of advanced technologies for climate change mitigation and adaptation is still very limited due to the unique characteristics of agriculture practices in Vietnam with both small-scale and traditional production patterns. Currently, there is no early warning system of climate change impacts to agriculture production. Some production practices address both mitigation and adaptation purposes (such as mangrove and forest development, models of SRI, and so on). Therefore, the incorporation of those purposes in the same models is necessary.

2.2.3 Financial Resources Mobilization

Compared to the objectives set up in the climate change action plan for the period 2011–15, actual achievements are still very limited. Two reasons may help justify this shortage of results: first, the development of the action plan did not involve al of the different stakeholders, which in turn lacked the participation of those stakeholders; secondly, most of the projects under the action plan are focused on developing and rehabilitating an adaptation infrastructure that requires a huge amount of investment while funding for the action plan is very small. This is mainly due to the low capacity of responsible agencies in financial mobilization (both internal and external). To enhance investment and sustain climate change funds role from the private sector becomes important.

Climate change response in agriculture requires consideration of overall agricultural priority arrangements together with understanding their relationship between agriculture and climate change, wherein the socioeconomic impacts are less understood than the biophysical impacts. Together with trade prioritization in agriculture, human and institutional abilities are required to be assisted by sustained funding criteria through investments in agricultural adaptations. Climate change mainstreaming into development plans enables us to tackle climate change impacts on agriculture and receive considerable funding for climate change adaptation.

2.3 RECOMMENDATIONS

A comprehensive policy system on climate change including stipulations and guidance on participation of the private sector and local communities together with mechanisms to create incentives and encourage them to participate in climate change mitigation and adaptation is essential. Besides, a clear policy on financial investment for long-term response to climate change is necessary. The integration of climate change in socioeconomic development plans should be institutionalized and concrete guidance on this integration should be developed. For development of the climate change action plan for the period 2016–20, a broad consultation and a clear list of priority activities should be developed. A science and technology program on climate change in the agriculture and rural development sectors is needed to create a sound foundation for effective response activities. A system to monitor and evaluate the effectiveness and impacts of different climate change measures is useful for long-term climate change response activities. It is necessary to set up an early warning system of climate change impacts on agriculture production and food security. Research to develop a model of climate smart agriculture that covers the three areas of climate change mitigation, adaptation, and economic growth seems suitable for Vietnam to obtain optimum benefits in climate change response.

Equally important is the strengthening capacity of relevant agencies of Vietnam in international negotiations and resources mobilization to implement climate change response activities. Climate change is opening opportunities for the simulation of global, multilateral, and bilateral cooperation through which such developing countries like Vietnam can have access to new mechanisms to receive financial support and technology transfer from developed countries. Climate change response in agriculture and rural development with the goal of ensuring food security and socioeconomic stability has long been a concern and has fostered active cooperation by many international organizations and partners.

As for the upcoming activities for climate change responses for the period 2016–20, capital investments in disaster prevention, mitigation, and climate adaptation requires focus. This should be mainstreamed into national and local development plans and strategies. Development of a warning system to reduce the risk of losing valuable crops is important while looking for ways to conserve traditional seed systems helps to adapt to climate change. This also increases farmers' self-reliance over planting materials while also entailing resistance over drought and salinity, as these varieties have been sustained for years through the process of continuous adaptation and acclimatization.

References

Dasgupta, S., Laplante, B., Meisner, C., Wheeler, D., Yan, J., 2007. The Impact of Sea Level Rise on Developing Countries: A Comparative Analysis. Policy Research Working Paper 4136, World Bank, Washington, DC.

Department of Crop Production, 2013a. Achievement in response to climate change in crop production sub-sector. Report for period 2008–2013, Department of Crop Production, Vietnam.

Department of Crop Production, 2013b. Reviewing the 2012 Rice Production, the 2013 Work plan and Promoting "Large-Scale Farming" Modality. Department of Crop Production, Vietnam.

Department of Livestock Production, 2013. Achievement in Response to Climate Change in Livestock Production Sub-sector Period 2008–2013. Department of Livestock Production, Vietnam.

Dickie, A., Streck, C., Roe, S., Zurek, M., Haupt, F., Dolginow, A., 2014. Strategies for mitigating climate change in agriculture: abridged Report. Climate Focus and California Environmental Associates, prepared with the support of the Climate and Land Use Alliance. Report and supplementary materials available at: www.agriculturalmitigation.org.

Directorate of Forestry, 2013. Achievement in Response to Climate Change in Forestry Sub-sector Period 2008–2013. Directorate of Forestry, Vietnam.

GoN, 2011. National Framework on Local Adaptation Plans for Action. Government of Nepal, Ministry of Environment, Singhdurbar.

IPCC, 2012. Managing the Risks of Extreme Events and Disasters to Advance Climate Change Adaptation. Cambridge University Press, New York.

MARD, 2009. Climate Change Adaptation Plan in Agriculture and Rural Development Sector of Vietnam. Ministry of Agriculture and Rural Development of Vietnam.

MARD, 2013. The Plan of Restructuring Agriculture and Rural Development Sector Towards Greater Added Value and Sustainable Development. Ministry of Agriculture and Rural Development, Vietnam.

MONRE (Ministry of Natural Resources and Environment), 2012. Climate Change, Sea Level Rise Scenarios for Vietnam. Ministry of Natural Resources and Environment, Hanoi. Retrieved from, http://www.preventionweb.net/files/11348_ClimateChangeSea LevelScenariosforVi.pdf.

Nguyen Viet Nam, et al., 2011. Impacts of climate change in Fishery and Aquaculture sector in Vietnam. In: Climate Change Impacts on Agriculture and Rural Development Sector in Vietnam and Response Measures. The Agricultural Publishing House, Hanoi, p. 155.

Tran Van The, et al., 2011. Impacts of climate change in Crop Production sub-sector in Vietnam. In: Climate Change Impacts on Agriculture and Rural Development Sector in Vietnam and Response Measures. The Agricultural Publishing House, Hanoi, p. 53.

Vu Tan Phuong, et al., 2011. Impacts of climate change in Forestry sub-sector in Vietnam. In: Climate Change Impacts on Agriculture and Rural Development Sector in Vietnam and Response Measures. The Agricultural Publishing House, Hanoi, p. 105.

CHAPTER

3

Assessing and Calculating a Climate Change Vulnerability Index for Agriculture Production in the Red River Delta, Vietnam

H.H. Duong, T. Thuc†, L. Ribbe‡*

*Vietnam Academy for Water Resources, Hanoi, Vietnam †Vietnam Institute of Meteorology, Hydrology and Climate Change, Hanoi, Vietnam ‡Cologne University, Cologne, Germany

3.1 BACKGROUND

Vietnam has been considered to be one of the countries that suffer from climate change. Agricultural production was assessed to be the most vulnerable sector to climate change impacts (Vietnam Ministry of Natural Resources and Environment, 2011). The impact of climate change on agricultural production and development in Vietnam includes (1) strong influence on agricultural land use; (2) changing the appropriateness of agricultural production with climate pattern; (3) natural disasters have more impacts on agriculture and tend to increase in frequency and intensity in the climate change context (Thang et al., 2010).

Although many investigations on the assessment of climate change impacts have been carried out (Duong, 2009), they have not much focused on vulnerability assessment, while this is very essential because it not only provides the basic information for proposing mitigation and adaptation measures but also is a foundation for establishing strategies, polices, and plans for a region, nation, community, or specific sector.

Currently, diverse methods and frameworks for assessing vulnerability to climate change have been developed and applied by different climate research institutes/organizations/agencies throughout the world (Duong, 2009). In Vietnam, however, the vulnerability assessment method is predominantly based on risk assessment and/or adopted from other countries; there is no unification of methods for vulnerability index calculation. Moreover, the

Redefining Diversity and Dynamics of Natural Resources Management in Asia Volume 2
http://dx.doi.org/10.1016/B978-0-12-805453-6.00003-6

research on vulnerability index in Vietnam has not focused much on the weight of the components that comprise the vulnerability index; if anything, it has been calculated as having equal weight as expert judgment weight.

It could be realized that climate change vulnerability assessment, in general, is still in its beginning stage and occupied with multiple challenges. The existing vulnerability assessment methods are not uniform in terms of concept, larger in assessment scale, and not paid much attention at the community level. There are many tools that support assessment and calculation of a vulnerability index that are applied all over the world; however, is difficult to be applied in Vietnam due to diverse complications related to supplementary use of a geographical information system (GIS), the web, or computer graphics. Moreover, these tools are applied mainly for vulnerability assessment at a large scale, such as at the region and country level with equal vulnerability component weights. The results are therefore not accurate as the vulnerability component weights differ significantly. Therefore, constructing a method and a supporting tool that are easy to understand, use, and visualize results in the possibility of application at the commune/district/province levels and is very necessary and urgent in Vietnam.

3.2 METHODOLOGY

3.2.1 The Concept of Vulnerability

The concept of vulnerability originated from the studies on natural disaster or food security and still is a controversial concept (Vincent, 2004). Therefore, the concept of vulnerability has been applied in different ways by different organizations such as the Intergovernmental Panel on Climate Change (IPCC), Organization for Economic Cooperation and Development (OECD), United Nations Development Program (UNDP), the Red Cross, and others. However, in terms of climate change research, the most popular concept was given by IPCC (2001): "... Vulnerability is a function of the character, magnitude, and rate of climate change and variation to which a system is exposed, its sensitivity, and its adaptive capacity"

Thus, vulnerability can be formulated as a function as

$$V = f(E,S,AC)$$

In the IPCC report, exposure (E) is defined as "the nature and degree to which a system is exposed to significant climatic variations"; sensitivity (S) is defined as "the degree to which a system is affected, either adversely or beneficially, by climate-related stimuli"; and adaptive capacity (AC) is defined as "the ability of a system to adjust to climate change (including climate variability and extremes), to moderate the potential damage from it, to take advantage of its opportunities, or to cope with its consequences."

3.2.2 Problem-Solving Diagram

Based on the vulnerability concept given by the IPCC, and diverse literature throughout Vietnam and all over the world, a vulnerability assessment framework that comprises an actual field situation is presented in Fig. 3.1.

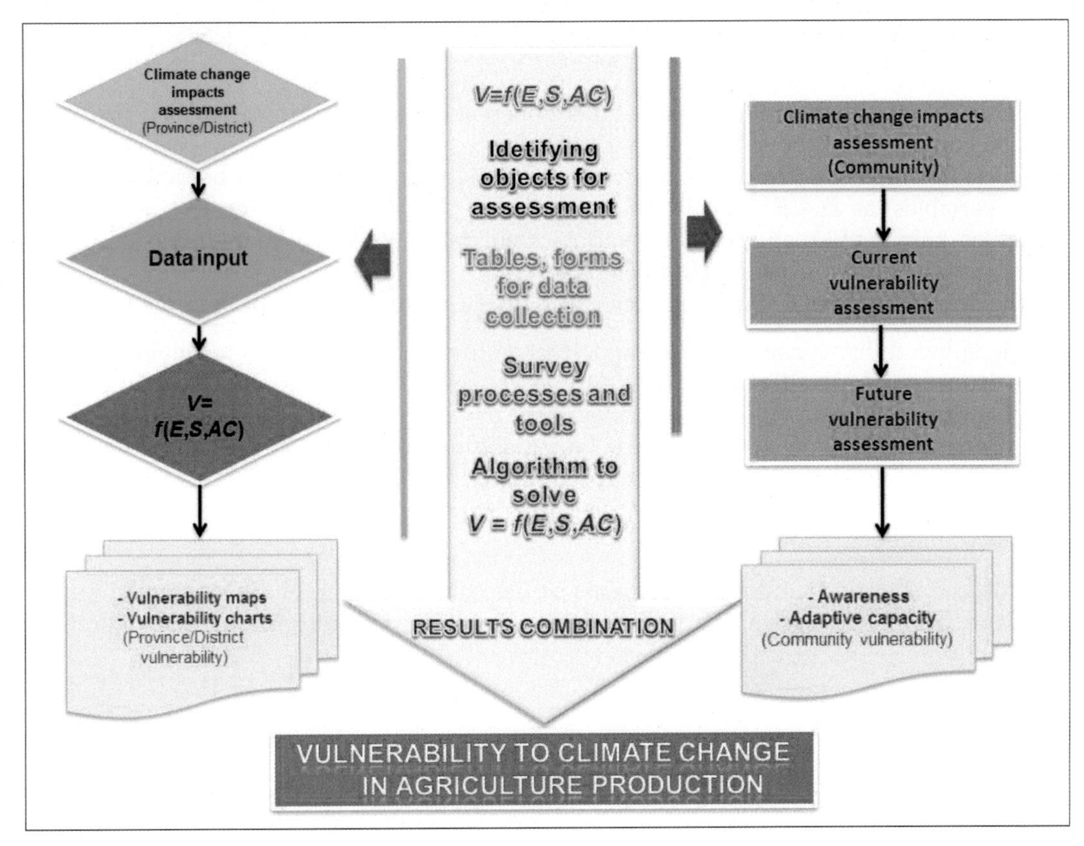

FIG. 3.1 Problem-solving diagram for the construction of a framework to assess vulnerability to climate change impacts on agriculture production.

According to Fig. 3.1, the proposed framework to assess climate change vulnerabilities on agriculture includes two main components:

- *Vulnerability assessment at the community level (commune/village):* This is to identify the vulnerabilities at the community level and then propose adaptation measures to climate change impacts and sea level rise. Community awareness and capacity on climate change understanding will be improved through surveys, assessments, and discussions. The main tools and methods used for the assessment at the community level are probabilistic risk assessment (PRA) tools, questionnaires, field trips, public hearings, and group discussions.
- *Vulnerability assessment at the district/province/region level:* This is to identify the vulnerability of different communes within a district/province/region and then determine the general vulnerability for that district/province/region. Different vulnerability levels will be determined by the construction of correlative vulnerability indexes and the results will be displayed by a vulnerability map for each area. The main support tools for assessment and calculation are climate vulnerability assessment support software (CVASS) and PRA tools, questionnaires, and field trips.

3.2.3 Theoretical Basis for Developing Climate Vulnerability Assessment Support Software

According to the theoretical basis as mentioned above, each main variable (E, S, and AC) will be comprised of the subvariables $E_1 \div E_n$, $S_1 \div S_n$, and $AC_1 \div AC_n$, respectively, and each subvariable *could* be comprise of different correlative components (can be defined as indicator variables) as $E_{11} \div E_{1n}$, $E_{n1} \div E_{nn}$, $S_{11} \div S_{1n}$, $S_{n1} \div S_{nn}$, và $AC_{11} \div AC_{1n}$, and $AC_{n1} \div AC_{nn}$. Main variables, subvariables and indicator variables will be outlined as shown in Fig. 3.2 and the calculation processes as presented in Fig. 3.4.

The development of CVASS has been based on the following concepts and methods.

- The vulnerability concept of IPCC and, thus, vulnerability includes three main variables as exposure (E), sensitivity (S), and adaptive capacity (AC).
- Quantitative climate vulnerability assessment by using the index construction method.
- Experience of climate change impacts on agriculture production in Vietnam.
- Availability and feasibility of required input data.
- Using the programming language C++ and integrating ArcGIS 9.1 for development of vulnerability maps and charts.

With the objectives and basic functions of the climate vulnerability assessment support software mentioned above, the software includes three modules: input, calculation, and display.

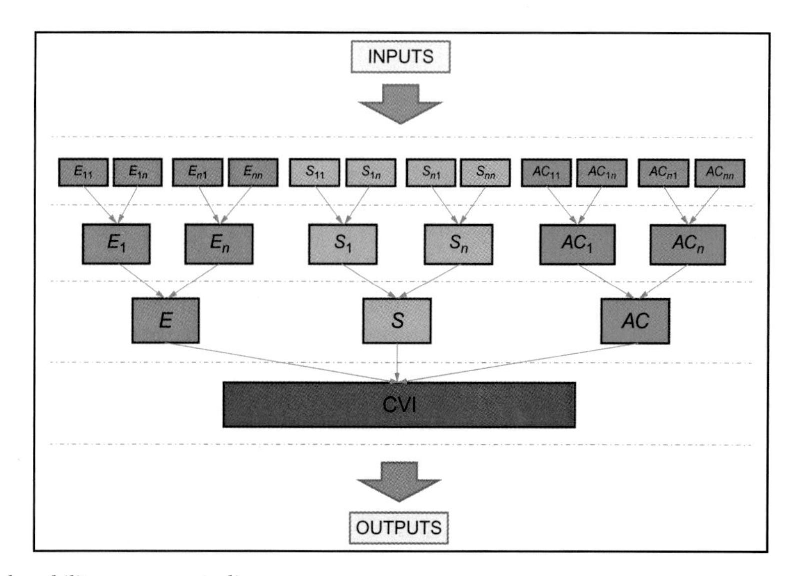

FIG. 3.2 Vulnerability components diagram.

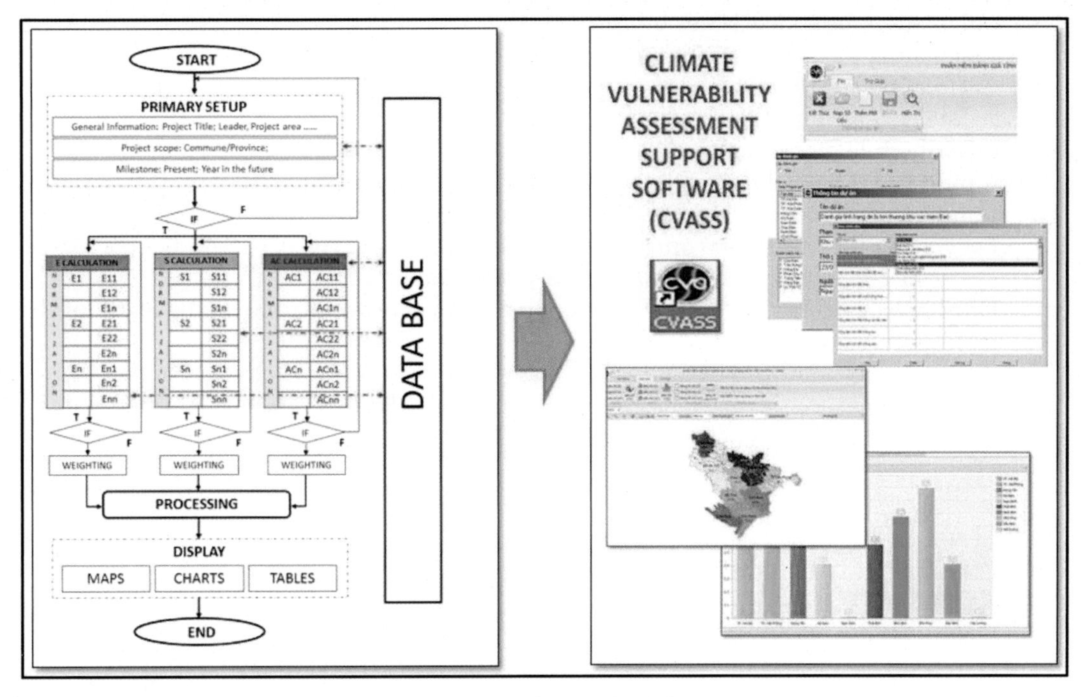

FIG. 3.3 Climate vulnerability assessment support software (CVASS).

CVASS includes three main modules: a data input module, a calculation module, and a display module. This software is best used by researchers and project staffs who are working with climate change vulnerability issues (Fig. 3.3).

- Input module

This module is allowed to provide the input data for calculating the components; hence, it estimates the sub- and main variables of *E*, *S*, *AC*, and CVI. The data used could be extracted from a built-in database and/or by direct inputting on the software interface. The input module basically consists of the following functions:

  - Entering data and general information on the project's name, scale, extent, and assessment time.
  - Displaying the data-entering interface to estimate the component values with two options as (1) call data from the built-in database and (2) direct inputting.
  - Each data will have its own data-inputting section.
  - Displaying error when wrong/incorrect/missing input.

- Calculation module

To calculate indexes of exposure (*E*), sensitivity (*S*), adaptation capacity (*AC*), and climate change vulnerability index (CVI), the following formulas will be used:

  - Normalization (Eq. 1);
  - Weight calculation (Eq. 2);

- Calculating the subvariables index (Eq. 3);
- Calculating the main variables index (Eq. 4);
- Calculating CVI (Eq. 5).

(See details of formulas in Fig. 3.4.)

- Display module

This module is allowed to display the output calculated in many forms such as map, table, graph, and/or the assessment result. The display module consists of the following functions:

- Displaying the calculation outputs of each index E, S and AC of the study area for every specific time in the forms of table, graph and map;
- Displaying calculated CVI of the study area for every specific time in the forms of table, graph and map;
- Illustrating comparison of CVI or E, S and AC under the forms of maps of the study area at every specific time, and presenting the degree difference through the color in each sub-area and among sub-areas;

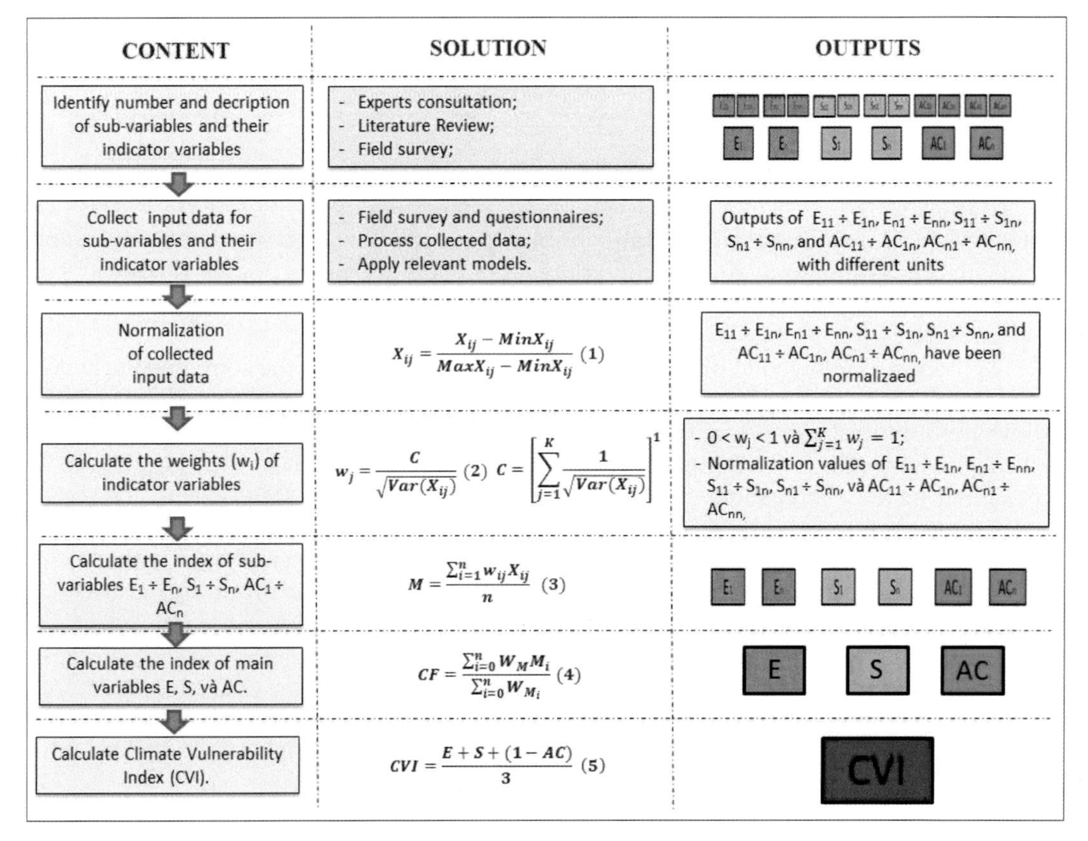

FIG. 3.4 Vulnerability calculation process.

- Displaying the detailed information of the study area following each scale, extent and time, which originally set up;
- Zooming in, zooming out and panning the maps; and exporting the maps in the format of JPEG, JPG, bitmap, etc.

Based on the contents and equations mentioned in Fig. 3.4, the process is to identify and calculate a climate vulnerability index as well as the indexes of exposure, sensitivity, and adaptive capacity, and is done as follows:

- The first and most important step is to identify the number of indicator variables ($E_{11} \div E_{1n}$, $E_{n1} \div E_{nn}$, $S_{11} \div S_{1n}$, $S_{n1} \div S_{nn}$, và $AC_{11} \div AC_{1n}$, $AC_{n1} \div AC_{nn}$). In theory, a greater number of indicator variables will make the results calculation more accurate. This step will be carried out through a literature review and expert consultation.
- It is very easy to recognize that identified and collected indicator variables will have different units; therefore, to apply the index method they will be normalized and the methodology used in UNDP's Human Development Index (HDI) (UNDP, 2006) is followed to normalize them (Eq. 1).
- Currently, there are three main methods to calculate weights: (1) method with equal weights; (2) method with unequal weights, which is divided into two types (expert judgment, and the Iyengar and Sudarshan method (Duong, 2009)); and (3) multivariate statistical techniques. Within the scale of this chapter, availability and feasibility of required input data, and conditions of case study provinces, Iyengar and Sudarshan's method has been applied to calculate weights. (Eq. 2).
- As mentioned above, each main variable (*E*, *S*, and *AC*) will have many subvariables ($E_1 \div E_n$, $S_1 \div S_n$, and $AC_1 \div AC_n$, respectively) and these subvariables will be calculated using Eq. (3) with weights calculated by the previous step.
- After having the subvariables, the main variables (*E*, *S*, and *AC*) will be calculated using Eq. (4).
- Finally, Eq. (5) is used to calculate the CVI.

3.3 RESULTS

3.3.1 Results of Constructing the Vulnerability Assessment Framework on Agriculture Production

The detailed content of each step follows (Fig. 3.5).

- Step 1: Preparation
  - Collecting the secondary data and documents.
  - Defining the assessment scope.
  - Selecting the climate change scenarios (including sea level rising).
  - Screening the main climate change impacts.
- Step 2: Field assessment

The approach focuses on two levels: the community (commune) level and the district/province/region level. Thus, the content of Step 2 is as follows:

- Assessing the vulnerability at the community level.
- Collecting the input data of the vulnerable components (Tables 3.1–3.3).

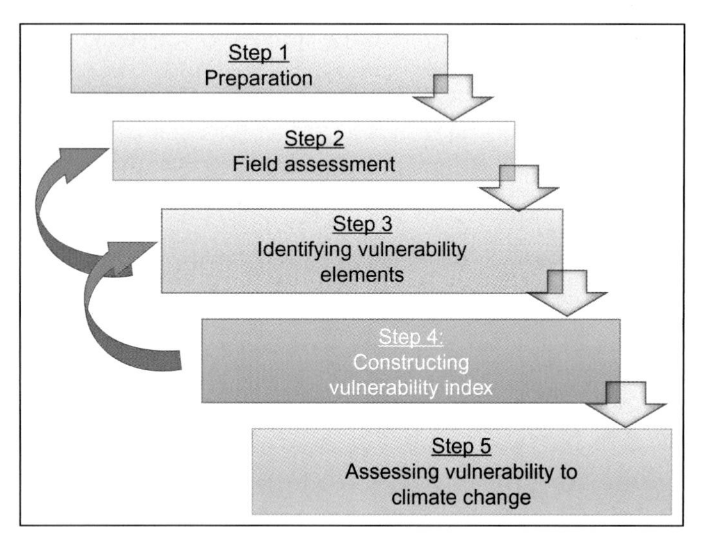

FIG. 3.5 The steps of the framework for assessing vulnerability to climate change.

TABLE 3.1 The Required Input Components to Calculate the Exposure Index (E)

Component	Subcomponent	Unit	Data Source
Climatic extreme event (E_1)	Mean annual number of storms (E_{11})	Event	TK
	Mean annual number of floods (E_{12})	Event	TK
	Mean annual number of droughts (E_{13})	Event	TK
Damages by the hazards (E_2)	Agricultural land affected by the storms (E_{21})	ha	TK
	Agricultural land affected by the floods (E_{22})	ha	TK/TT
	Agricultural land affected by the droughts (E_{23})	ha	TK/TT
Changes in climatic variables (E_3)	Mean maximum monthly rainfall (E_{31})	mm	TK
	Mean maximum monthly rainfall (E_{32})	mm	TK
	Mean daily maximum temperature (E_{33})	T°	TK
	Mean daily minimum temperature (E_{34})	T°	TK
Sea level rising (E_4)	Agricultural land affected by the saline intrusion (E_{41})	ha	TK/TT
	Salinity (E_{42})	$^0/_{00}$	TK/TT

- Step 3: Defining the vulnerability factors
  - Defining the vulnerability factor means determining the exposure (impact degree) (E), sensitivity, and adaptation capacity (AC) according to the IPCC definition.
  - Factors E, S, and AC are determined through the data collected at Step 2; if not available, it needs to come back to Step 3 for collecting more.

TABLE 3.2 The Required Input Components to Calculate the Sensitivity Index (S)

Component	Subcomponent	Unit	Data Source
Land (S_1)	Land used for agriculture production (S_{11})	ha	TK
	Irrigated agriculture land (S_{12})	ha	TK
	Agricultural land that converted to the other types (S_{13})	ha	TK
	Degraded agricultural land (S_{14})	ha	TK
Water sources (S_2)	Water demand for agricultural production (S_{21})	m^3	TK/TT
	Water-use efficiency for agricultural production (S_{22})	%	TK/TT
Labor (S_3)	Total number of agricultural households (S_{31})	HH	TK
	Agricultural household density (S_{32})		TK
	Total number of agricultural households transformed to other production (S_{33})	HH	TK
	Total number of poor and marginally poor households (S_{34})	HH	TK
Income (S_4)	Total income from agricultural production (S_{41})	VND	TK/TT
	Total income from other sources (S_{42})	VND	TK/TT
	Number of households only living on agricultural production (S_{43})	VND	TK/TT

TABLE 3.3 The Required Input Components to Calculate the Adaptive Capacity Index (AC)

Component	Subcomponent	Unit	Data Source
Infrastructure (AC_1)	Rate of modernized irrigation and drainage system (AC_{11})	%	TK
	Rate of upgraded rural road system (AC_{12})	%	TK
	Rate of upgraded on-farm road system (AC_{13})	%	TK
	Rate of mechanization in agriculture (AC_{14})	%	TK
	Rate of state-grid electricity use (AC_{15})	%	TK
Economy (AC_2)	GDP (AC_{21})		TK
	Agricultural investment (AC_{22})	VND	TK
	Budget to recover agriculture after hazard/disaster (AC_{23})	VND	TK
	Agricultural climatic insurance (AC_{24})	VND	TK/TT
Society (AC_3)	Rate of working-age population (AC_{31})	%	TK
	Rate of university-degree population (AC_{32})	%	TK
	Number of agencies related to agriculture (AC_{33})	TC	TK
	Human Development Index (AC_{34})		TK/TT

Remark: TK = Statistic; TT = Estimate.

- Step 4: Constructing the vulnerability index
  - Process and analyze the data collected at Step 2.
  - Based on the estimate of E, S, and AC at Step 3, construct the vulnerability index.
  - Construct vulnerability maps/graphs.
- Step 5: Assessing the vulnerability
  - The assessment is based on the output of Step 2 and Step 4.
  - Propose measures to respond to climate change impacts.

3.4 RESULTS OF PILOT ASSESSMENT IN SOME PROVINCES IN THE RED RIVER DELTA, VIETNAM

Due to climate change impacts and extreme climatic events, the Red River Delta is divided into two subareas as follows:

- Area 1: Suffered from the impacts of climate change and sea level rise, including the Nam Dinh, Hai Phong, Ninh Binh, and Thai Binh provinces.
- Area 2: Suffered from the impacts of climate change onlym including the Hai Duong, Vinh Phuc, Bac Ninh, and Hanoi city provinces.

Based on the natural conditions of each subarea, the impacts of climate change and sea level rise in each province/city as well as the availability and feasibility of required input data, the following provinces were selected for pilot assessment using a developed framework and tool:

- Area 1: Nam Dinh Province (communes of Giao Xuan and Giao Lac, Xuan Thuy District), and Hai Phong City (Tan Trao Commune, Kien Thuy District).
- Area 2: Ha Nam Province (Lien Son Commune, Kim Bang District) and Hai Duong City (communes of Tien Tien and Phuong Hoang, Thanh Ha District).

3.5 DISCUSSION ON RESULTS

3.5.1 The Effectiveness of a Developed Framework

To estimate the effectiveness of a developed framework, a set of questionnaires has been developed with its content to estimate the effectiveness of each part of a developed framework. This will be verified through data collection, discussion content and method, and organizing assessment. These questionnaires were delivered to all members participating in the assessment including staff, local partners, and officials. The estimation result is synthesized as follows:

Results of weight calculation (Table 3.4)

TABLE 3.4 Results of Weight Calculation

Weight	E_{11}	E_{12}	E_{13}	E_{21}	E_{22}	E_{23}	E_{31}	E_{32}	E_{33}	E_{34}	E_{41}	E_{42}
The weights of indicator variables of exposure index												
Present	0.020	0.026	0.027	0.033	0.026	0.031	0.029	0.026	0.030	0.021	0.023	0.024
2030	0.027	0.027	0.020	0.032	0.027	0.027	0.028	0.027	0.028	0.029	0.023	0.022

Weight	S_{11}	S_{12}	S_{13}	S_{14}	S_{21}	S_{22}	S_{31}	S_{32}	S_{33}	S_{34}	S_{41}	S_{42}	S_{43}
The weights of indicator variables of sensitivity index													
Present	0.032	0.032	0.029	0.029	0.032	0.028	0.028	0.029	0.025	0.029	0.030	0.029	0.031
2030	0.030	0.030	0.029	0.028	0.027	0.027	0.031	0.031	0.026	0.028	0.030	0.028	0.029

Weight	AC_{11}	AC_{12}	AC_{13}	AC_{14}	AC_{15}	AC_{21}	AC_{22}	AC_{31}	AC_{32}	AC_{33}	AC_{34}
The weights of indicator variables of adaptive capacity index											
Present	0.025	0.029	0.025	0.031	0.027	0.032	0.027	0.028	0.025	0.023	0.029
2030	0.030	0.030	0.035	0.026	0.030	0.031	0.025	0.025	0.028	0.027	0.025

Results of constructing vulnerability maps at the provincial level (Fig. 3.6)

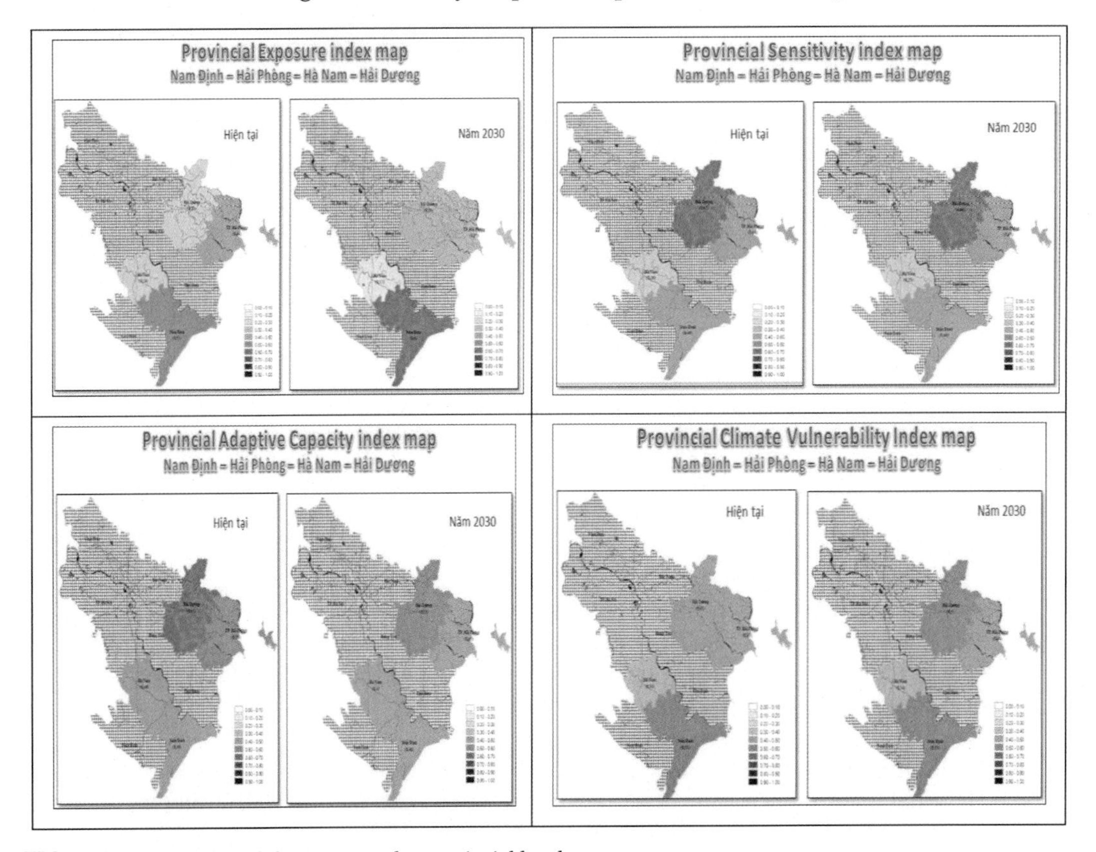

FIG. 3.6 Set of vulnerability maps at the provincial level.

Results of constructing vulnerability charts at the community level (Fig. 3.7)

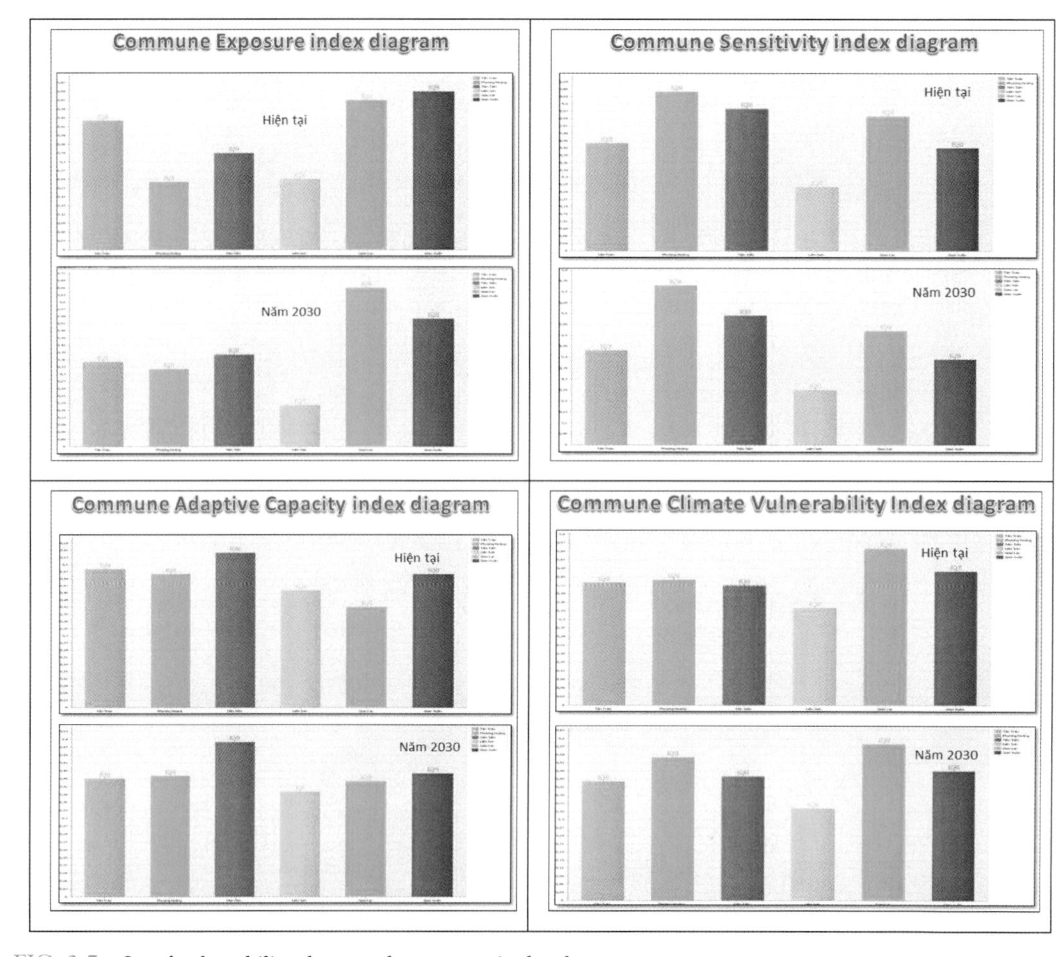

FIG. 3.7 Set of vulnerability charts at the community level.

- Data collection part: 80% of the interviewees estimated that data collection tables were designed logically and reasonably and had enough required data; 20% of the respondents estimated that there were some unnecessary data and the focus should be more on data related to climate change and agriculture.
- Dicussion content: 90% of the interviewees estimated that disussion content was appropriate; 10% of the respondents said this was unappropriate.
- Discussion method: 85% of the interviewees agreed with the proposed topics, 9% disagreed and suggested to add more discussion on present natural disasters, while 6% of the respondents had no comments.
- Approximately 100% of the interviewees estimated that the organizing assessment group was reasonable with necessary preparation. However, capacity of individual members

is not equal in terms of climate change knowledge and they should be be trained before participating in the assessment.
- Approximately, 100% of the interviewees estimated that awareness and knowledge on climate change improved after the assessment, especially for local staff.

3.5.2 The Effectiveness of CVASS

- The interface of CVASS is easy to understand and apply and its flexibility in calculation makes it especially able to add or delete the indicator variables of exposure/sensitivity/adaptive capacity.
- Data input is simple and convenient to use.
- CVASS is able to calculate unequal weights of the indicator variables of exposure/sensitivity/adaptive capacity.
- CVASS is able to calculate the indexes of exposure, sensitivity, adaptive capacity, as well as vulnerability index.
- High visual display results.

3.5.3 The Result of Constructing Vulnerability Maps and Diagrams

- The vulnerabily maps and diagrams clearly presented the most vulnerable communes and provinces.
- The vulnerability maps and diagrams presented impacts of climate change in each pilot province. For example, it can be seen very clearly through the exposure map/diagram that the coastal provinces (Nam Dinh and Hai Phong) suffered more from the impacts of climate change and sea level rise than did the inland provinces (Ha Nam and Hai Duong).
- The exposure, sesitivity, adaptive capacity, and vulnerability map shows that they are complied with the vulnerability concept of the IPCC. This means the vulnerability index depends on all three factors: exposure, sesitivity, and adaptive capacity. Therefore, these maps and diagrams prove that the coastal provinces with much more climate change impacts had high adaptive capacity with a lower vulnerability index than inland provinces with less impacts and low adaptive capacity.
- The results shown in the maps/diagrams are very high in visual information and will be the basis to comprise the vulnerability level among assessed communes/provinces. Based on the vulnerability comparisons, mitigation and adaptive measures, plans, policies, visions, and strategies for agriculture and socioeconomic development can be proposed to reduce the vulnerability to climate change.

3.6 CONCLUSION

It is very clear that most vulnerability assessments applied different methods/frameworks for different climate research organizations/countries. Currently, there is no unified method/framework for assessing vulnerability to climate change in Vietnam, while the vulnerability assessment has not focused on the calculation of the weights for each components of the vulnerability index, especially unequal weights.

This research provides a theoretical and scientific basis for constructing a framework and a tool for general vulnerability assessment and can be applied for agriculture production in particular. The proposed framework is based mainly on the vulnerability concept of the IPCC, along with the method for calculating the vulnerability index. This framework has been piloted in some provinces of the Red River Delta, whereby the input given is through the result obtained through data collection and locality assessment using climate change scenarios for each province. Local policies and strategies to respond to climate change impacts can then be incorporated accordingly. Besides, through the assessment activities, local communities were trained on the basic knowledge and awareness on climate change and vulnerabilites. Moreover, local people were able to propose the measures to responding to the current and future climate change impacts through their indigenous knowledge.

To support the vulnerability assessment, a tool (CVASS) has been developed. The calculation results of CVASS are reliable as it integrates unequal weight calculations. The results of calculating weight for the individual CVI index proved the weight of these components to be different in both theory and in practice. The outputs of CVASS are vulnerability maps and diagrams that highlight diverse vulnerability levels among different pilot provinces.

The results of testing the effectiveness of the framework and tool have proved to have high feasibility in practice; however, some issues should be paid more attention. Field survey and assessment should be carefully planned so that the principles of participatory rapid assessment are complied with. Prior to undertaking field surveys, climate change introductory materials could be handed over to the community to ensure that the measures developed to respond to climate change at the community level is feasible and realistic. The data collected at the community level needs to be cross-checked to ensure data accuracy. The indicator variables of exposure, sensitivity, and adaptive capacity needs to be based on actual field conditions backed up by science. In cases with no statistical data available, relevant supplemental models could be used to calculate indicator variables before using CVASS.

References

IPCC, 2001. Climate Change 2001: Impacts, Adaptation and Vulnerability. A report of working Group II of the Intergovernmental Panel on Climate Change. Intergovernmental Panel on Climate Change.

Ministry of Natural Resources and Environment, 2011. Vietnam National Strategy on Climate Change. Vietnam Publishing House of Natural Resources, Environment and Cartography. Hanoi, December, 2011.

Duong, H.H., 2009. Literature review on the methodology to assess the vulnerabilities caused by climate change. Vietnam J. Water Res. Sci. Tech. 22, 101–104.

Thang, N.V., Hieu, N.T., Thuc, T., Huong, P.T.L., Hieu, N.T., Lan, N.T., Thang, V.V., 2010. Climate Change and Its Impacts in Vietnam. Hanoi Science and Technology Publishing House, Hanoi.

UNDP, 2006. Beyond Scarcity: Power, Poverty and the Global Water Crisis. Human development report. United Nations Development Programme, New York, USA.

Vincent, K., 2004. Creating an Index of Social Vulnerability to Climate Change for Africa. Working paper no. 56. Tyndal Centre for Climate Research.

SECTION III

PAYMENT FROM ECOSYSTEM SERVICES

CHAPTER

4

Cash-Based Versus Water-Based Payment for Environmental Services in the Uplands of Northern Vietnam: Potential Farmers' Participation Using Farm Modeling

D. Jourdain[*,¶], E. Boere[†], M. van den Berg[†], D.D. Quang[‡], C.P. Thanh[§], F. Affholder[¶]*

[*]Asian Institute of Technology, Bangkok, Thailand [†]Wageningen University, Wageningen, The Netherlands [‡]Northern Mountainous Agriculture and Forestry Science Institute (NOMAFSI), Phu To, Vietnam [§]Thainguyen University of Economics and Business Administration, Thainguyen (TUEBA), Vietnam [¶]CIRAD UR SCA, Montpellier, France

4.1 INTRODUCTION

The mountainous and forested areas of northern Vietnam serve as biodiversity reservoirs and have important watershed regulating functions with effects that go well beyond their boundaries. Starting in the late 1990s, land redistribution, market liberalization, and population growth have been driving agricultural area expansion and increasing cropping intensity (Pandey, 2006; Tai, 2006; Folving and Christensen, 2007). The induced land-use changes pose a threat to the provision of environmental services (ES) (eg, Valentin et al., 2008).

Payments for environmental services (PES) is an innovative approach for improving natural resources management (Ferraro and Kiss, 2002; Gómez-Baggethun et al., 2010). PES programs are studied and introduced for diverse issues such as the protection of forested area, the enhancement of watershed protection, the stimulation of carbon sequestration, and for the beautification of landscapes (Landell-Mills and Poras, 2002). Early promoters envisioned that PES programs would transform conservation practices from costly requirements imposed on

Redefining Diversity and Dynamics of Natural Resources Management in Asia
Volume 2
http://dx.doi.org/10.1016/B978-0-12-805453-6.00004-8

landowners to potential sources of revenue conditioned to conservation targets (Kinzig et al., 2011). In particular, when negative externalities are present between well-identified groups, individuals and communities would be financially motivated to engage in mutually beneficial agreements regarding resource management (Wunder, 2005).

This chapter investigates two alternative PES programs tailored to reestablish natural or productive forests in the uplands of northern Vietnam. In the first PES program, farmers are proposed to set aside some of their cultivated sloping land and receive monetary payments for maintaining their newly forested land (Wunder et al., 2005; Hoang et al., 2008; UN-REDD, 2010). It pays a flat rate for all types of farmers and for all types of retired land to minimize transaction and monitoring costs. The rationale for these payments is that farmers who retire agricultural land are providing additional ES and should be remunerated for doing so by those who benefit from the ES (Bulte et al., 2008; Lipper et al., 2008). The government serves as intermediary between the actual users (downstream households) and the suppliers (farm households).

An alternative PES program could be tailored to the specific settings of farmers of these mountainous areas, where an important share of the agricultural land has been privately attributed and farmers cannot expand their agricultural land by deforestation of common lands (in contrast with earlier studies such as Angelsen, 1999; Tachibana et al., 2001). While most PES analyses seem to concentrate on pure rain-fed agriculture (Engel et al., 2008; Quintero et al., 2009), a sizable portion of farmers in the mountainous areas of Vietnam are practicing some form of *composite swiddening*: an agroecosystem that combines upland rotating crop/fallow plots and downstream permanent wet rice fields into a single household resource system (Lam et al., 2004; Nguyen et al., 2008; Vien et al., 2009). This diverse system offers an alternative to the standard financial (Bui and Hong, 2006) or food compensations (Uchida et al., 2005; Gauvin et al., 2010) often considered in PES. The main idea underlying the alternative PES program is to compensate production losses resulting from setting aside land by helping farmers to increase the productivity of their remaining agricultural land. In a context of imperfect markets where farmers are concerned about food security, such a program may be more attractive than the standard PES using financial rewards. Through the construction of terraces and linking these to existing water bodies some sloping lands can be converted to plots that can be irrigated at least during part of the year. The literature indicates that this phenomenon has already occurred elsewhere in response to increased population pressure (Boserup, 1981; Krautkraemer, 1994). Yet, while the possibilities of converting land into terraces have not been exhausted, the costs of linking additional terraces to water bodies are prohibitive for individual farmers. Therefore, the second alternative PES program involves compensating farmers for conversion of agricultural land to forests by bearing the cost of converting part of their sloping rain-fed land into terraces with access to irrigation.

The second alternative PES program fulfills the three requirements laid out by Pagiola et al. (2005) on targeting poor farmers. First, poor farmers generally cultivate land with low agricultural production but high potential for ES provision, and are therefore in the "right place" to participate. Second, our program allows farmers to increase (food) production on their remaining land and thus directly compensates them for production losses, which makes it less biased toward large farmers. Third, despite large differences in the size of farms, all farmers in the study area have long-term cultivation rights for the land with potential for ES provision and are therefore able to participate (Bulte et al., 2008). In contrast, set-aside/cash-based PES programs are mostly attractive to larger landholders who have a lower opportunity cost of land and are more likely to be willing and able to participate. For small landholders, setting aside land reduces the already scarce land available for food production and thus increases their food insecurity and financial instability (eg, Jourdain et al., 2009).

Our main objective is to compare the cash-based with the irrigated-terrace programs in terms of target population: Which farmers are most likely to be able and willing to participate in the two programs and what are the consequences of their participation on their revenues?

Because the second program is hypothetical, we used a simulation approach. Using mathematical programming, we developed a set of farm models corresponding to typical farms of a mountainous district of northern Vietnam where currently no PES is in operation. We simulated the level of participation of different types of farms in the two types of PES programs. For each specific PES, we analyzed participation, measured by the area of land converted into forest land, the cost of the program, and its impacts on land use and household revenues, at the individual farm and village level. The results suggest that the cash-based program would be biased against the smallest landholders of the region, while the irrigated-terrace program could create a win-win situation where the smallest and poorest farmers of the area would benefit most.

4.2 STUDY AREA

The study was carried out in the mountainous zone of the district of Van Chan (Yen Bai Province, northwestern region of Vietnam). Seventy-six percent of the district land has slopes greater than 15% and agriculture is found at altitudes ranging from 200 to 1000 masl. The climate is driven by the monsoon regime. The winter season, from January to March, is cold and humid and characterized by low rainfall and persistent drizzle. The summer season, from May to October, is hot and rainy with 80% of yearly rain falling in just 5 months. The two short interseasons, April and November, are characterized by low rainfall, strong sunshine, and low humidity.

Households' current access to land and water is the result of the last redistribution of land formerly managed by agricultural cooperatives to individual households. However, as in other provinces, this process has not been carried out consistently (Castella and Quang, 2002; Chính, 2008; Sikor and Müller, 2009). Although the general idea was to maintain equity among the former members of cooperatives, various redistribution rules have been implemented leading to an important differentiation of farming systems (Jourdain et al., 2011b): a significant portion of the households has poor access to irrigated lowlands and has to rely on the more fragile and less productive sloping lands to produce the food they need; for these households, this results in difficulties to diversify their livelihood and increased poverty (Jourdain et al., 2011a). The most important crops cultivated in the region are irrigated rice on the irrigated lowland, and upland rice, tea, cinnamon, cardamom, maize, and cassava on the dry uplands. Tea, cinnamon, and cardamom are the main cash crops. Other crops are staple crops that are sold on markets only by households able to produce more than their consumption. In the bottom valleys, irrigated rice can be cultivated during two distinct seasons: the summer season and, when irrigation water is sufficient, the winter season. Two successive upland crops may be cultivated between March and November.

Farming households of the study area face major market constraints. Opportunities to engage in local off-farm work are limited and the ethnicity of the inhabitants increases the transaction costs and limits the opportunities to find jobs outside of their villages due to language barriers. Market restrictions are not limited to the labor market; for all crops besides the main staple rice crop, markets have not been fully established. A good example of this is the market for ginger, which is quickly satiated resulting in highly fluctuating prices. As a result, risk-averse farmers devote only a limited amount of their resources to the cultivation of ginger. Moreover, most farmers have very limited access to formal credit. Lending and borrowing does occur, but either at high interest rates, or among friends and relatives. The combination of limited access to markets and large transaction costs leads most farmers to aim for self-sufficiency in their production.

Three main types of land were identified: bottom valleys, rain-fed sloping lands, and terraces established on the sloping lands. In the bottom valleys and on the terraces, plots were differentiated by their access to water during the two cropping seasons. On plots that can be irrigated during both cropping seasons, two irrigated crops can be cultivated each year. Because water flows are very low during the winter season, some plots are only irrigated during the summer season. The rain-fed sloping zone is divided into areas where upland annual crops may be cultivated between March and November, and areas dedicated to perennial crops.

4.3 DATA AND DESCRIPTIVE STATISTICS

Data were collected using two separate rounds of field work in the Van Chan District. First, a rapid farm household survey was conducted to understand the organization of agricultural activities and to identify types of farm households based on their resource endowments (Jourdain et al., 2011a). A typology was developed to differentiate farm groups according to household size, labor force, and the cultivated areas per land type. Six types of farms were identified (Jourdain et al., 2011a):

- Water scarce land scarce (WSLS) farms are characterized by a small landholding, no access to irrigation water, and low human capital. Minimum food requirements can hardly be met by household production, even if they cultivate almost all of their land with staple crops. As labor is limited and mobilized for the production of upland rice, off-farm activities are limited. However, wage employment is an important source of monetary income of the household.
- Water scarce land rich (WSLR) farms have a very low amount of irrigated lowland and have not constructed terraces. However, they have large areas of rain-fed sloping land. As a result, they generate a marketable surplus: large sloping areas offset their lack of access to water.
- Off-farm work (OFFW) farms involve a small group of households endowed with a large workforce and relatively good access to off-farm markets. They mostly have access to irrigation water only during the summer season and therefore have a low amount of irrigated lowland per head. Because they do not possess enough sloping land to compensate for the lack of access to water, they have largely turned toward off-farm work.

- Terraces and labor rich (TERLAB) farms are managed by large households who do not have access to paddies. Available labor and sloping land permitting, they converted large areas of sloping land into terraces. However, only a third of these terraces receive water during winter. In addition, a large number of workers also spend a significant amount of time on tea production and off-farm work.
- Paddy rich (PARI) farms cultivate large areas of paddy land, which receives a lot of water during both seasons, allowing for a sizeable rice production. Staple and perennial crops are cultivated on the slopes. No terraces were constructed in the sloping zones.
- Terraces and upland (TERUPL) farms cultivate large areas of uplands. They have also converted large parts of this upland into terraces that have access to water throughout the year. While attaining food self-sufficiency, these households can diversify their production with cash crops, either annual crops such as maize or cassava or perennial crops like tea.

A second round of interviews was then conducted to gather the particulars of crop technology and market access through in-depth interviews with farmers. For the interviews, we selected villages within the district of Van Chan that were contrasting in terms of access to markets and availability of irrigation (Table 4.1). Within these villages, we selected 45 farm households. We used a stratified random procedure taking into account poverty status and amount of paddy and terrace land available to ascertain sufficient diversity in the sample. Between September and December 2009, we interviewed the households using a semistructured questionnaire including questions about household characteristics, crop activities, assets, off-farm and nonfarm activities, food balances, and crop cultivation. Many questions involved the details of crop production during the agricultural year 2008–9: crops cultivated, activities undertaken, inputs used, and yields obtained on a field basis. To cover all these aspects, three interviews with each farm household were conducted, which also gave us the chance to check with farmers' data that appeared unclear or incorrect.

Based on the questions on crop production we constructed a cropping calendar consisting of the main crops and their labor activities (Fig. 4.1). The most important crops observed were irrigated rice (one and two cycles) on paddy and terrace land; upland rice, maize, peanuts, cassava, soybeans, sweet potatoes, and ginger on sloping land; and tea, cardamom, and cinnamon on perennial land.

TABLE 4.1 Characterization of Selected Villages

Village (Commune)	Irrigation Availability	Proximity to Main District
Pang Cang (Suoi Giang)	No bottom valley, few terraces (SU)	Near
Giang Cay (Nam Lanh)	Large bottom valley (WIN+SU), few terraces (SU)	Far
Ban Tun (Tu Le)	Small bottom valley (WIN+SU), large terraced area (WIN+SU)	Near
Nam Chau (Nam Bung)	Small bottom valley (WIN+SU), few terraces (SU)	Far

SU, water available during summer only; *WIN+SU*, water available during winter and summer seasons.

Land type	Crop	Jan	Feb	Mar	Apr	May	Jun	Jul	Aug	Sep	Oct	Nov	Dec
		T1	T2	T3	T4		T5 / T6	T6	T7				T1
Irrigated Lowland	Irrigated rice Two cycles	PL	TR	WE		HA							PL
							PL TR	WE			HA		
	Irrigated rice One cycle			PL		PL	TR	WE			HA		
Upland	Upland rice					PL SE			WE	WE	HA		
	Maize Two cycles			PL	SE			WE HA					
									PL SE			WE HA	
	Maize - Peanuts			PL	SE			WE HA					
									PL SE		WE	HA	
	Cassava	HA	HA	SE	HA	HA		HA		HA	HA		HA
	Soybean Two cycles			PL SE		WE HA							
								PL SE		WE HA			
	Sweet Potato			PL SE		WE	HA						
	Ginger			PL	PL SE		WE		WE			HA	
Perennial	Tea			WE	HA		HA	HA		HA			
	Cardamom				WE			WE			HA		
	Cinnamon			WE	HA							HA	

FIG. 4.1 Cropping calendar of the main crops and their labor requirements per type of land. *PL*, planting; *TR*, transplanting; *SE*, seeding; *WE*, weeding; *HA*, harvest.

4.4 METHODOLOGY

We used a farm household modeling (FHM) approach that explicitly models the objectives, resources, possible activities, and the socioeconomic environment of the farm household and has been used extensively over the past decades to assess the effects of policy measures on farm households (Kruseman and Bade, 1998; Van den Berg et al., 2007; Laborte et al., 2009).

4.4.1 Farm Household Model

4.4.1.1 General Structure

The whole farm model is built using mathematical programming (Hazell and Norton, 1986) and developed on a general algebraic modeling system (GAMS) platform (Rosenthal, 2007). To account for the diversity of farming systems in the area, we developed six versions of the model: one for each of the farm types described above. While the overall model structure and available crop technologies are the same for each farm type, the models differ in the parameters specifying household members, labor, land differentiated by access to water during the two growing seasons, intensity of crop cultivation, and initial capital endowment.

The models were designed to reproduce the behavior of farm households selecting from a set of cropping systems and off-farm activities to fulfill their objectives within the boundaries of their constraints. Livestock production was not included, as it was only of limited importance for the surveyed farmers and unlikely to substantially affect land allocation decisions. Modeled households' land resources were considered fixed, as land transactions, either sale or rent, were not observed during the surveys. Farmers were assumed to maximize discretionary income; that is, their income after satisfying basic food requirements. Constraints relate to crop technology, the availability of resources—land, water, labor, and cash—and market access. The simulation horizon for the household model was one year based on the data of the cropping year 2008–9.

Basic food requirements were expressed as a minimum amount of rice to be consumed daily, with a possible substitution of maize and cassava. Maize is a second choice in terms of staple food, and farm households were assumed to consume maize only when they cannot produce or buy sufficient rice for household consumption. Households can sell and buy farm products, but there are shallow market opportunities. This leads to the inclusion of a price band between the farm-gate price and the consumer (retail) price of the crops produced. This price band, expressed in percentages of market price, was used as a calibration parameter. Shallow market opportunities will induce farm households to give priority to production for home consumption and try to reach food self-sufficiency before aiming to produce for the market (de Janvry et al., 1991).

We defined 25 cropping systems based on the main combinations of crops observed during the survey periods. We defined 14 single-crop systems, 7 double-crop systems of which four are not rice-based, and 4 rotation systems of which one is not rice-based (Table 4.2). While in the real world the rotation spans over several years, we modeled these as annualized composite cropping systems. As irrigated crops are cultivated with various input intensities, we defined cropping systems with low, medium, and high input use for the lowlands. Similarly, upland maize can be cultivated with low or high input intensity. The "high input" labels should be interpreted with care as we intended to differentiate the mountainous cropping systems on a local scale of intensity. "High input" cropping systems of the mountainous areas use much lower amounts of chemical inputs than the intensive cropping systems found in the Red River and Mekong Delta areas of Vietnam.

The cropping systems were defined by a set of technical coefficients including yields and input requirements. As crop cultivation is highly seasonal, we defined seven time periods based on the main labor patterns in crop cultivation identified in the second round of surveys and calculated labor and other input requirements for each time period. The model describes which cropping systems can be established on what type of land and puts a constraint on the amount per type of land.

As the opportunities to sell cash crops are often restricted, these crops are allowed a restricted percentage of land to be cultivated based on survey observations.

In each time period, the household is assumed to allocate its available labor between the different cropping activities and off-farm opportunities. If family labor is not sufficient to fulfill all tasks needed on the farm, additional labor can be contracted. In the case of a surplus of family labor, family members can search for a job off-farm. However, we assumed that there were transaction costs for hiring and selling labor, representing the expenses of finding off-farm work, and making the transition to the job.

Like labor, cash is balanced for each period in the model. Cash is generated through sales of crop products and engagement in off-farm employment. Expenses arise from minimum cash living expenses, the purchase of staple food crops, inputs for crops cultivated, and wages paid for hired labor. Households can borrow cash but are facing a fixed interest rate per period. Initial cash and food endowments are farm-type specific and were obtained through calibration of the model.

TABLE 4.2 List of Cropping Systems Available to Farmers for Each Season and Agroecological Compartment

Cropping System	Crops	Input Level	Winter	Summer	IRRPAD_1	IRRPAD_2	IRRTER_1	IRRTER_2	SLOPING	PERMCASH
CA	Cassava	Low		X					x	
CI	Cinnamon									X
CM	Cardamom	Low								X
GI	Ginger			X					x	
IR1IR2H	Irrigated rice	High	x	X		x		X		
IR1IR2L	Irrigated rice	Low	x	X		x		X		
IR1IR2M	Irrigated rice	Medium	x	X		x		X		
IR2H	Irrigated rice	High		X	x	x	x	X		
IR2L	Irrigated rice	Low		X	x	x	x	X		
IR2M	Irrigated rice	Medium		X	x	x	x	X		
MA1H	Maize	High	x						x	
MA1L	Maize	Low	x						x	
MA1MA2H	Maize	Low	x	X					x	
MA1MA2L	Maize	High	x	X					x	
MA1PE	Maize; peanuts								x	

PE	Peanuts						x	
ROT1	Maize (3 years)—cassava (3 years)			X			x	
ROT2	Upland rice (1 year)—fallow (1 year)			X			x	
ROT3	Upland rice (2 years)—maize (3 years)—cassava (3 years)—fallow (3 years)			X			x	
ROT4	Upland rice (2 years)—fallow (2 years)			X			x	
SB1	Soybean	Low	x				x	
SB1SB2	Soybean (two seasons)		x	X			x	
SB2	Soybean (summer)			X			x	
SP	Sweet potato			X			x	
TE	Tea							X

IRRPAD_2, irrigated paddies receiving water during spring and summer; *IRRPAD_1*, irrigated paddies receiving water during summer only; *IRRTER_2*, terraces on sloping land receiving water during spring and summer; *IRRTER_1*, terraces on sloping land receiving water during summer only; *SLOPING*, sloping rain-fed land; *PERMCASH*, permanent cash crops (tea, cardamom, etc.).

4.4.2 Model Calibration and Validation

For the calibration process, we selected six farms deemed sufficiently representative of each typological group among the farms surveyed during the first round. The parameters used for the calibration were the maximum percentage of staple food that can be obtained through the market, price bands, and the maximum percentage of family labor able to find off-farm employment in each period. The models were then validated with six additional farms among the farms surveyed during the second round. Given the purpose of the study, particular attention was paid to the capacity of the model to predict the areas under the different cropping systems in each land type for the different farm types. The quality of these predictions of the model was assessed using the percentage absolute deviation[1] (PAD) proposed by Hazell and Norton (1986) and used by Yiridoe et al. (2006). Once calibrated and validated, we ran base-run simulations for average farms of each group as defined in Table 4.3.

4.4.3 PES Simulation Scenarios

We simulated scenarios in which we introduced the option to participate in one of the PES programs. We defined two types of programs, tagged "Payments for Forest" (PFF), and "Terraces for Forest" (TFF). All programs involved setting aside sloping land for reforestation.

TABLE 4.3 Main Typological Groups and Their Modeled Parameters

	Water Scarce Land Scarce (WSLS)	Water Scarce Land Rich (WSLR)	Off-farm Work (OFFW)	Terraces and Labor Rich (TERLAB)	Paddy Rich (PARI)	Terrace and Upland (TERUPL)
HH size	4	5	7	8	6	6
HH workforce	2	2	5	5	3.5	3.5
Land available in each zone per farm household (m^2)						
IRRPAD_2	0	0	200	200	1500	100
IRRPAD_1	0	0	1500	250	3000	250
IRRTER_2	0	0	0	300	0	1000
IRRTER_1	200	500	800	2100	0	3000
SLOPING	6500	4000	7800	8000	15,000	16,000
PERMCASH	600	6000	200	9000	4500	5000

IRRPAD_2, irrigated paddies receiving water during spring and summer; *IRRPAD_1*, irrigated paddies receiving water during summer only; *IRRTER_2*, terraces on sloping land receiving water during spring and summer; *IRRTER_1*, terraces on sloping land receiving water during summer only; *PERMCASH*, permanent cash crops (tea, etc.).

[1] Percentage absolute deviation is calculated according to the following formula: $PAD = (\sum_i |X_i - X_0|)/(\sum_i X_i)$, where X_i is the modeled area of activity *i*; X_0 is the observed area of activity *i*; for PAD the closer to zero, the better the prediction quality.

During each simulation, the farmers got the option to participate in a specific program, characterized by the structure of its benefits. The core element of TFF programs is the conversion of sloping land into terraces with access to irrigation. For each area A of land converted into forests, an additional $A \times tff$ area of owned sloping land is converted into terraces with access to water during the summer period. *tff*, the terraces for forest ratio, is a characteristic of the program and was parametrically increased from 0.15 to 0.9. In addition, farmers receive annual payments for the maintenance of the forest that were parametrically increased from US$0 to US$500 per hectare. For PFF programs, the cash payment is the only compensation and ranges from US$300 to US$700 ha/year. The amount of sloping land that can be converted into terraces is limited to 50% of the farms' sloping land. This ceiling of 50% of the land convertible into terraces may seem high, but the terraces considered would receive water during the summer season only. In that season, there is much less water constraints, so establishment of terraces is more limited by topography than by potential access to water.

We assumed that the additional work to maintain the new forest land is spread evenly throughout the year and may conflict with other cropping or off-farm activities. Finally, we made the realistic assumption, at least in the short term, that no forest products either in the form of food or cashable products can be extracted from the new forested areas. Hence, in our simulations the only possible benefits from the forest were the PES compensations.

We assumed that all cash compensations are paid during the last accounting period, such that possible household cash constraints during the cropping season will not affect the decision to participate. This was made explicit as most subsidy programs are known to be slow in disbursing money to farms, among others, because controls of farm claims are often needed.

A full mathematical description of the model and of the simulations can be provided to interested readers.

4.5 RESULTS

4.5.1 Model Calibration and Validation and Base Run

Satisfactory similarity between simulated and actual sampled farms plans was obtained by introducing a price band of 15% in the product market, and differentially limiting food and off-farm market opportunities. With a PAD of 14.5%, the model was found to adequately simulate the sets of cropping systems and their allocation in the different agroecological zones of the farms according to the model assessment criteria.

Once calibrated and validated, we ran a base-run simulation for average farms of each group (Fig. 4.2). All modeled farm households allocate all their irrigated land to rice with high inputs. This suggests very few potential cash and labor constraints for the adoption of the most intensive irrigated rice cropping system. In case of cash constraints, external inputs are allocated first to land most responsive to external inputs. Possible cash constraints are also alleviated by off-farm labor incomes, especially for land-scarce (WSLS, OFFW) farms.

Labor and input intensive cropping systems in the lowlands do not prevent cultivation in the uplands. With very small landholdings, land and food sufficiency constraints are more stringent than cash and labor constraints. Rice needs force the two poorest groups of farm households to mainly use the rice-based rotations on the slopes. In particular, WSLS farms are growing part of their sloping land with short cassava-rice rotations. Other farm types use

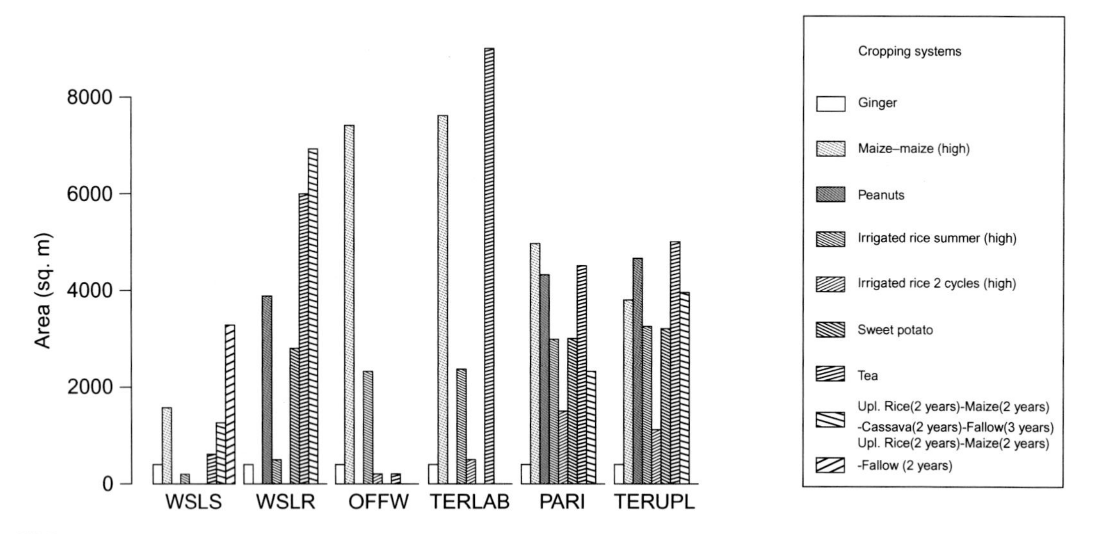

FIG. 4.2 Cropping systems selected by the different simulated farms. *WSLS*, water short land short; *WSLR*, water scare land rich; *OFFW*, off-farm; *TERLAB*, terraces and labor rich; *PARI*, paddy rich; *TERUPL*, terraces and upland.

TABLE 4.4 Simulated Income Indicators for the Different Farm Types (Base Scenario)

Income Indicators	WSLS	WSLR	OFFW	TERLAB	PARI	TERUPL
Household cash revenue (USD/year)	552.1	1331	2540.2	2659	2459.4	2481.7
Household total revenue (USD/year)	830.6	1637.8	2853.5	3017	2727.9	2750.2
% cash income coming from off-farm activities	73.4	27.4	78	50.1	35.5	35.2
% total revenue coming from off-farm activities	48.8	22.3	69.5	44.2	32	31.8
Household size (head)	4	5	7	8	6	6
Total revenue per head (USD/day)	0.6	0.9	1.1	1	1.2	1.3

1 USD = 18,000 VND (2010); *WSLS*, water scarce land scarce; *WSLR*, water scarce land rich; *OFFW*, off-farm; *TERLAB*, terraces and labor rich; *PARI*, paddy rich; *TERUPL*, terraces and upland.

more intensive cropping systems based on maize and cassava on the sloping areas. In the base run, cash and total revenues were very unevenly distributed ranging from US$0.60 to US$1.30 person/day (Table 4.4).

4.5.2 Simulation Results: Which Farms Participate and Why?

4.5.2.1 Who Is Participating in What Program?

Participation in the various programs was measured as the percentage of the sloping land converted to forest land (Figs. 4.3 and 4.4).

Participation in the PFF program (Fig. 4.3) starts at an annual payment per hectare of sloping land converted of US$400/ha for WSLR, TERUPL, and PARI farms. OFFW and TERLAB farms participate only when payments rise to US$550/ha. Finally, WSLS farms never participated in

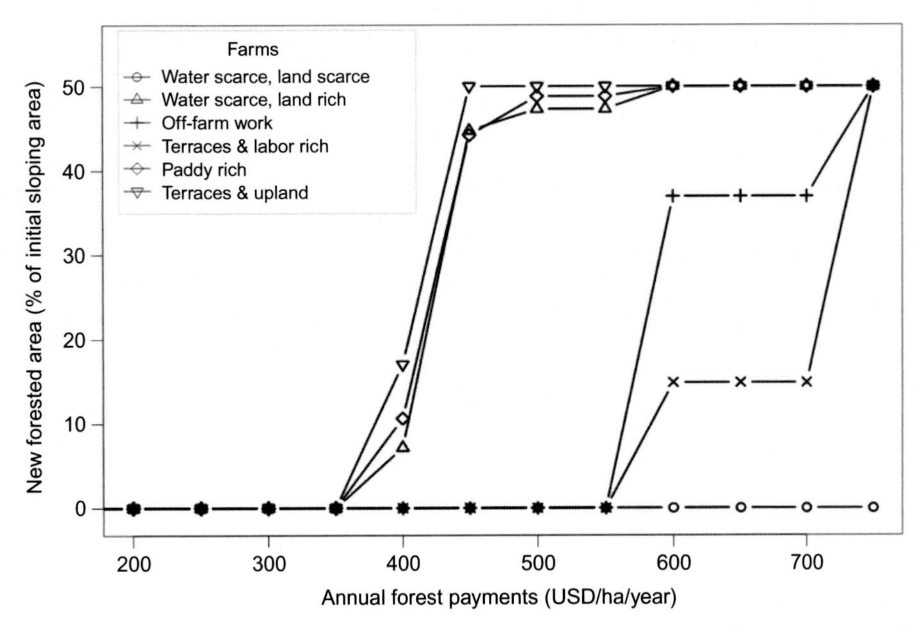

FIG. 4.3 Simulated percentages of sloping land converted to forest under the PFF differentiated by the annual payments.

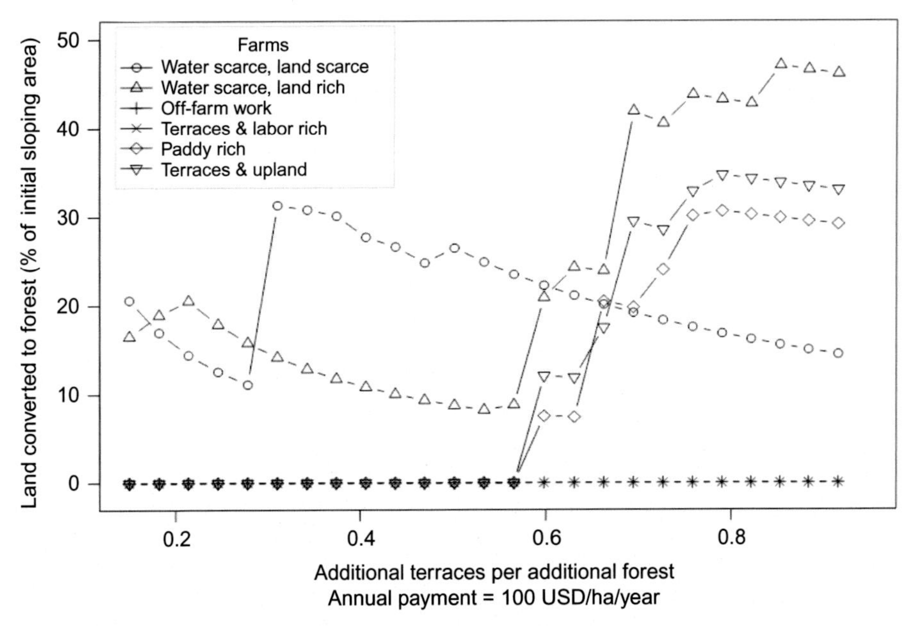

FIG. 4.4 Simulated percentages of sloping land converted to forest as a function of *tff* (terraces for forest ratio) with an annual payment of US$100/ha for forestry.

the proposed programs. On the practical side, annual payments exceeding US$500/ha are very unlikely, and one can anticipate that PFF programs would attract mainly *land rich farms*.

Four patterns of participation in TFF programs can be identified (Fig. 4.4): (1) WSLS farms are converting land for all *tff* levels; however, the percentage of conversion remains below 35%, even when the annual compensation is set as high as US$100/ha/year; at a *tff* around 0.3, forest conversion peaks to decrease slightly at higher levels, as farmers can get the same amount of terraces with less forest conversion. (2) WSLR farms are also adopting for all levels of *tff*. At low *tffs*, participation rises with increasing *tff*, but then decreases slightly when a desired amount of terraces is obtained. As *tff* gets higher than 0.7, the percentage of adoption raises sharply, getting close the maximum participation of 50 percent at *tffs* close to 1. (3) Farms with large areas of irrigated land (TERUPL and PARI) do not adopt for most of the range of the *tff* ratio, but then switch rapidly as the *tff* gets higher than 0.6–0.7. (4) Finally, the farms with large quantities of labor (OFFW and TERLAB) never convert sloping land into forests. These different patterns of adoption suggest that different farm types face different types of constraints for participating in the proposed programs.

As expected, more land is converted to forests at a given *tff* ratio when annual payments are higher, as reflected by the shifts of the participation curves to the left in Fig. 4.4. The shift is more apparent for PARI, TERUPL, and WSLR farms; that is, for farmers with relatively abundant land.

The differences in adoption patterns between TFF and PFF indicate that the two programs are addressing two different farm types: TFF is likely to attract land and water scarce farms while PFF is more likely to attract the more wealthy land-rich farms. Our results are in line with those of Pagiola et al. (2005), who concludes that monetary PES-schemes are biased toward richer farms. Below we will investigate what types of farm constraints causes this bias toward either rich or poor farms.

4.5.2.2 Constraints to Participation in PFF and TFF Programs

LAND AND WATER CONSTRAINTS

Examination of land and water constraints gives a first tangible explanation for the differentiated amount of land supplied for forestry. For TFF programs, we approximated the economic impact of converting 0.16 ha of sloping land into terraced land for every hectare of sloping land set aside using the opportunity costs to different types of farms of the different land types involved: forest, terraces irrigable during the summer season, and sloping land obtained from the base-run model (Table 4.5). WSLS farms have the highest opportunity cost of sloping land. Therefore, their participation to TFF programs can be explained by their very high opportunity cost for irrigable terraces (ie, for irrigation water): having an additional unit of terrace clearly outweighs the loss of the six units of sloping land to forest. WSLR farms have the lowest opportunity cost of sloping land, so they can easily give up some sloping land, especially because their opportunity cost for irrigable terraces is also relatively high. For other types of farms, the gains of adding new irrigable terraces are not sufficient to offset the loss of sloping land to forest at the set ratio, resulting in a negative impact. For PFF programs, opportunity cost of sloping land for the base-run model also gives important information regarding the different types of farmers' willingness to participate. Land rich farms (WSLR, PARI, TERUPL) have the lowest opportunity cost for sloping land and are likely to participate in PFF schemes with lower annual payments than other farm types.

TABLE 4.5 Simulated Opportunity Costs of Land and Cost Impact Measured Per ha of Land Converted to Forest (US$/ha)

Farm Group	Forest	Terraces (Summer)	Sloping	Impact per ha of New Forest
Water scarce land scarce	111	16 212	1 057	+1 580
Water scarce land rich	47	2 475	308	+101
Off-farm	98	1 219	527	−314
Terraces and labor	98	1 219	527	−314
Paddy rich	64	919	381	−227
Terraces and upland	64	895	360	−207

TFF program with a *tff*=0.16; that is, 1/6 units of new terraces is constructed when one unit of land is retired to forest; opportunity costs of forest are calculated for an annual payment of US$100 per ha of new forest, and a labor requirement for forest maintenance of 80 working days per year per ha. The impact is calculated per unit of additional forest as $OC_{Forest}+1/6\ OC_{Terraces(summer)}-7/6\ OC_{sloping}$.

FOOD CONSTRAINTS

In the base-run and PFF simulations, food constraints are binding for water-scarce farms (WSLS and WSLR). In particular, part of the farmers' diet is composed of maize. Food constraints have no effect on the other farm types. When participating in the TFF program, simulated water-scarce farms abandon the use of maize for food. They increase their maize production sharply, but all produce is sold to the market. Rice consumption requirements are met through increased home production allowed by better access to irrigation. Therefore, participation in TFF had a clear impact on water-scarce farms by allowing them to avoid rice shortages. Beyond the straight economic calculations, this is likely to be a very attractive feature for those households in terms of welfare, because of the strong preferences for rice as staple food in the study area.

4.5.2.3 *Impact on Cropping Patterns*

Despite the apparent complex changes, some general conclusions can be drawn. When farms are participating in the TFF, increased irrigated area allows a higher rice production and alleviates the minimum rice production constraint. As a result, farmers are abandoning the rice rotations they were practicing in the remaining sloping area. These rotations are replaced by intensive continuous cropping systems (eg, continuous maize cultivation). As demand for maize is increasing in Vietnam for the production of animal feedstuff, a scenario where increased maize production saturates the market and induces a fall in maize prices is unlikely. Therefore, the overall impact of the TFF program is an intensification of the sloping areas that are not set aside for forestry.

The effects of PFF programs are more diverse. WSLS farms do not adopt PFF programs even for high compensations, so they are not changing their cropping patterns. When adopting PFF programs, WSLR farms increase the percentage of their land under upland rice-based rotations. This is mainly due to the food self-sufficiency constraints (also related to the high price band) and their reduced area of sloping land. Other farms (PARI, TERUPL) increase the proportions of continuous cropping as they would under the TFF program. Therefore, the overall impact of the PFF program is also an intensification of the sloping areas that are not set aside for forestry, either by increasing the areas of continuous cropping systems, or by increasing the share of sloping land under upland rice-based rotation.

4.5.2.4 Impact on Farm Revenues

Participation in PES programs changes farm revenues. The change includes the difference in revenue obtained by altering activities of the farm and the annual compensation for the maintenance of the newly forested areas. The changes are extremely different across the farm types.

The farms that would benefit most from adoption of the TFF program are those currently having the lowest revenues (WSLS and WSLR). For WSLS the change in revenue would range between around 10% to 30% of their current revenue (for programs with US$100/ha/year of compensation for forestry work). The change in revenue for WSLR farms could reach 12% of their current revenues. However, their initial revenues are higher than WSLS farms. Other farms would see only a small impact on their revenues (<5%). Given these small benefits and the possible transaction costs to participation, it is likely that they would be reluctant to participate. Therefore, the TFF program is very likely to attract mainly farms with the lowest initial revenues.

The farms that would benefit most from adoption of the PFF program are the relatively large farm types (WSLR, TERUPL, and PARI). Interestingly, the WSLR farm (having the second lowest level of income) would benefit the most from this program. However, it should be noted that even a payment of US$600/ha/year would only increase their annual revenues by 10%.

4.6 DISCUSSION AND CONCLUSIONS

The main objective of this chapter was an ex-ante analysis of two alternative PES potential programs: TFF (water-based) and PFF (cash-based). Careful examination of the potential effects of PES programs in a mountainous district of Vietnam, based on a typological and farm modeling approach, confirms that cash-based land diversion types of PES are unlikely to attract small landholders. On the other hand, we have shown that helping farms to develop their terraced areas could significantly increase participation and revenues of small landholders. Hence, the existing reforestation programs, which are biased against the smallest landholders, could be transformed into win-win programs that increase the forested areas and reduce inequalities within mountainous rural communities of northern Vietnam. In terms of equity, our alternative program directly targets the poor households, as these households are in the "right place," able and the most willing to participate in the program (Pagiola et al., 2005). However, because the smallest landholders are more likely to respond, large transaction costs should be expected as many farms will bring small additional areas of forest and this should probably be investigated further.

Although the model was developed for a specific district, the analysis brings us several conclusions that are likely to be applicable in many other areas of Southeast Asia. The composite swiddening farming systems are present in various forms across that region, and households aiming for food self-sufficiency represent an important share of the rice-producing households. As we modeled the decision to participate in PES programs, we could identify the specific constraints to participation expressed theoretically elsewhere (Pagiola et al., 2005).

The TFF program laid out in this study fulfills most of the elements that define PES as explained by Wunder (2005). First, there is a clear service defined by setting aside sloping land for reforestation and consequently maintaining this newly forested land. This service is hypothesized to maintain/improve ES provision. Second, the service is bought by a service buyer. In our case, these are the households downstream, with an intermediate responsible for the design and monitoring of the PES program. Third, the service provision is voluntary. Based on the typology of different groups of potential service providers we analyzed the participation rate at which each of the groups would be voluntarily willing to set aside some of their sloping land. Fourth, there is a service supplier. These are the farm households in the uplands who set-aside their sloping land. In return for service provision, they receive a combination of more irrigated land and monetary payments per hectare of land set-aside.

The final condition might be a problem with TFF programs: the payment is made conditional upon service provision. After irrigation has been secured, the incentive for conditionality of the ES supplier is reduced to the amount of the yearly monetary payment. So, although the farmers would already have incentives to participate without receiving those yearly payments, selection of programs with some annual payments for forest maintenance may be warranted. Otherwise, incentives arise to revert back newly established forests into agricultural land after completion of the terraces. However, because irrigated terraces are more productive than sloping land, participation in a program reduces the opportunity cost of sloping land, and hence reduces the incentives to convert back initially set-aside forests. Yet, this alone is not enough to eliminate incentives to revert forest back to cultivated land. For example, for WSLS farmers the opportunity cost of sloping land diminishes from US$1057/ha to US$568/ha with $tff = 0.3$ and a payment of US$20/year/ha (payment level chosen as it was observed in other projects in Vietnam; eg, Peter et al. (2009) and To et al. (2012)). However, as forest is growing, households are likely to be able to extract nontimber products that were not taken into account in our model, making the opportunity cost of forest more important than established here, and reducing further the incentives to convert forest into cultivated land. Moreover, conversion of forests into agricultural land is costly and will add up to the disincentives to convert forest back into agricultural land.

In our household approach, we have assumed that off-farm work opportunities are not affected by introduction of the PES programs. However, under scenarios of setting land aside as considered here, smallholders being highly dependent on off-farm activities (especially on farm work contracted by larger landholders) could be affected by a reduction of off-farm work opportunities (Bulte et al., 2008; Zilberman et al., 2008). Our analysis showed that this is more likely to happen under the PFF program than under the TFF program; under TFF programs smallholders are converting some land into terraces requiring more on-farm labor, and making them less dependent on off-farm work. On the other hand, with PFF larger landholders are the ones converting cultivated lands into forestland, thereby reducing labor requirements for agriculture. Therefore, by considering an unchanged labor market under the TFF program, we are more likely to underestimate the positive effects of the TFF programs and the potential negative effects of the PFF program on small landholders (WSLS and OFFW).

Finally, the possible extension of these programs will be limited to the areas where farmers are direct owners of their lands, and where the possibilities of developing industrial crops such as rubber are low. Management of mountainous land by state forest enterprises (SFEs) remains important in some areas of Vietnam and would probably lead to different types of participation. In particular, SFEs may be reluctant to distribute more land to individual farmers if they foresee potential future gains in participating in PES programs (McElwee, 2012).

On another note, we haven't been able to address the overall effect of the program in terms of environmental services provided (Tomich et al., 2004). In particular, farms that participate in the programs are likely to intensify their cropping systems. While there is little concern over erosion created by the newly established terraces, there is a risk that the transition to continuous and intensive cropping systems like maize is likely to increase erosion on the remaining cropped land (Valentin et al., 2008). To analyze the different biophysical effects and provisions of ES, this purely economic model should be extended to properly account for the following changes in ES provisions: (1) the construction of terraces will reduce erosion and this would have effects on the lower parts of the catchments and would also serve the uplanders keeping the soil fertility capital in the upper-catchment instead of letting it leak to the lower part; (2) erosive rotations such as rice-cassava-fallow cycles would be abandoned by participants (this should reduce erosion); (3) forest areas will be reestablished providing all the associated ES; (4) continuous cultivation of maize-cassava will replace old rotations and could possibly be more erosive, and more polluting. Linking our purely economic models to real territories would be necessary to assess those aspects. We would then need to see the joint distribution of the farms and the potential ES. A combination of the earlier approaches of Shively and Coxhead (2004), and Antle and Valdivia (2006) working on the distribution of farms in the territory with works by Quintero et al. (2009) relying more on landscape models would be interesting to pursue.

Finally, we have developed a model based on some restrictive behavioral hypothesis. Acceptance of PES programs may be linked to other variables that have not been considered here. As such, economic experiments to test the validity of our model results should be conducted. This could take the form of real auctions (Jack et al., 2009; Jindal et al., 2011; Narloch et al., 2013), or of choice experiments (Jaeck and Lifran, 2013) to further test the potential interest of such programs.

References

Angelsen, A., 1999. Agricultural expansion and deforestation: modelling the impact of population, market forces and property rights. J. Dev. Econ. 58, 185–218.

Antle, J.M., Valdivia, R.O., 2006. Modelling the supply of ecosystem services from agriculture: a minimum-data approach. Aust. J. Agric. Resour. Econ. 50, 1–15.

Boserup, E., 1981. Population and Technological Change: A Study of Long-Term Trends. University of Chicago Press, Chicago, USA.

Bui, D.T., Hong, B.N., 2006. Payments for Environmental Services in Vietnam: Assessing an Economic Approach to Sustainable Forest Management. EEPSEA research report. Economy and Environment Program for Southeast Asia (EEPSEA), Singapore, p. 55.

Bulte, E.H., Lipper, L., Stringer, R., Zilberman, D., 2008. Payments for ecosystem services and poverty reduction: concepts, issues, and empirical perspectives. Environ. Dev. Econ. 13.

Castella, J.C., Quang, D.D., 2002. Doi moi in the mountains. In: Land use changes and farmers' livelihood strategies in Bac Kan Province, Viet Nam. The Agricultural Publishing House, Hanoi.

Chính, N.V., 2008. From swidden cultivation to fixed farming and settlement: Effects of sedentarization policies among the Khmu in Vietnam. J. Vietnamese Studies 3, 44–80.
de Janvry, A., Fafchamps, M., Sadoulet, E., 1991. Peasant household behavior with missing markets: some paradoxes explained. Econ. J. 101, 1400–1417.
Engel, S., Pagiola, S., Wunder, S., 2008. Designing payments for environmental services in theory and practice: an overview of the issues. Ecol. Econ. 65, 663–674.
Ferraro, P.J., Kiss, A., 2002. Direct payments to conserve biodiversity. Science 298, 1718–1719.
Folving, R., Christensen, H., 2007. Farming system changes in the Vietnamese uplands using fallow length and farmers' adoption of Sloping Agricultural Land Technologies as indicators of environmental sustainability. Danish J. Geogr. 107, 43–58.
Gauvin, C., Uchida, E., Rozelle, S., Xu, J., Zhan, J., 2010. Cost-effectiveness of payments for ecosystem services with dual goals of environment and poverty alleviation. Environ. Manage. 45, 488–501.
Gómez-Baggethun, E., de Groot, R., Lomas, P.L., Montes, C., 2010. The history of ecosystem services in economic theory and practice: from early notions to markets and payment schemes. Ecol. Econ. 69, 1209–1218.
Hazell, P.B.R., Norton, R.D., 1986. Mathematical Programming for Economic Analysis in Agriculture. Macmillan Publishing Company, New York.
Hoang, M.H., Van Noordwijk, M., Pham, T.T. (Eds.), 2008. Payment for Environmental Services: Experiences and Lessons in Vietnam. VNA Publishing House, Hanoi.
Jack, B.K., Leimona, B., Ferraro, P.J., 2009. A revealed preference approach to estimating supply curves for ecosystem services: use of auctions to set payments for soil erosion control in Indonesia. Conserv. Biol. 23, 359–367.
Jaeck, M., Lifran, R., 2014. Farmers' preferences for production practices: a choice experiment study in the rhone river delta. J. Agric. Econ. 65, 112–130.
Jindal, R., Kerr, J.M., Ferraro, P.J., Swallow, B.M., 2011. Social dimensions of procurement auctions for environmental service contracts: evaluating tradeoffs between cost-effectiveness and participation by the poor in rural Tanzania. Land Use Policy 31, 71–80.
Jourdain, D., Tai, D.A., Quang, D.D., Pandey, S., 2009. Payments for environmental services in upper-catchments of Vietnam: will it help the poorest? Int. J. Commons 3, 64–81.
Jourdain, D., Quang, D.D., Van Cuong, T.P., Jamin, J.-Y., 2011a. Différenciation des exploitations agricoles dans les petits bassins versants de montagne au Nord du Vietnam: le rôle clé de l'accès à l'eau? Cah. Agric. 20, 48–59.
Jourdain, D., Rakotofiringa, A., Quang, D.D., Valony, M.J., Vidal, R., Jamin, J.Y., 2011b. Irrigation water management in the northern mountains of vietnam: towards more autonomy? Cah. Agric. 20, 78–84.
Kinzig, A.P., Perrings, C., Chapin, F.S., Polasky, S., Smith, V.K., Tilman, D., Turner, B.L., 2011. Paying for ecosystem services-promise and peril. Science 334, 603–604.
Krautkraemer, J.A., 1994. Population growth, soil fertility, and agricultural intensification. J. Dev. Econ. 44, 403–428.
Kruseman, G., Bade, J., 1998. Agrarian policies for sustainable land use: bio-economic modelling to assess the effectiveness of policy instruments. Agr. Syst. 58, 465–481.
Laborte, A.G., Schipper, R.A., van Ittersum, M.K., van den Berg, M.M., van Keulen, H., Prins, A.G., Hossain, M., 2009. Farmers' welfare, food production and the environment: a model-based assessment of the effects of new technologies in the northern Philippines. NJAS 54, 345–372.
Lam, N.T., Patanothai, A., Rambo, A.T., 2004. Recent changes in the composite swidden farming system of a Da Bac Tay ethnic minority community in Vietnam's northern mountain region. J. Southeast Asian Stud. 42, 273–293.
Landell-Mills, N., Porras, I.T., 2002. Silver bullet or fools' gold? A global review of markets for forest environmental services and their impact on the poor, Instruments for sustainable private sector forestry series. In: International Institute for Environment and Development (IIED), London, p. 272.
Lipper, L.M., Sakuyama, T., Stringer, R., Zilberman, D. (Eds.), 2008. Payment for Environmental Services in Agricultural Landscapes: Economic Policies and Poverty Reduction in Developing Countries. Springer-Verlag Inc & FAO, New York.
McElwee, P.D., 2012. Payments for environmental services as neoliberal market-based forest conservation in Vietnam: Panacea or problem? Geoforum 43, 412–426.
Narloch, U., Pascual, U., Drucker, A.G., 2013. How to achieve fairness in payments for ecosystem services? Insights from agrobiodiversity conservation auctions. Land Use Policy 35, 107–118.

Nguyen, V.D., Tran, D.V., Nguyen, T.L., Tran, M.T., Cadisch, G., 2008. Analysis of the sustainability within the composite swidden agroecosystem in northern Vietnam: 1. Partial nutrient balances and recovery times of upland fields. Agr. Ecosyst. Environ. 128, 37–51.

Pagiola, S., Arcenas, A., Platais, G., 2005. Can payments for environmental services help reduce poverty? An exploration of the issues and the evidence to date from Latin America. World Dev. 33, 237–253.

Pandey, S., 2006. Land degradation in sloping uplands: Economic drivers and strategies for promoting sustainable land use. In: NAFRI (Ed.), Sustainable Sloping Lands and Watershed Management: Linking Research to Strengthen Upland Policies and Practices. Lao PDR, Luang Prabang, pp. 517–527.

Peter, J., Nguyen, C.T., Nguyen, T.B.T., 2009. The Pilot Payment for Forest Environmental Services in Lam Dong Province. FSSP Newsletter, Hanoi. pp. 11–13.

Quintero, M., Wunder, S., Estrada, R.D., 2009. For services rendered? Modeling hydrology and livelihoods in Andean payments for environmental services schemes. For. Ecol. Manage. 258, 1871–1880.

Rosenthal, R.E., 2007. GAMS: A User's Guide. GAMS Development Corporation, Washington, D.C.

Shively, G., Coxhead, I., 2004. Conducting economic policy at a landscape scale: examples from a Philippines watershed. Agr. Ecosyst. Environ 104, 159–170.

Sikor, T., Müller, D., 2009. The limits of state-led land reform: an introduction. World Dev. 37, 1307–1316.

Tachibana, T., Nguyen, T.M., Otsuka, K., 2001. Agricultural intensification versus extensification: a case study of deforestation in the northern-hill region of Vietnam. J. Environ. Econ. Manag. 41, 44–69.

Tai, D.A., 2006. Micro-level perspectives of rural livelihoods in mountain regions of northern Vietnam. In: Doppler, W., Praneetvatakul, S., Mungkung, N., Sattarasart, A., Kitchaicharoen, J., Thongthap, C., Lentes, P., Tai, D.A., Gruninger, M., Weber, K.E. (Eds.), Resources and Livelihood in Mountain Areas of South East Asia. Margraf Publishers, Weikersheim (Germany), pp. 333–375.

To, P.X., Dressler, W.H., Mahanty, S., Pham, T.T., Zingerli, C., 2012. The prospects for payment for ecosystem services (PES) in Vietnam: a look at three payment schemes. Hum. Ecol. 40, 237–249.

Tomich, T.P., Thomas, D.E., van Noordwijk, M., 2004. Environmental services and land use change in Southeast Asia: from recognition to regulation or reward? Agr Ecosyst Environ 104, 229–244.

Uchida, E., Xu, J., Rozelle, S., 2005. "Grain for Green": cost-effectiveness and sustainability of China's Conservation Set-Aside Program. Land Econ. 81, 247–264.

UN-REDD, 2010. Design of a REDD-Compliant Benefit Distribution System for Vietnam. UN-REDD Program, Hanoi, p. 191.

Valentin, C., Agus, F., Alamban, R., Boosaner, A., Bricquet, J.P., Chaplot, V., de Guzman, T., de Rouw, A., Janeau, J.L., Orange, D., Phachomphonh, K., Do Duy, P., Podwojewski, P., Ribolzi, O., Silvera, N., Subagyono, K., Thiébaux, J.P., Tran Duc, T., Vadari, T., 2008. Runoff and sediment losses from 27 upland catchments in Southeast Asia: impact of rapid land use changes and conservation practices. Agric. Ecosyst. Environ. 128, 225–238.

Van den Berg, M.M., Hengsdijk, H., Wolf, J., Van Ittersum, M.K., Guanghuo, W., Roetter, R.P., 2007. The impact of increasing farm size and mechanisation on rural income and rice production in Zhejiang province, China. Agricultural Systems 94, 841–850.

Vien, T.D., Rambo, A.T., Lam, N.T. (Eds.), 2009. Farming with Fire and Water: The Human Ecology of a Composite Swiddening Community in Vietnam's Northern Mountain. Kyoto University Press, Kyoto, Japan.

Wunder, S., 2005. Payments for Environmental Services: Some Nuts and Bolts. Center for International Forestry Research, Bogor, Indonesia. p. 32.

Wunder, S., The, B.D., Ibarra, E., 2005. Payment is Good, Control Is Better: Why Payments for Forest Environmental Services in Vietnam Have so far Remained Incipient. CIFOR, Bogor, Indonesia.

Yiridoe, E.K., Langyintuo, A.S., Dogbe, W., 2006. Economics of the impact of alternative rice cropping systems on subsistence farming: whole-farm analysis in northern Ghana. Agric. Syst. 91, 102–121.

Zilberman, D., Lipper, L., McCarthy, N., 2008. When could payments for environmental services benefit the poor? Environ. Dev. Econ. 13, 255–278.

CHAPTER

5

A Voluntary Model of Payment for Environmental Services: Lessons From Ba Be District, Bac Kan Province of Vietnam

T.D. Vien, C.T. Son*, N.T.T. Dung†, N.T. Lam**

***Vietnam National University of Agriculture, Hanoi, Vietnam †Office of Natural Resource and Environment, Gia Lam, Hanoi, Vietnam**

5.1 INTRODUCTION

Environmental services (ES), also known as ecosystem services, are the benefits people obtain from ecosystems (MA, Millennium Ecosystem Assessment, 2003, 2005). Today, however, many ecosystems and the services they provide are under increasing pressure, due to a rapid increase in the global population and economy. Extended research has confirmed that nearly two-thirds of valuable ecosystem services around the world are in decline (MA, Millennium Ecosystem Assessment, 2005). The fact is that "the benefits reaped from our engineering of the planet have been achieved by running down natural capital assets" (MA, Millennium Ecosystem Assessment, 2005), whereas the decline in these assets is often bigger than the optimal benefits that we can obtain, as shown by the existence of many forms of market failures, such as the presence of external effects, imperfect property rights, and lack of knowledge and information about ecosystems (Tietenberg, 2006). These issues not only trigger degradation of natural ecosystems, but also are an injustice for ES providers.

Payment for environmental services (PES) is an incentive-based mechanism under which the users or beneficiary of an ecosystem service make a payment to ES providers (Engel and Palmer, 2008). PES is hence understood as an economic instrument that allows ES users to pay ES suppliers for the maintenance of these services. In addition to this, PES contributes to creating a sustainable financial mechanism for protecting natural resources and the environment. Therefore, PES has been receiving increased attention from the world community over the

Redefining Diversity and Dynamics of Natural Resources Management in Asia
Volume 2
http://dx.doi.org/10.1016/B978-0-12-805453-6.00005-X

past decade. Numbers of PES programs have been successfully implemented throughout the world at different levels, such as the National Payment for ES Program in Costa Rica (Pagiola, 2008) and Mexico (Munoz-Pina et al., 2008); agricultural environmental plans in Europe and North of America (Baylis et al., 2008; Claassen et al., 2008; Dobbs and Pretty, 2008); the priority and encourage conservation policies (Hardner and Rice, 2002; Niesten et al., 2004); and the carbon stock program (Smith and Scherr, 2002).

In Vietnam, pilot PES projects have been implemented in Lam Dong and Son La provinces since 2008. Lessons from these cases were reviewed and have become a solid foundation for Decree No. 99/ND-CP about payments for forest ES (PFES). Decree No. 99 is the initial legal document that plays a significant role for PES implementation in Vietnam. Through the promulgation of this decree, Vietnam has become the first country in Asia to institutionalize a national policy on PFES (Pham et al., 2013). However, most PES mechanisms ran through the national programs, which were managed by governmental institutions. A voluntary PES model is, however, very limited.

In Ba Be District, Bac Kan Province in 2013, through the support of the Pro-Poor Partnership for Agroforestry Development Project (3PAD), a mechanism of a voluntary PES model has been established. This is considered as the first voluntary PES model in Vietnam (Ba Be People's Committee, 2014). In this model, service receivers are the beneficiaries of ES comprising homestay entrepreneurs, boating communities in Ba Be Lake, and Nam Mau Communes (people living downstream). Service holders in PES including all the villagers of Duong Hamlet of Hoang Tri Commune (people living upstream) protect the forest and water source at Ba Be Lake. The purpose of this study is to identify benefits and drawbacks of the model, which will help to design mechanisms of PES in the future.

5.2 THE STUDY AREA

This study was carried out in two hamlets in Ta Lengriver's watershed, namely Duong Hamlet of Hoang Tri Commune situated upstream and Pac Ngoi Hamlet belonging to Nam Mau Commune downstream.

Duong Hamlet has 29 ethnic minority households that include 24 households of Tay, who have been residing in the hamlet for many generations; and 5 households of Dao, who has been migrated to the hamlet from Cao Bang since 1970. The main local livelihood patterns include paddy rice, swidden cultivation, livestock, and forest use. The hamlet currently has 10 poor households[1] (34.48%) and 19 near-poor households[2] (65.52%). As it is situated upstream, Duong Hamlet helps downstream communities including Pac Ngoi Hamlet in terms of ES. The quality of forests in Duong Hamlet strongly impacts the quality and quantity of water in Nam Mau Commune and Ba Be Lake. In 2013, 550 ha of Duong Hamlet was occupied by forests, which includes 180 ha of protection forest, 350 ha of production forest, and 20 ha of rehabilitation forest (Fig. 5.1).

[1] The households who have income less than 400,000 VND/person/month, equivalent to US$ 18.4/person/month.

[2] The households who have income from 401,000 to 520,000 VND/person/month, equivalent to US$ 18.45 to US$ 23.92/person/month.

General information of Duong hamlet	
Number of household (household)	29
Population (Person)	137
Forest area (ha)	550
+ Protection forest	*180*
+ Production forest	*350*
+ Rehabiliation forest	*20*
Poor household (%)	34.48
Near-poor household (%)	65.52
Ethnic group	Tày, Dao
- *Tày (%)*	*82.76*
- *Dao (%)*	*17.24*

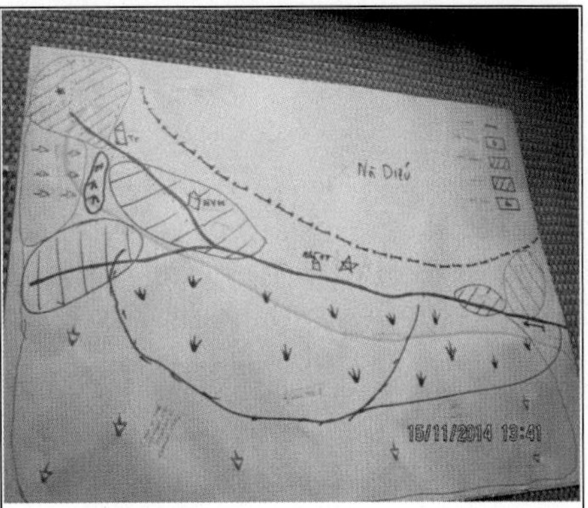

FIG. 5.1 Sketch diagram of Duong Hamlet. *Source: Field Survey, 2014*

Pac Ngoi Hamlet is located at the lakeside of Ba Be Lake. The hamlet provisions for homestay and canoe activities to visitors, who visit Ba Be National Park. In 2013, the hamlet had 22 homestays and a canoe union with 86 canoes for serving tourists. The tourism entrepreneurs have volunteered to contribute the financial resources to support upstream watershed forest protection of Duong Hamlet.

5.3 METHODS

Both secondary and primary data were used for the study. Secondary data was collected from various departments in Ba Be District. These include Ba Be District People's Committee, Department of Agriculture, and Rural Development, Nam Mau and Hoang Tri Commune People's Committee, and the office of the 3PAD project. The following methods were employed to collect primary data in the field.

5.3.1 Key Informant Interview

A number of key informants were interviewed, including the vice chairman of Ba Be District People's Committee, the head of the Department of Agriculture and Rural Development, chairman of Nam Mau Commune, 3PAD project officers in Nam Mau and Hoang Tri communes, and the head of Duong Hamlet. These interviews were designed to collect information about the historical time line of the voluntary PES model, stakeholder involvement, payment mechanisms, as well as stakeholders' perception about advantages and disadvantages of the model.

5.3.2 Group Discussion

A group discussion was arranged with nine male and female villagers in Duong Hamlet from different age groups. The group discussion helped to collect general information about the Ta Leng River watershed, forest protection activities in the hamlet and its historical time line, as well as the local livelihood patterns. Villagers also asked to draw a sketch map of Duong Hamlet, which shows residential areas and the location of main resources.

5.4 RESULTS AND DISCUSSIONS

5.4.1 Stakeholders

The study has identified three key stakeholders who are directly involved in the voluntary PES model in Ba Be District: ES providers, ES consumers, and moderators. These are explained in more details in the following section.

5.4.1.1 ES Providers

The Duong Hamlet owns 180 ha of protection forest. This forest area plays a significant role in maintaining watershed and natural landscape of Ba Be Lake. The forest area is maintained by villagers of Duong Hamlet; thus, they are considered to be ES providers. The forest resource in Duong Hamlet is currently under high pressure, because of the dependence of the local villagers on forest products and illegal logging from neighboring districts (eg, Cho Don District); meanwhile the financial support from the government for forest protection is very limited. To maintain the forest, protection activities must be paid for with equal benefits that people might obtain from forest product extraction or converting to forestland use (Pagiola and Platais, 2007). However, the financial support for forest protection activities from the government is very limited; thus, it is necessary to generate financial support from beneficiaries of the forest services.

5.4.1.2 ES Buyers

Homestay and canoe entrepreneurs (Nam Mau Canoe Union) and Ba Be National Park volunteered to contribute the financial resources to support Duong's villagers to protect watershed forest, environment, and water sources to maintain a clean water supply to Ba Be Lake. In this case, these actors are considered as ES consumers. They are aware of the impacts of Ba Be landscape on their business activities; thus, they have the motivation to make spontaneous payments for watershed forest and environmental protection in Duong Hamlet. This type of model is believed to obtain high efficiency (Pagiola and Platais, 2007) because the payment is in terms of the need and awareness of consumers. Researchers have also asserted that in the voluntary PES, ES receivers, and providers have opportunities to negotiate. ES buyers also have authority to evaluate the quality of ES, to define the rate of payment, or to decide to continue or stop paying (Coase, 1960).

5.4.1.3 *Moderators*

The 3PAD project, the Administrative Board of Ba Be National Park, and security police of Nam Mau Commune play an important role as intermediaries to connect the ES providers and ES receivers. Particularly, the administrative board of the 3PAD project promotes and encourages Duong's villagers, along with homestay entrepreneurs and boatmen to participate in the PES model. Whereas, the Administrative Board of Ba Be National Park and the security police of Nam Mau Commune collect and transfer the funds to service providers. The relationship among the previously mentioned stakeholders is presented in Fig. 5.2.

5.4.2 Payment Mechanism

Fig. 5.3 represents the payment mechanism along with involved stakeholders of the PES model in Ban Be District.

As mentioned earlier, ES receivers including the Nam Mau Canoe Union, have committed to contribute 2% of their profit generated through tourism activities (1% contributed by the canoe union management board and the remainder contributed by its members). The payment was transferred to the Administrative Board of Ba Be National Park. Similarly, homestay entrepreneurs in Ba Be Lake, Nam Mau Commune, committed to extract 4000 VND (US$ 0.18) per guest per night. The quantity of guests is determined according to the temporary residence registration at the police station of Nam Mau Commune. The contributions were then transferred to the Ba Be Forest Management Board. The board committed Duong Hamlet to

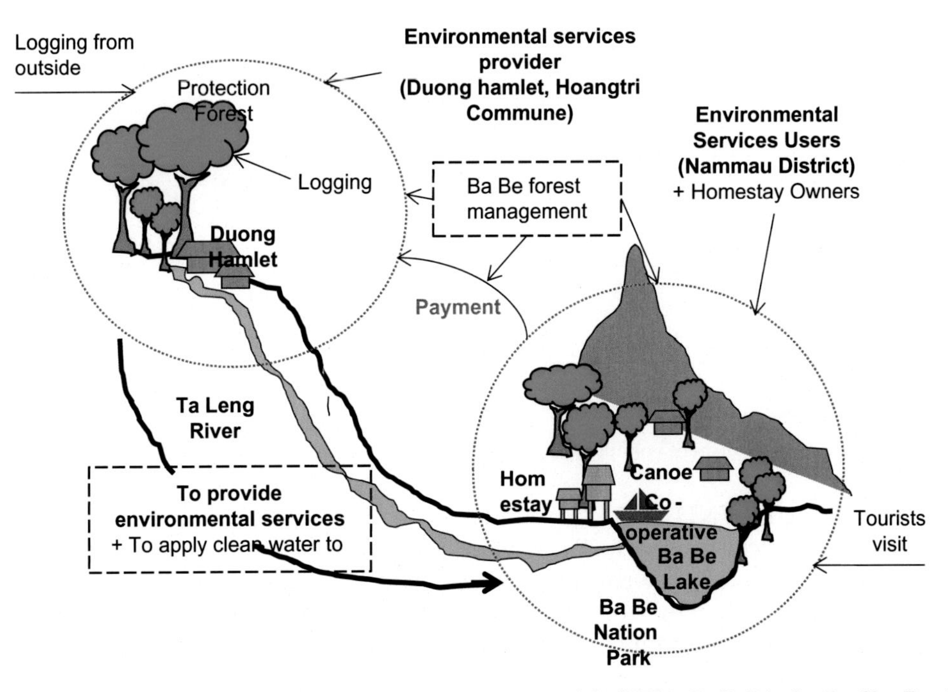

FIG. 5.2 Relationships between stakeholders in the voluntary model of PES in Ba Be District, Bac Kan Province.

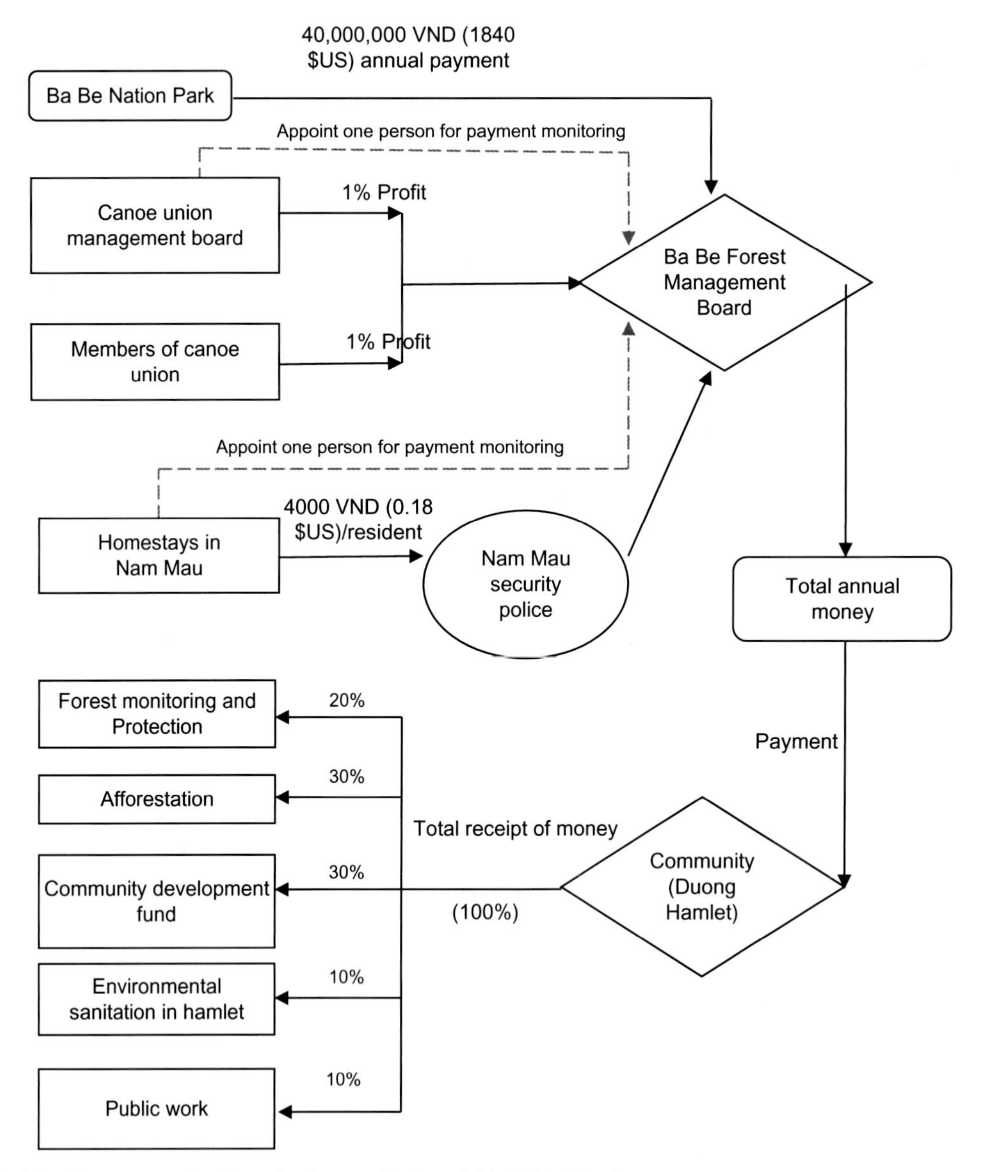

FIG. 5.3 Payment mechanism of voluntary PES model in Ba Be District.

supporting 40,000,000 VND (US$ 1840) annually. This amount of money is deducted through the national park's ticket sales. The Ba Be Forest Management Board takes the overall responsibility of gathering and transferring the contributions to Duong Hamlet.

After receiving the money from the Ba Be Forest Management Board, service providers (Duong Hamlet) have to commit to using the money in accordance with the regulation. The total received funds were used in accordance with the following ensuring purposes: 20% allocated for forest monitoring and protection activities; 30% allocated for afforestation activities; 30% allocated for the community development fund; 10% allocated for environmental sanitation activities around the hamlet; and 10% allocated for public services such as schools, roads, and community houses. These expenses must be reported to ES receivers as a financial disclosure.

The Ba Be Forest Management Board, along with the security police of Nam Mau Commune and 3PAD officers, play a moderator role to connect the ES providers and ES receivers. In addition, the Nam Mau Canoe Union and the homestay entrepreneurs nominate a representative each for monitoring the payment process.

The previously mentioned payment mechanism was established through direct negotiation between ES receivers and ES providers. This is done in compliance with economic theory of Coase, which considers the problem in relation to social cost (Coase, 1960). To proceed with the payment, transparency of ownership and lower operation cost are the basic conditions applied. In this study, ownership of watershed forest is clearly defined by the Duong Hamlet community. The operation cost was low because of small-scale implementations with two participating communities. Furthermore, the operation cost was mainly supported by the 3PAD project. This assistance certainly reduced the cost of implementation; however, it potentially led to unstable conditions in the long run, especially when the 3PAD project no longer exists. Similar to other PES models, a cash payment method was also applied in the model. The rate of payment is based on ES receivers' income; thus the profit generated by ES receivers from their business has an enormous impact on the sustainability of the PES model.

The study found that there was no clause in the contract between ES receivers and ES providers to address the responsibilities and specific punishments for those who break the contract. Normally, the contract must be guaranteed by a third party, such as local government. Although the 3PAD project and the Administrative Board of Ba Be National Park act as the third parties, these organizations do not have enough jurisdiction to solve and judge a possible conflict situation among stakeholders.

5.4.3 Outcomes and Weakness of the Model

After 1 year of implementation, the total received payment from the Nam Mau Canoe Union and homestay entrepreneurs was 26,000,000 VND (US$ 1196). This amount of money was paid in two installments to the head of Duong Hamlet in Jul. 2013 and Dec. 2013 (13,000,000 VND, equivalent to US$ 598 each time). However, the Ba Be Forest Management Board did not pay their committed amount of 40,000,000 VND to Duong Hamlet.

After receiving the payment, villagers of Duong Hamlet used the money as per their commitment with the ES buyers on the following activities:

- Forest monitoring: Earlier, forest monitoring was regularly and voluntarily conducted by Duong Hamlet villagers once a month. After receiving the payment, forest monitoring increased to three times a month. A total of 50,000 VND was provided to each patrolman during each patrolling.
- Sanitation: A head of Duong Hamlet in 2013 revealed that local residents carried an environmental sanitation program within the watershed area. Each household in the hamlet appointed an individual to join the program. This individual was awarded 50,000 VND per day for their participation.

- Community development fund: Approximately 30% of the total payment (7,800,000 VND, equivalent to US$ 358.80) was given to the Women's Union within the hamlet. This organization lends money to the poor households with at 0% interest rate for a period of 12 months. After the money is paid back, it is further shared among other poor households.
- Forest protection and plantation and improving public services: According to the head of the hamlet, expenditures from forest protection and plantations were used for improving public services. However, there is a little evidence that these activities are actually carried out in the hamlet.

All the previously mentioned expenditures must be approved at community meetings. However, the outcomes from Duong Hamlet's activities were not documented nor presented in actual reports. According to the head of the hamlet, since 2014 Duong Hamlet did not receive any payments through PES. To the contrary, the ES receivers claimed that there was a lack of transparency in the PES process, which made them stop paying for ES for Duong Hamlet.

Causes of Failures

Through consultations and discussions with stakeholders in the PES model in Ba Be District, the following is revealed to be the reasons for failure:

- *Duong Hamlet Community:* The head of Duong Hamlet responded that most of the people in the hamlet were unaware of the reasons for not receiving the payment, though he claimed that people still carried on with forest protection activities as per the commitment. He also added that in 2013, there was a staff from the 3PAD project who would come regularly to encourage the villagers. However, after the 3PAD project phased out in 2014 there was no connection among participating villagers and other involved stakeholders. As such, the villagers had less access to required information. Available information shows villagers, despite having the passion to continue the PES, lack the competency to negotiate directly with ES receivers.
- *ES receivers:* According to the available information from the Department of Agriculture and Rural Development, ES receivers stopped the payment for not receiving financial disclosure and activity reports. The staff working under the Ba Be National Park claimed that, although committed activity worth 40,000,000 VND (US$ 1840) was carried out it had not been paid for due to the disagreement among stakeholders. The sudden break in payment had created many difficulties for ES providers. However, the contract had no clause that would address the responsibilities and specific punishments for stakeholders for breaking the regulations. The limitation of the PES model in Ba Be is a common issue for PES programs in Vietnam (Vietnam Forest Protection and Development Fund, VNFF, 2014).
- *Intermediary*: Hoang Tri and Nam Mau People's Committee claimed that, they only notified the ES program through documents without being directly involved in the project. The Administrative Board of Ba Be National Part and security police of Nam Mau Commune stated that they had finished the money transaction processes from the 3PAD project and ES receivers. When the 3PAD project wrapped up in 2014, the support from the intermediaries was not maintained; thus, the intermediaries were not involved in the

PES process. Obviously, the financial assistance played an important role in maintaining the motivations for the intermediaries.

Evidently, the failures of the PES model of Ba Be mostly have arisen from the internal partners. As a consequence of the failures arising due to a lack of competencies, the villagers were not able to conduct financial disclosure and submit activity reports, which led to the loss of trust from ES receivers and resulted in a breakdown of the PES model. In addition, the intermediaries did not play a role as promoters after they stopped receiving benefits from the model. This is common in a spontaneous PES mechanism, whereby the ES receiver loses their trust in ES providers, concomitantly with the termination of the PES contract (Pagiola and Platais, 2007).

5.4.4 Lessons Learned From the PES Model in Ba Be District

5.4.4.1 Weakness

The model has the following weaknesses.

DETERMINING INAPPROPRIATE ES RECEIVERS

As mentioned earlier, the ES receivers in the PES model were boatmen and homestay entrepreneurs. However, they are not the direct users of ES. The direct ES users probably are tourists and Ba Be National Park. Therefore, if the ES receiver spontaneously signs the contract, the payment would be acceptable. Nevertheless, with regard to the theory of market competiveness, many errors could be found in the model. For examples, different tourists might use different modes of transportation such as bicycles, motorcycles, and cars for traveling in Ba Be instead of using boats. Similarly, tourists may choose alternative accommodations instead of homestays or even move out of Ba Be National Park in a day. In other words, the tourists do not have to pay for ES if they do not stay in the homestays. These examples prove there are inequalities among tourists in paying for ES; consequently, it causes losses in payments for ES providers.

DETERMINING INAPPROPRIATE RATE OF PAYMENT

According to economic principles, the price of a product must be based on the cost of production (Coase, 1960). Thus, if we perceive ES of Duong Hamlet is a product, the price of this product must be determined according to its cost and quality. In reality, the rate of payment is based on the profit of ES buyers (homestay businessmen and boatmen); thus, it was much lower than the alternative income that villagers earn from forest products. For example, the normal income of a bamboo shoot picker ranges from 100,000 VND to 150,000 VND (US$ 4.60 to US$ 6.90) per day and a timber transporter is paid 200,000 VND (US$ 9.20) per day. Nevertheless, the forest patrolmen were each paid 50,000 VND (US$ 2.30) per day; much lower than the income they might earn in other jobs. The rate of payment, which is lower than other opportunity costs, is one vital cause leading to the unsustainability of PES (Vietnam Forest Protection and Development Fund, VNFF, 2014).

In addition, the PES model in Ba Be generates the funds for payment from three sources: tourist tickets, boatmen, and homestay entrepreneurs. However, the actual buyers of ES are tourists because the ES cost is added to the price for accommodation or transportation. It means the tourists have to pay three times for the ES if they use three services: accommodation, tourist tickets, and a boat. This issue holds inequality for the tourists.

UNSUSTAINABLE OPERATION MECHANISM

The PES model in Ba Be lacks sustainable promoters. The 3PAD project plays an important role in promoting stakeholders and also provides additional funds to support involvement of Ba Be National Park and security police of Nam Mau Commune. However, this is short-term support and gets terminated when the 3PAD project wraps up. The staffs of Ba Be National Park and the police of Nam Mau Commune lack motivation to be sustainable promoters. Their participation in the PES process ended when they stopped receiving support funds for intermediaries from the 3PAD project for their involvement.

LACK OF STAKEHOLDERS' COMPETENCIES

- *ES providers:* All of the ES providers are ethnic minorities (Tay and Dao people), with low education and a lack of competencies in terms of financial management, writing reports, and maintaining official documents. These weaknesses led to difficulties for villagers in presenting and reporting the activities to ES receivers. The unclear and deficient information of the PES process from villagers caused the breakdown of the PES model in Ba Be.
- *ES receivers:* The criteria to evaluate the quality of the ES was not provided and not established. Thus, it is difficult for ES buyers to check and take over the quality of services that they receive; for instance, the quality of forest protection, forest plantation, and watershed protection. Unclear benefits of ES had a vital effect in stopping the payment (Pagiola and Platais, 2007).
- *Intermediaries:* 3PAD is a research project among organizations; the administrative board of Ba Be National Park is an organization in charge of resource management; and the police station of Nam Mau Commune is a security organization. All three of these intermediaries lack jurisdictions to solve and judge possible conflicts among stakeholders. Thus, to solve the conflicts among stakeholders in accordance with the legal documents, the participation of local government is important (Vietnam Forest Protection and Development Fund, VNFF, 2014). Ba Be District People's Committee, Nam Mau Commune, and Hoang Tri Commune should participate in this PES model at Ba Be. The conflicts between Duong Hamlet and ES receivers in Ba Be, which led to the end of the PES model, might be resolved if the local institutions are actively involved.

5.4.5 Learned Lessons

There are many valuable lessons from the PES model of Ba Be, which are presented below.

- *Defining the ES receivers and ES providers:* To clearly define ES receivers and ES providers, it is important to identify actual consumers and providers of the service (Wunder, 2005).
- *Pricing the service:* The payment should be at least equal to, or larger than, the production cost of the service (Pagiola and Platais, 2007). The equality is a warranty for the sustainability of PES.

- *Improving competencies of stakeholders:* The important competencies that should be provided for ES providers, especially the ethnic minority groups, include writing reports, negotiation skills, and project management. The ES receivers should be provided with criteria established to evaluate the quality of the ES to determine the appropriate price of the received service. The core competencies of intermediaries are skills for solving conflicts, promoting, and connecting stakeholders. Improving the competencies of stakeholders is an effective solution to ensure the sustainability of PES programs (Elisabeth et al., 2014).
- *Payment mechanism:* The operation mechanism must be flexible and should ensure benefits to all stakeholders. Particularly, the providers must receive enough committed payment equal to the production cost of the service. The buyers must have the right to reduce the price and delay or stop paying for the service if the equality of the service is lower than committed to in the contract. The responsibilities of the intermediaries must also be attached to the models. A part of the payment could be allocated to pay for the intermediaries. This solution has been widely used in Bac Kan Province. For example, a PES program of some hydropowers extracted 5–10% of total payments for general management activities (Forest Protection and Development in Bac Kan province of Vietnam, 2015).

5.5 CONCLUSION

The PES model of Ba Be District is the first model of voluntary PES in Vietnam. The model was established in 2013 and had some positive outcomes as it had defined ES receivers and ES providers as well as a mechanism for payment. However, the model had many weaknesses due to inappropriately defined ES receivers and payment rates, an unsuitable operation mechanism, and lack of stakeholders' competencies.

References

Baylis, K., Peplow, S., Rausser, G., Siimon, L., 2008. Agri-environmental policies in the EU and United State: a comparison. Ecol. Econ. 65, 753–764.

Ba Be People's Committee, 2014. Report of Payments for environmental services in Duong hamlet of Ba Be district in 2013 [in Vietnamese]. Bac Kan, Vietnam.

Claassen, R., Cattaneo, R., Jonhansson, R., 2008. Cost-effective design of agri-environmental payment program: US experience in theory and practice. Ecol. Econ. 65, 737–752.

Coase, R.H., 1960. The problem of social cost. J. Law Econ. 3, 1–44.

Dobbs, T.L., Pretty, J., 2008. Case study of agri-environmental payment: the United Kingdom. Ecol. Econ. 65, 765–775.

Elisabeth, S., Bac, D.V., Hoan, D.T., Rebecca, T., Delia, C., Lan, D.N., 2014. Recommendations for capacity building programme on payment for environmental services for stakeholders in Vietnam. Int. J. Agric. Rural. Dev. 1, 101–109 (in Vietnamese).

Engel, S., Palmer, C., 2008. Payment for environmental services as an alternative to logging under weak property rights: the case of Indonesia. Ecol. Econ. 65, 799–809.

Forest Protection and Development in Bac Kan Province of Vietnam, 2015. Action Planning for Payment of Forest Environmental Services for Forest Owners in Bac Kan Province in 2015. People's Committee of Bac Kan Province, Vietnam.

Hardner, J., Rice, R., 2002. Rethinking green consumerism. Sci. Am. 286 (5), 88–95.

MA (Millennium Ecosystem Assessment), 2003. Ecosystem and Human Well-Being a Framework for Assessment: Millennium Ecosystem Assessment Report. Island Press, Washington, DC.

MA (Millennium Ecosystem Assessment), 2005. Ecosystem and Human Well-Being: Synthesis: Millennium Ecosystem Assessment Report. Island Press, Washington, DC.

Munoz-Pina, C., Guevara, A., Torres, J.M., Brana, J., 2008. Paying for hydrological services of Mexico's forest: analysis, negotiations and results. Ecol. Econ. 65, 725–736.

Niesten, E.T., Rice, R.E., Ratay, S.M., Paratore, K., 2004. Commodities and Conservation: The Need for Greater Habitat Protection in the Tropics. Conservation International, Washington, DC.

Pagiola, S., 2008. Payment for environmental services in Costa Rica. Ecol. Econ. 65, 712–724.

Pagiola, S., Platais, G., 2007. Payments for Environmental Services: From Theory to Practice. World Bank, Washington, DC.

Pham, T.T., Bennet, K., Vu, T.P., Brunner, J., Le, N.D., Nguyen, D.T., 2013. Payment for forest environmental services in Vietnam: from policy to reality: Special Subject Report No. 98. CIFOR, Vietnam.

Smith, J., Scherr, S.J., 2002. Forest Carbon and Local Livelihoods: Assessment of Opportunities and Policy Recommendations: CIFOR Occasional Report No. 37. CIFOR, Bogor.

Tietenberg, T., 2006. Environmental and Natural Resource Economics, sixth ed. Addision Wesley, Boston.

Vietnam Forest Protection and Development Fund (VNFF), 2014. Payment for Forest Environmental Services in Vietnam: Finding from Three Years of Implementation. VNFF, Vietnam.

Wunder, S., 2005. Payments for Environmental Services: Some Nuts and Bolts: Occasional Report No. 42. CIFOR, Bogor, Indonesia.

SECTION IV

LAND-USE PLANNING

CHAPTER

6

Land-Cover and Land-Use Transitions in Northern Vietnam From the Early 1990s to 2012

S.J. Leisz

Colorado State University, Fort Collins, CO, United States

6.1 INTRODUCTION

One of the most obvious and significant changes to take place in Vietnam since the onset of *doi moi* is the landscape-level modifications that are taking place in all regions of Vietnam. In the north, these changes are taking place in lowland and delta areas and across the midland and upland hinterlands of Hanoi and farther afield in the rural areas of the region. In the lowland and delta areas, urban centers are expanding into previously agriculturally dominated villages and communes. These villages and communes are acquiring urban characteristics and some are developing into periurban centers attached to older urban areas while others are organically taking on urban characteristics in a process that has been referred to as rural urbanization or invisible urbanization (Douglass et al., 2002; DiGregorio et al., 2003). This situation is similar to urban and periurban growth patterns that are seen in other parts of Southeast Asia and East Asia (McGee, 1991; Rigg, 2001; Rigg et al., 2008). These changes are reflected in changing land-use patterns, which are due to the dominant livelihoods in the lowlands and deltas changing, and also in the corresponding changes in the land cover found in these areas.

Midland and upland areas are also experiencing land-use and land-cover changes. In these areas the changes are driven by livelihood transitions, but the transitions are not as directly influenced by nearby urban areas and foreign direct investment as are the changes in the lowland and delta regions. Instead an argument can be made that midland and upland livelihood changes are influenced by what has recently been termed *teleconnections* (Seto et al., 2012). Teleconnections refer to actions that take place in distant locations, often the lowland and delta regions, or even farther afield, and have repercussions reaching into very distant, often rural, regions. The result of the teleconnections in northern Vietnam is changing livelihood

Redefining Diversity and Dynamics of Natural Resources Management in Asia
Volume 2
http://dx.doi.org/10.1016/B978-0-12-805453-6.00006-1

patterns in the midlands and uplands, which lead to changes in where people live, what their major activities are, changes in how they use their natural resource base, and modifications in the way they use their land areas. All of these result in the land-cover changes found in the midlands and uplands.

This chapter is an overview of the way that land use and land cover has changed and is changing in the lowland and delta regions, midlands, and uplands of northern Vietnam. The drivers of these changes and possible impacts are also reviewed and discussed. Northern Vietnam is here defined as stretching from the southern border of Nghe An Province to the border with China.

6.2 ORIGIN OF THE CHANGES IN LAND USE AND LAND COVER

From the late 1970s until the present, government efforts to influence land use and land cover in the lowland, midland, and highland areas of the country have focused on the introduction of laws and regulations related to how land and forest are managed. In the 1960s, 1970s, and into the 1980s land and forest management took the form of state-run agricultural cooperatives and forest enterprises. By the end of the 1970s these policies had created a situation where there were chronic shortfalls in agricultural production and serious deforestation and degradation of forest areas in the uplands (De Koninck, 1999). In response to these situations Decree 100 was implemented in 1981. This decree allowed farmers to use land in return for a fixed amount of the crop produced on it. In the late 1980s *doi moi*, or economic reforms, was instituted throughout the country. As part of these reforms, in 1988 Resolution 10 was passed. Resolution 10 emphasized the importance of private property rights, further dismantled the cooperative system, and allowed individuals to have the right to use a plot of land for a period of 15 years. After the 15-year term the land was supposed to revert to the state and it was then possible that the state would redistribute it to another person or family. The 1993 Land Law expanded private use rights for agricultural land and allowed for the transfer of individual land rights from one person or family to another person or family under certain conditions (Government of the Socialist Republic of Việt Nam, 1993). In 2003 a new land law was passed. The 2003 Land Law allowed for a person or household to hold land for 20 years if it is being used for annual crop farming and for 50 years if the land is used for perennial crop farming (Government of the Socialist Republic of Việt Nam, 2003).

The result of land allocation in the lowlands was the dissolution of cooperatives and distribution of land to households for private use for a limited period of time. The implementation of the land law is credited as one of the main drivers of increased agricultural production in Vietnam since the mid-1980s (Gomiero et al., 2000; Đỗ and Iyer, 2003). In the uplands from 1943 through the mid-1990s forests were degraded and the forest cover area reduced. Overall it is estimated that forest cover decreased from 43% to 16% of the country (De Koninck, 1999). During the late 1980s and early 1990s forestland ownership and management were both reformed. In 1993, the government extended the 1988 Land Law into the uplands (Ahlback, 1995; MAFI, 1993; Sikor, 1995; De Koninck, 1999) allocating forestland to upland farmers in the same manner as agricultural land. At this time, the government also implemented the Law for the Protection and Development of Forests.

This law formalized the process for forestland allocation and based it on the willingness of households to plant trees (Government of the Socialist Republic of Việt Nam, 1994; Gomiero et al., 2000). Decree No. 85/1999/ND-CP in 1999 (Socialist Republic of Việt Nam, 1999) modified the 1993 Land Law and the 2003 Land Law further revised the allocation of forestland, specifically production and protection forestland, so that it could be allocated to an individual or household for use for 50 years (Government of the Socialist Republic of Việt Nam, 2003).

6.3 LAND-USE AND LAND-COVER CHANGES IN LOWLAND DELTA AREAS

Lowland and delta areas have seen locally dramatic changes since the beginning of *doi moi*, but overall the land cover and land use of these areas has remained constant. During the cooperative period the lowland areas outside of urban areas were devoted to agriculture, specifically to growing irrigated rice. Given the influx of foreign direct investment and the growth of urban areas in the deltas and lowlands, as well as the infrastructure associated with this growth, it is reasonable to assume that large amounts of lowland and delta rice areas would have been converted from rice land/agricultural use to land uses and land covers associated with expanding urbanicity. However, a review of land-use statistics and land cover in the lowland and delta areas does not show large changes. Instead, while residential areas are being built in the immediate vicinity of urban areas, and urban areas are expanding, as can be seen with the expansion of Hanoi, and industrial parks have been built in the lowland and delta areas, what is mostly seen over the larger area is the densification of the inhabited areas within village boundaries, as these areas are built up and take on more "urban" characteristics (DiGregorio et al., 2003), and the continuation of large areas devoted to agriculture, specifically to irrigated rice growing (McPherson, 2012).

The main reason for this is that it is difficult for people to convert agricultural land devoted to rice growing to other land uses. The government of Vietnam defines food security as secure and adequate production of rice (McPherson, 2012). Recognizing this the 1993 Land Law purposely made it difficult to convert rice-growing land to other land uses. The law allowed conversion of rice-growing land to other purposes only if the area was less than 2 ha and the provincial authorities approved the land-use change. If the area was larger than 2 ha, the prime minister's approval was needed (Government of the Socialist Republic of Việt Nam, 1993). The 2003 Land Law has similar provisions, prohibiting the conversion of agricultural land used for rice growing to other forms of use unless the relevant government officials approved it (see Article 74 2003 Land Law, Government of the Socialist Republic of Việt Nam, 2003). These prohibitions were reaffirmed by Decree 69 of August 2009, Articles 6 and 10 (Socialist Republic of Việt Nam, 2009). The enforcement of these restrictions regarding the use of land associated with land devoted to rice growing has had the repercussion that while some areas outside of established urban areas have been converted to industrial parks, most periurban areas have been densified, filling gaps in their residential areas by building up these areas. The same is true for village areas not connected directly with urban areas. This pattern of land use is leading to "rural urbanization" or "invisible urbanization" (McGee, 1991; Douglass et al., 2002; DiGregorio et al., 2003).

"Rural urbanization" takes place when communities that were previously classified as "rural" villages, and where people made their living mainly from agricultural pursuits, start to take on characteristics of urban areas. These characteristics include inhabitants' livelihoods changing from agricultural-based to artisanal or industrial-based, regardless of whether the livelihood activity means that they travel to a formally recognized urban area or whether the livelihood is practiced within the community; changes in housing style; changes in the way open space in the "village" is managed (eg, traditional ponds may be filled, roads paved and expanded, traditional buildings knocked down and replaced by modern high rises, etc.). This process means that the previous agricultural village changes to a dense built-up area often covered with larger buildings and buildings that are high rises, and with little public open space. Open green space and village ponds are replaced by concrete and thus land-cover changes. Overall this process is seen in much of the Red River Delta (Douglass et al., 2002; DiGregorio et al., 2003) and there is some evidence that this is happening in other lowland areas in northern Vietnam.

While the lowland areas devoted to rice agriculture have not experienced land-use or land-cover change over large areas, even while the nature and the quality of the land use and land cover may have changed, land use and land cover in the coastal areas in northern Vietnam have changed significantly. The most significant change that has occurred in these areas is the land-cover change from mangrove dominated shorelines to shorelines dominated by shrimp aquaculture (Hien et al., 2006; Dat and Yoshino, 2012). It is estimated that between 63% and 70% of the mangroves in some parts of northern Vietnam has been lost and that most of this has been replaced by shrimp aquaculture (Hien et al., 2006; Beland et al., 2006). This change has repercussions for the resiliency of the shoreline in the face of extreme weather, as well as biodiversity implications and other long-term ecological impacts (Hai-Hoa et al., 2010). Conversely, shrimp aquaculture has been lauded for its economic benefits for the area (Beland et al., 2006). Thus, in the case of the coastal areas, both land cover and land use has changed since the early 1990s.

6.4 LAND-USE AND LAND-COVER CHANGES IN MIDLAND AND UPLAND AREAS

In the much of the northern uplands the effect of the implementation of the land laws since 1988 is more ambiguous than the results in the lowlands. Prior to *doi moi* natural resource use in the highlands and midlands was dominated by agricultural systems and logging. The agricultural systems practiced were irrigated rice growing, which was usually found in valley bottoms, but also found in some terrace systems, most notably in Lai Chau (Sapa), and swidden systems, also referred to as shifting cultivation, and composite swidden systems (Rambo, 1998), which utilize both irrigated rice in valley bottoms and swidden fields on hillsides. Complementary to these agricultural systems, small-scale animal husbandry was practiced, dominated by free-ranging pigs and cattle, with some tended water buffalo also being raised (Viên et al., 2006; Leisz et al., 2011). Prior to *doi moi*, forestry, which mainly consisted of logging, was organized and operated by state forest enterprises.

In the early 1990s *doi moi* found its way into the midlands and uplands through the land laws, the implementation of which is often credited with changing the land-use systems and ultimately the land cover found in these areas. The allocation of individual rights to forest and agricultural land in the uplands under the new forest and land laws is aimed to address forest loss. Individual households were given rights to agricultural and forestland. This allocation of "private property rights" also had the goal of sedentarizing the households in cases where households and hamlets still practiced nomadic swidden (Thành et al., 1995; Đức, 2003; Castella et al., 2006; Tran, 2007) and fixing in one place the cultivation of fields where rotational swidden was practiced (Lambin and Mefroidt, 2010). At the same time as the implementation of the new land and forest laws, logging was banned in special use and natural forests in the northern mountains and the amount of logging activity was greatly reduced (Tottrup, 2002; McElwee, 2004). The resulting changes in upland land-use have followed a number of different trajectories.

6.4.1 Land-Use Change in the Midland and Upland Areas

The first trajectory is one of status quo. In this situation, despite the implementation and enforcement of the land laws since the early 1990s, swidden is still practiced in many places. Swidden agriculture manages land by alternating the use of the land between fields, which are cleared and cultivated with crops for one, two, or more years, depending upon the type of swidden practiced, and leaving land in fallow. Depending upon different factors, such as the number of times the land has been cleared and planted in crops, the length of time between clearing/crop growing, the soil type, soil structure, and soil fertility, the fallow land may have grass-dominated land cover, shrub-and bush-dominated land cover, small tree-dominated land cover, and even tree cover structurally similar to primary forest. Since the 1990s there are some parts of the uplands, specifically areas in the provinces of Nghe An, Thanh Hoa, and Hoa Binh (Viên et al., 2006; personal observation June 2009, April 2015), as well as areas in other northern provinces where swidden is still practiced and where the fallow length has, on average, stayed the same (Leisz et al., 2007; Leisz, 2009). There are many other districts within the northern part of Vietnam where swidden is still found, but fallow length has been reduced. This has been reported for areas of Nghe An (Leisz et al., 2007; Jakobsen et al., 2007), Hoa Binh (Fox et al., 2000, 2001; Leisz et al., 2007), Son La (Vu, 2007; Vu et al., 2013), and other parts of the northern uplands and midlands (Fox and Vogler, 2005; Leisz et al., 2005).

A second trajectory that has been documented is the shifting of land use from swidden cultivation systems to permanent cover of various tree crops. In the uplands some farmers who previously practiced swidden agriculture have replaced some or all of their swidden fields with tree crops. In some areas this has led to new plantings or expansion of rubber trees (Thiha et al., 2007), coffee trees (Vu, 2007), or fruit trees (Vu, 2007). In these cases fallow lands have been taken out of the swidden system, decreasing the area where swidden is practiced within the farmer's land or the farmer has converted all of the land to tree crops, removing swidden from the household's livelihood system entirely. A similar land-use change trajectory is also seen in midland areas, a region where lowland delta transitions to uplands. In much of the midlands annual cropland has been converted to perennial tree crops. In most of these cases the trees that have been planted are fruit trees and the crop is sold in the expanding lowland markets.

A third trajectory that has been documented is the transition of swidden areas to permanent cropland. These croplands are dominated by maize, fodder crops, or legumes. In parts of the northern mountains hybrid maize has been promoted as part of an agricultural package. The maize is part of a commodity chain leading down to lowland feedlots that are producing large amounts of beef for an expanding urban market. Maize production has increased in the mountains and in many areas has replaced the swidden systems that were found there (Leisz et al., 2005, 2007; Vu, 2007). Annual production of legumes, peanuts, and other fodder crops on permanent fields has also replaced swidden fields in some places (Vu, 2007).

A fourth trajectory is the replacement of swidden with plantation timber. This change has been in response to government projects such as the 323 program and the 5 million hectares program, both of which encouraged and subsidized the planting of tree crops on previously swidden agriculture land. The result of this has been the expansion of plantation timber in some upland provinces (Vu, 2007; Zingerli et al., 2002). Parallel to this change, though on a much smaller scale, has been the expansion of protected forest areas, which have been reported in some provinces (Zingerli et al. 2002).

A fifth trajectory of land-use change is the increasing use of fallow areas as pastureland for cattle. This change in the use of fallow land has been observed in Nghe An Province (Leisz et al., 2007, 2011). The land-use change is in response to demand in lowland cities and other areas for increased beef in diets. While swidden systems are still present in the uplands in these areas, the use of fallow areas as pasture has expanded and the size of cultivated fields within the swidden land areas has decreased.

6.4.2 Land-Cover Change in the Uplands and Midlands

Each land-use change trajectory from dominant swidden systems to other land-use systems has an implication for how land cover has changed. Table 6.1 illustrates the relationship between land uses in the uplands and midlands and different types of land covers. In general where different tree crops, tree plantations, or protected forests replaced swidden systems, there has been a change in the land cover from a mosaic of land covers made up of cultivated areas, grass-dominated areas, bush-dominated areas, and tree-dominated areas that are scattered over a village area to a more monolithic land cover for the same village area. In places where permanent field crops, such as maize, legumes, groundnuts, or other fodder crops

TABLE 6.1 Relationship Between Areas of Swidden/Fallow Systems Prior to *doi moi* and Land Cover in the Uplands and Midlands of Northern Vietnam

Initial Land-Use (Pre-Early 1990s)	**Specific Land-Cover Associated With Land-Use**	**Land-Use After *doi moi* (Late 1990s to Present)**	**Land-Cover Associated With Land-Use**
Swidden/fallow system	Mosaic of cultivated land and land-cover associated with fallow lands: grass, grass and bush, bush, bush and small trees, medium to large trees, bamboo	Swidden	Same as pre-*doi moi*
		Permanent tree crops	Tree cover
		Permanent upland crops	Cultivated land
		Plantation timber or production forest or protected forest	Tree cover
		Pasture for animal husbandry	Mixed grass and bush-dominated areas

have replaced swidden systems, the land cover has become dominated either by permanently cultivated land, or has changed to a "binary" land cover where part of the village area is permanently cropped fields and part is maturing tree cover. In areas where swidden systems still exist, but animal husbandry is playing a more central role in the livelihood system, the influence of increased numbers of cattle on the landscape is perpetuating a mosaic of land covers on the area, but the period where the land cover is dominated by grass and bush is extended, as cattle graze the area and keep tree cover from reestablishing itself.

6.5 DISCUSSION AND CONCLUSION

The pattern of land-use and land-cover change in northern Vietnam is different for the lowlands, the midlands, and the uplands of the region. In the lowlands, despite the introduction of private property through the new land laws during the post-*doi moi* era, the way that irrigated rice land can be managed is still closely regulated by the government. This has created a situation in the periurban lowland areas where land-use change from agricultural to urban uses and the corresponding land-cover changes have not happened to the extent that many would expect in a country that is experiencing a rural-urban shift and agricultural-industrialization shift as Vietnam experienced from 1990 to 2012 (McPherson, 2012).

In the midlands and uplands the change in land uses from swidden-fallow dominated land-use systems to tree crop and permanent cropping systems means that the land-cover mosaics made up of cultivated fields, grass-dominated fields, grass and bush fields, bush-dominated fields, bamboo-dominated fields, and tree covered fields that are found within the fallow areas of swidden-fallow systems have been disappearing. The land cover that has replaced these fields is either large areas dominated by monocrops of trees, such as tea, coffee, rubber, or plantation timber (production forests), or tree-dominated protected forests, which have less temporal diversity in the ages of trees; or binary land cover of cropped agriculture fields interspaced with tree covered fields over an area. If animal husbandry has become more central within a swidden-fallow system, a land use that leads to more pasture dominated land cover, rather than a mixture of grass and bush, dominates these areas.

The land-use transitions and associated land-cover changes in the midlands and uplands can have implications for carbon sequestration and biodiversity in these areas. As researchers investigate the issue, it has been noted that contrary to what is popularly held, swidden-fallow systems, especially if they include long-term fallow areas, do a good job of sequestering carbon when the whole landscape of a region is considered (Bruun et al., 2009; Van Vliet et al., 2012; Ziegler et al., 2012). This is precisely because of the land-cover mosaic found over the swidden system landscapes and the large role that soil plays in sequestering carbon. The mosaic of land-cover types associated with swidden-fallow systems are thought to interact with the soil of an area in such a way that the carbon remains sequestered in it, compared to how soil is managed in other land uses (Bruun et al., 2009).

The land-use transitions and associated land-cover changes in the midlands and uplands also have implications for the biodiversity of these areas. Rerkasem et al. (2009) and Ziegler et al. (2011) detail the implications of changing swidden-fallow systems to other land uses and their associated land covers. In all of the cases, except for when the land-use change is to protected natural forest, as the land-cover simplifies the biodiversity of the area decreases.

Overall, the literature suggests that land-use and land-cover changes in the midlands and uplands have been significant since the onset of *doi moi*. Indications also are that these changes are leading to less carbon sequestration and less biodiversity. Furthermore, the changes in the lowland areas are not as great as one would expect given the level of urban growth and industrialization that is taking place. Whether these patterns continue into the next decades or change, as both urban and rural development continues in northern Vietnam, is an interesting question.

References

Ahlback, A.J., 1995. On Forestry in Vietnam, the New Reforestation Strategy and Assistance. UNDP/FAO Project VIE/92/022, Hànội.

Beland, M., Goita, K., Bonn, F., Pham, T.T.H., 2006. Assessment of land-cover changes related to shrimp aquaculture using remote sensing data: a case study in the Giao Thuy District, Vietnam. Int. J. Remote Sens. 27 (8), 1491–1510.

Bruun, T.B., de Neergaard, A., Lawrence, D., Ziegler, A., 2009. Environmental consequences of the demise in swidden agriculture in Southeast Asia: carbon storage and soil quality. Hum. Ecol. 37, 375–388.

Castella, J.C., Boissau, S., Thanh, N.H., Novosad, P., 2006. Impact of forestland allocation on land use in a mountainous province of Việtnam. Land Use Policy 23, 147–160.

Dat, P.T., Yoshino, K., 2012. Mangrove analysis using ALOS imagery in Hai Phong City, Vietnam. In: Remote Sensing of the Marine Environment II Book Series. Proceedings of SPIE, Vol. 8525.

De Koninck, R., 1999. Deforestation in Viet Nam. IDRC Resources Books, Ottawa.

DiGregorio, M., Leisz, S.J., Vogler, J., 2003. The invisible urban transition: rural urbanization in the Red River Delta. In: Paper Presented for the 7th International Congress of the Asian Planning Schools Association, Hanoi, 12–14 September.

Đỗ, Q.-T., Iyer, L., 2003. Land Rights and Economic Development: Evidence from Viet Nam. World Bank, Development Research Group, Washington, DC. Policy Research Working Paper 3120.

Douglass, M., DiGregorio, M., Pichaya, V., Boonchuen, P., Brunner, M., Bunjamin, W., Forster, D., Handler, S., Komalasari, R., Taniguchi, K., 2002. The Urban Transition in Vietnam. UNCHS, Fukuoka, Japan. p. 290.

Đức, T., 2003. The Farm Economy in Việt Nam. The Gioi Publishers, Hanoi.

Fox, J., Vogler, J., 2005. Land use and land cover change in montane mainland Southeast Asia. Environ. Manag. 36, 3394–3403.

Fox, J., Truong, D.M., Rambo, A.T., Tuyen, N.P., Cuc, L.T., Leisz, S., 2000. Shifting cultivation: a new paradigm for managing tropical forests. Bioscience 50 (6), 521–528.

Fox, J., Leisz, S., Truong, D.M., Rambo, A.T., Tuyen, N.P., Cuc, L.T., 2001. Shifting cultivation without deforestation: a case study in the mountains of northwestern Vietnam. In: Millington, A.C., Walsh, S.J., Osborne, P.E. (Eds.), Applications of GIS and Remote Sensing in Biogeography and Ecology. Kluwer Academic, Boston, pp. 289–308.

Gomiero, T., Pettenella, D., Giang, Triều Phan, Maurizio, P., 2000. Vietnamese uplands: environmental and socio-economic perspective of forest land allocation and deforestation process. Environ. Dev. Sustain. 2, 119–142.

Government of the Socialist Republic of Việt Nam, 1993. Allocation of Agricultural Land, Decree No. 64-CP, September. Official Gazette, Hànội.

Government of the Socialist Republic of Việt Nam, 2003. Land Law. National Political Publisher, Ha Noi. 2004.

Hai-Hoa, N., Pullar, D., Duke, N.C., McAlpine, C., Hien, N.T., Johansen, K., 2010. Historic shoreline changes: an indicator of coastal vulnerability for human land-use and development in Kien Giang, Vietnam. In: Proceedings of Asian Association on Remote Sensing.

Hien, P.T.T., Beland, M., Bonn, F., Goita, K., Dubois, J.M., Cu, P.V., 2006. Assessment of land-cover changes related to shrimp farming in two districts of northern Vietnam using multitemporal Landsat data. In: Global Developments in Environmental Earth Observation from Space. 25th Symposium of the European-Association-of-Remote-Sensing-Laboratories (EARSeL), 6–11 June 2005. Univ Porto, Oporto, Portugal, pp. 845–853.

Jakobsen, J., Rasmussen, K., Leisz, S., Folving, R., Quang, N.V., 2007. The effects of land tenure policy on rural livelihoods and food sufficiency in the upland village of Que, north central Vietnam. Agric. Syst. 94 (2), 309–319.

Lambin, E.F., Mefroidt, P., 2010. Land use transitions: socio-ecological feedback versus socio-economic change. Land Use Policy 27, 108–118.

Leisz, S.J., 2009. Dynamics of land cover and land use changes in the upper Ca river basin of Nghe An, Vietnam. Tonan Ajia Kenkyu (Southeast Asian Studies) 47 (3), 287–308.

Leisz, S.J., thi Thu Ha, N., thi Bich Yen, N., Lam, N.T., Vien, T.D., 2005. Developing a methodology for identifying, mapping and potentially monitoring the distribution of general farming system types in Vietnam's northern mountain region. Agric. Syst. 85, 340–363.

Leisz, S.J., Rasmussen, K., Olesen, J.E., Vien, T.D., Elberling, B., Jacobsen, J., Christiansen, L., 2007. The impacts of local farming system development trajectories on greenhouse gas emissions in the northern mountains of Vietnam. Reg. Environ. Chang. 7 (4), 187–208.

Leisz, S.J., Ginzburg, R.F., Lam, N.T., Vien, T.D., Rasmussen, K., 2011. Geographical settings, government policies, and market forces in the uplands of Nghe An. In: Sikor, T., Tuyen, N.P., Sowerine, J., Romm, J. (Eds.), Upland Transformations: Opening Boundaries in Vietnam. National University of Singapore Press, Singapore, pp. 115–145.

MAFI (Ministry of Agriculture and Forest Industry), 1993. Bare lands in Việt Nam: The Existing Situation and the Improving and Using Orientation up to the Year 2000. MAFI, Hànội.

McElwee, P., 2004. You say illegal, I say legal: the relationship between 'illegal' logging and land tenure, poverty, and forest use rights in Vietnam. J. Sustain. For. 19, 97–135.

McGee, T.G., 1991. The emergence of Desakota in Asia: expanding a hypothesis. In: Ginsberg, N., Koppel, B., McGee, T.G. (Eds.), The Extended Metropolis. University of Hawaii Press, Honolulu, pp. 3–25.

McPherson, M.F., 2012. Land policy in Vietnam: challenges and prospects for constructive change. J. Macromark. 32 (1), 137–146.

Rambo, A.T., 1998. The composite swiddenning agroecosystem of the Tày Ethnic Minority of the Northwestern Mountains of Vietnam. In: Patanothai, A. (Ed.), Land Degradation and Agricultural Sustainability: Case Studies from Southeast and East Asia, Regional Secretariat The Southeast Asian Universities Agroecosystem Network (SUAN). Khon Kaen University Press, Khon Kaen, pp. 43–64.

Rerkasem, K., Lawrence, D., Padoch, C., Schmidt-Vogt, D., Ziegler, A.D., Bruun, T.B., 2009. Consequences of swidden transitions for crop and fallow biodiversity in Southeast Asia. Hum. Ecol. 37, 347–360.

Rigg, J., 2001. Embracing the global in Thailand: activism and pragmatism in an era of deagrarianization. World Dev. 29 (6), 945–960.

Rigg, J., Veeravongs, S., Veeravongs, L., Rohitarachoon, P., 2008. Reconfiguring rural spaces and remaking rural lives in central Thailand. J. Southeast Asian Stud. 39 (3), 355–381.

Seto, K., Reenberg, A., Boone, C.G., Fragkias, M., Haase, D., Langanke, T., Marcotullio, P., Munroe, D.K., Olah, B., Simon, D., 2012. Urban teleconnections and sustainability. Proc. Natl. Acad. Sci. U. S. A. 109 (20), 7687–7692.

Sikor, T., 1995. Decree 327 and the restoration of barren land in the Vietnamese highlands. In: Rambo, A.T., Reed, R.R., Cuc, L.T., DiGregorio, M.R. (Eds.), The Challenges of Highland Development in Vietnam. East-West Center, Honolulu.

Socialist Republic of Việt Nam, 2009. In: Decree 69/2009/CP Addendum on Land Use Planning, Land Price, Land Recall, Compensation, Support and Relocation. The Government of Socialist Republic of Vietnam, Hanoi. August 13.

Socialist Republic of Vietnam, 1994. Resolution no 02/CP. In: Trần, D. (Ed.), The Farm Economy in Việt Nam. The Gioi Publishers, Hano. 2003.

Socialist Republic of Vietnam, 1999. Decree No. 85/1999/ND-CP. In: Trần, D. (Ed.), The Farm Economy in Việt Nam. The Gioi Publishers, Hano. 2003.

Thành, N.L., Mễ, V.V., Tiêm, N.V., 1995. Land classification and land allocation of forest land in Vietnam – A meeting of the national and local perspective. Forest, Trees and People Newsletter 25, 31–36.

Thiha, Webb, E.L., Honda, K., 2007. Biophysical and policy drivers of land use-cover change in a Central Vietnamese District. Environ. Conserv. 33, 2164–2172.

Tottrup, C., 2002. Deforestation in the upper Ca river basin in North Central Vietnam – a remote sensing and GIS perspective. In: Geographica Hafniensia C12. Institute of Geography, University of Copenhagen, Copenhagen, Denmark.

Tran, T.H.T., 2007. The Impact of the renovation policies on the livelihoods of upland forest-dependent people in Northern Vietnam. In: Conference on Critical Transitions in the Mekong Region, 29–31 January 2007, Chiang Mai, Thailand.

Van Vliet, N., Mertz, O., Heinimann, A., Langanke, T., Adams, C., Messerli, P., Leisz, S., Pascual, U., Schmook, B., Schmidt-Vogt, D., Castella, J.C., Jorgensen, L., Birch-Thompson, T., Hett, C., Bech-Bruun, T., Ickowitz, A., Vu, K.C., Yasuyuki, K., Fox, J., Dressler, W., Pacoch, C., Ziegler, A.D., 2012. Trends, drivers and impacts of changes in swidden cultivation in tropical forest-agriculture frontiers: a global assessment. Glob. Environ. Chang. 22 (2), 418–429.

Viên, T.D., Leisz, S.J., Lâm, N.T., Rambo, A.T., 2006. Using traditional swidden agriculture to enhance rural livelihoods in Vietnam's uplands. Mt. Res. Dev. 26 (3), 192–196.

Vu, K.C., 2007. Land Use Change in the Suoi Muoi Catchment, Vietnam: Disentangling the Role of Natural and Cultural Factors. PhD Dissertation, KatholiekeUniversiteit Leuven, Belgium.

Vu, K.C., Rompaey, A.V., Govers, G., Vanacker, V., Schmook, B., Hieu, N., 2013. Land transitions in Northwest Vietnam: an integrated analysis of biophysical and socio-cultural factors. Hum. Ecol. 41, 37–50.

Ziegler, A.D., Fox, J.M., Webb, E.L., Padoch, C., Leisz, S., Cramb, R.A., Mertz, O., Bruun, T.T., Vien, T.D., 2011. Recognizing contemporary roles of swidden agriculture in transforming landscapes of Southeast Asia. Conserv. Biol. 25 (4), 846–848.

Ziegler, A.D., Phelps, J., Yuen, J.Q., Webb, E.L., Lawrence, D., Fox, J.M., Bruun, T.B., Leisz, S.J., Ryan, C., Mertz, O., Dressler, W., Pascual, U., Padoch, C., Koh, L.P., 2012. Carbon outcomes of major land-cover transitions in SE Asia: great uncertainties and REDD+ policy implications. Glob. Chang. Biol. 18 (10), 3087–3099.

Zingerli, C., Castella, J.C., Manh, P.H., Cu, P.V., 2002. Contesting policies: rural development versus biodiversity conservation in the Ba Be National Park area. In: Castella, J.C., Quang, D.D. (Eds.), Doi Moi in the Mountains. Land Use Changes and Farming Systems Differentiations in Bac Kan Province, Vietnam. Agricultural Publishing House, Hanoi, Vietnam, pp. 249–275.

CHAPTER

7

The Role of Land-Use Planning on Socioeconomic Development in Mai Chau District, Vietnam

D.V. Nha

Vietnam National University of Agriculture, Hanoi, Vietnam

7.1 CONCEPTS OF LAND-USE PLANNING

FAO/UNEP (1997) defined land and land resources as an area of the Earth's terrestrial surface, encompassing all attributes of the biosphere immediately above or below this surface, including those of the near-surface climate, the soil and terrain forms, the surface hydrology (including shallow lakes, rivers, marshes, and swamps), the near-surface sedimentary layers and associated groundwater and geohydrological reserve, the plant and animal populations, the human settlement pattern and physical results of past and present human activity (terracing, water storage or drainage structures, roads, buildings, etc.). According to the FAO and UNEP (1999), the basic functions of land in supporting human and other terrestrial ecosystems can be summarized as follows:

- A store of wealth for individuals, groups, or a community.
- Production of food, fiber, fuel, or other biotic materials for human use.
- Provision of biological habitats for plants, animals, and microorganisms.
- Codeterminant in global energy balance and the global hydrological cycle, which considers both a source and a sink for greenhouse gases.
- Regulation of the storage and flow of surface water and groundwater.
- Storehouse of minerals and raw materials for human use.
- A buffer, filter, or modifier for chemical pollutants.
- Provision of physical space for settlements, industry, and recreation.
- Storage and protection of evidence from the historical or prehistorical record (fossils, evidence of past climate, archeological remains, etc.).
- Enabling or hindering the movement of animals, plants, and people from one area to another.

Redefining Diversity and Dynamics of Natural Resources Management in Asia Volume 2

http://dx.doi.org/10.1016/B978-0-12-805453-6.00007-3

Based on the vital role of land, understanding and using it becomes the crucial responsibility of land users, authorities, scientists, and planners. From all appearances, people had used land resource naturally for their purposes to exist. Therefore, land use is a product of interaction between the biophysical and human driving forces (Weng, 2010). The ways of using land resources have been changed from rudimentary to modern methods throughout different periods of time. In actual development, an emerging issue is how to use land resources to meet the needs of present and future generations. Moreover, land use is characterized by the arrangements, activities, and through the people to produce, change, or maintain a certain land-cover type (FAO, 2005). Land use defined in this way establishes a direct linkage between land cover and the people's actions in their environment. Thus, it can be defined as the human use of land that involves both the manner of including biophysical attributes of land and purposes for use of land (Weng, 2010). In using land, the pressure on land has been gradually increased by different causes (The National Land Use Planning Commission, 1998) as follows: a growing number of conflicts between the different land users; insecurity of land use and tenure; poor development of land markets; deforestation; and increasing migrations of people and livestock.

Land-use planning (LUP) is a potential solution for sustainable use of land in the long term by optimizing the effective use of land resources. According to Crowley et al. (1975) LUP is defined as planning for the allocation of activities to land areas to benefit human. The discipline involves three sets of tasks:

- Forecasting requirements or demands for goods and services.
- Estimating the supply of land available to produce the goods and services (in terms of amount, location, quality, suitability, or capability).
- Evaluating, implementing, and monitoring the alternative management and control strategies.

The aim of LUP is, therefore, to meet the needs of land users and development, including urban (residential, commercial, industrial, institutional), transportation, agriculture, forestry, mining, and outdoor recreation. For instance, in the 1970s people in the United States had found the best way to use land for meeting different needs, such as goods and services through LUP. In addition, LUP had significantly contributed to the land management and land-use strategies in the long run (Crowley et al., 1975). Thus, the relationship between land availability and land requirements in an area has been mentioned in both present and future contexts.

To disseminate LUP worldwide, the Food and Agriculture Organization of the United Nations (FAO) (1993) defined it as: "Land use planning is the systematic assessment of land and water potential, alternatives for land use and economic and social conditions in order to select and adopt the best land use options." The purpose is to select and take into practice the land uses that will best meet the needs of the land users as well as safeguarding resources for future. The need for change, the need for improvement of management, and the need for a quite different pattern of land use are dictated by changing circumstances.

This definition was, therefore, accepted for its comprehensive contents, including its economic, social, and environmental components. In particular, the needs of people for land resources in LUP are permanently met both in the present and the future, so the participation of local people is important (Huy, 2009).

To understand and clearly apply LUP for sustainable development, FAO and UNEP (1999) proposed a definition of LUP as: "Land-use (or Land Resources) Planning is a systematic and iterative procedure carried out in order to create an enabling environment for sustainable development of land resources which meets people's needs and demands. It assesses the physical, socio-economic, institutional and legal potentials and constraints with respect to an optimal and sustainable use of land resources, and empowers people to make decisions about how to allocate those resources."

The Working Group on Integrated Land Use Planning defined LUP as: "Land use planning is an iterative process based on the dialogue amongst all stakeholders aiming at the negotiation and decision for a sustainable form of land use in rural areas as well as initiating and monitoring its implementation" (Amler et al., 1999).

In short, LUP is understood as a solution in the future for sustainable development and land resources use in the economic, social, and environmental dimensions. In addition, LUP has also been directly related to public administration and policies (Bristow, 1981; Puginier, 2002). However, understanding and using these definitions of LUP in practice are a concern for planners so as to avoid some misunderstandings that lead to making inefficient LUP in the future.

7.2 METHODS OF LUP

Some approaches and methods of LUP have been actually proposed and applied in different regions to grow land user's income and pursue sustainable development. As the numbers of people continue to increase, there is an increasingly urgent need to match land types and land uses in the most rational way possible (FAO, 1993). Some of the methods will be explained below.

7.2.1 FAO Approach and Guidelines for LUP

The FAO approach and guidelines helps those involved in the planning process, including development, management, and conservation of rural development. LUP described in FAO's, 1993 guidelines contains 10 steps (see Fig. 7.1): (1) establish goals and terms of reference; (2) organize the work; (3) analyze the problems; (4) identify opportunities for change; (5) evaluate land suitability; (6) appraise the alternatives: environmental, economic and social analysis; (7) choose the best option; (8) prepare the land-use plan; (9) implement the plan; (10) monitor and revise the plan.

FAO (1993) indicated the aim of LUP is to make the best use of limited resources by:

- Assessing the present and future needs and systematically evaluating the land's ability to supply them.
- Identifying and resolving conflicts between competing uses, between the needs of individuals and those of the community, and between the needs of the present generation and those of future generations.
- Seeking the sustainable options and choosing those that best identified needs.
- Planning to bring about desired changes.
- Learning from experience.

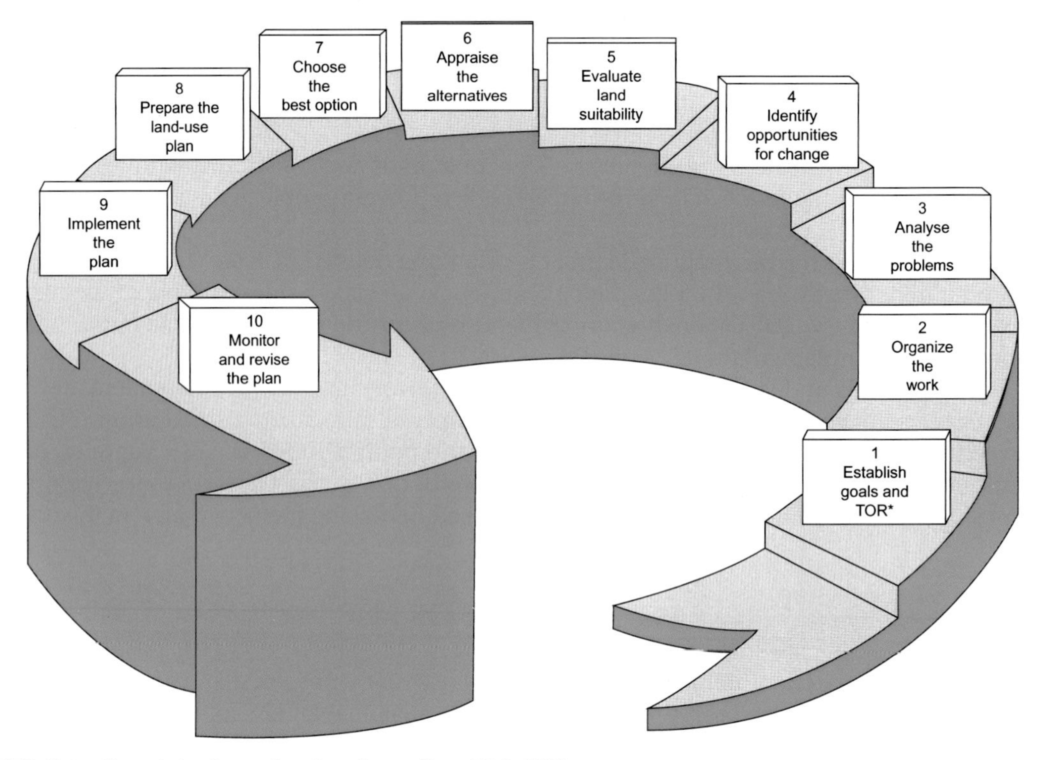

FIG. 7.1 Steps in land-use planning. *Source: From FAO, 1993.*

According to Fig. 7.1, the contents of land evaluation are described from step 5 to step 7. Land evaluation is mainly the analysis of data about land containing soils, climate, vegetation, an so forth, in terms of realistic alternatives for improving the use of the land. It is focused upon the land itself, its properties, functions, and potentials (FAO, 2007). The primary aim of land evaluation is to determine the suitability of land for alternative, actual, or potential land uses that are relevant to the area under consideration (Loi, 2008). Based on the results of the land evaluation, some land-use scenarios for the future are proposed and implemented in specific areas. However, some conflicts between different stakeholders are not clearly solved in the LUP process, so the decision makers need to trade off between different conflicting goals. According to Huy (2009), FAO's Guidelines for Land Use Planning states that three different levels at which LUP can be applied are national, district, and local. These different levels of LUP are relevant to the levels of government at which decisions about land use are made.

A technique used to solve these conflicts is multi-criteria evaluation (MCE), which emerged during the 1970s (Trung, 2006). This method includes six components: (1) a goal; (2) the decision maker; (3) evaluation criteria; (4) decision alternatives; (5) a set of uncontrollable variables; and (6) a set of outcomes. According to Trung (2006), the integrating the FAO approach with MCE for LUP includes six steps: (1) biophysical land evaluation; (2) socioeconomic assessment; (3) environmental assessment; (4) standardization; (5) calculation of suitability scores; and (6) scenarios analysis.

The advantages of the FAO-MCE approach is that it allows integration between biophysical land evaluation and socioeconomic and environmental assessment. Scenarios analysis can help decision makers trade-off different goals in long-term sustainable development. However, the drawbacks of the FAO-MCE approach are less realistic with large land mapping units (LMUs) and the possibly subjective justification and standardization of chosen criteria (Trung, 2006).

7.2.2 Participatory Land-Use Planning

People's participation in rural development was formulated in the mid-1970s. Participation of rural people in the institutions that govern their lives is a basic human right (UN, 2009). The FAO launched the people's participation program (PPP), then PPP has implemented pilot projects throughout the developing world in an attempt to test and develop an operational method of people's participation for incorporation in larger rural development schemes (FAO, 1990). According to Trung (2006), in a recent year, participatory land-use planning (PLUP) has gained international recognition as an important tool for achieving sustainable development and resource management by local communities. Based on this approach, several studies were carried out in various regions such as, China, Thailand, India, and Sri Lanka (Albecht et al., 1996). There were also some case studies carried out in Vietnam, Lao, Cambodia, and Thailand (Trung et al., 2004).

The PLUP approach focuses on the capacities and needs of the local people, so that sustainable land resources management can be achieved when the local people participate in management. The approach is promising in improving farmer awareness of environmental problems and solutions, as well as in linking local and scientific knowledge (Fagerström et al., 2003). According to (Fagerström et al., 2003), there are three steps in this approach: (1) researchers learning about local conditions; (2) the analyses of land use by local farmers; and (3) the feedback of the farmer-researcher in different land-use scenarios and potential effects on erosion and the household economy. Accordingly, a PLUP workshop is organized in a village to investigate the land-use problems, their causes, effects, and possible solutions. The workshop confirms integrating scientific and local knowledge to concrete options for sustainable land use that fits local realities and aspirations (Hessel et al., 2009). PLUP tries to identify land-use options that are accepted by stakeholders and satisfy the needs of all parties involved. Consequently, land users have to agree with the purposes of LUP, so they have to realize their responsibilities for sustainable land use and development.

In brief, PLUP uses a bottom-up approach to achieve the purposes of land use for the future. The land-use options are contributed from each land user, so PLUP is very useful to apply in small areas like villages and communes (Trung, 2006). In contrast, the difficulties of PLUP are related to conducting it in a large area and using the abilities of the local people involved.

7.2.3 Land-Use Planning and Analysis System

The land-use planning and analysis system (LUPAS) methodology was developed under the systems research network for ecoregional land use planning in tropical Asia (SysNet) project (1996–2000). The SysNet is a system of research networks in South and Southeast

Asia, established for the development and evaluation of methodologies to enhance strategic land-use policies (Trung, 2006, p. 5).

Actually, LUPAS has three main methodological components: (1) land evaluation; (2) scenario construction; and (3) multiple goal linear programming (Laborte et al., 2002). Similar to other methodologies, land evaluation and scenario are the basic components of LUP based on the biophysical and socioeconomic conditions in the specific area. In addition, multiple goal linear programming is the computerized component assisting planners, experts, and authority agents in setting up the targets for long-term development. LUPAS is a computerized decision support system based on the interactive multiple goal linear programming approach. It can be applied for scenario analysis of a complex problem like conflicts in land use (Roetter et al., 2005).

Furthermore, LUPAS consists of four main parts: (1) resource balance and land evaluation; (2) yield estimation; (3) input-output estimation; and (4) interaction multiple goal linear programming (Fig. 7.2). It can be used to point out the constraints of development deriving from the resources available, such as labor resources, capital limitations, and natural resources. Based on the development targets, goal restrictions are formed in LUPAS, such as minimum rice production for food security. Based on these, different land-use scenarios can be analyzed and optional land use can be selected to implement in a specific area. However, when development constraints and goals restrictions are changed, the land-use scenarios are changed accordingly. The LUPAS approach is still a top-down approach, so the participation of stakeholders is limited, even though they participated in setting development constraints and goal restrictions. Estimation of input-output depends on the secondary data. Thus, the precise results of models depends on government policy and ecological situations (Trung, 2006, p. 64).

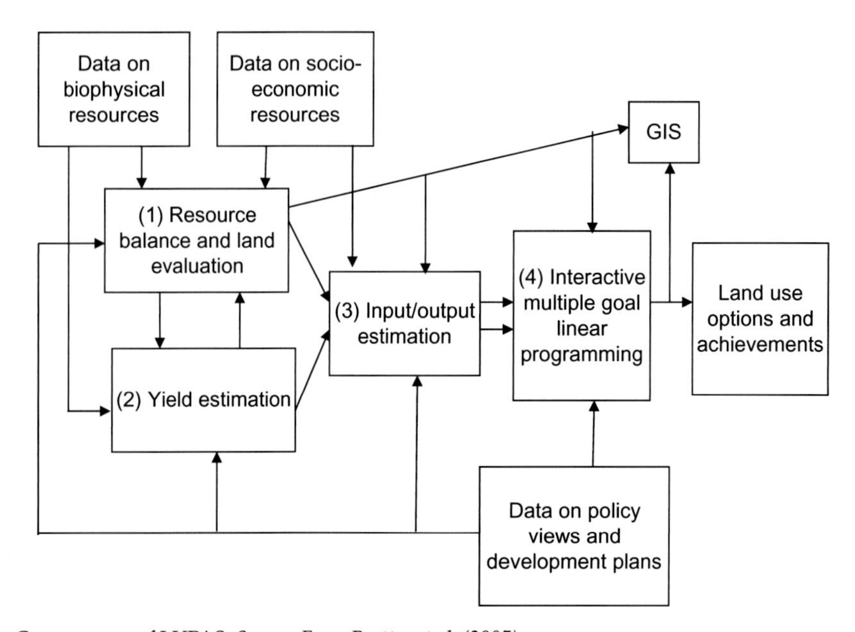

FIG. 7.2 Components of LUPAS. *Source: From Roetter et al. (2005).*

7.3 OVERVIEW OF LUP IN VIETNAM

LUP and plans in Vietnam is one of the 13 contents of State management on land (Article 6, Land Law, 2003) (The Socialist Republic of Vietnam, 2003). It is an administrative process to manage land resources. LUP divides and allocates land for specific purposes and development among different sectors. Not only it is the only spatial plan in the country, but there are also urban development plans, agriculture development plans, forest planning, and many more. However, LUP is, in theory, the overriding spatial plan that covers all land and is also the legal basis for any types of land use. It is developed every 10 years (planning) and 5 years (plan) for all administrative levels as well as for high-tech zones and special economic zones (The Socialist Republic of Vietnam, 2003; SEMLA, 2009). There were two main periods of development regarding to LUP, as discussed next.

In the period of 1975–86, all of Vietnam had a centrally planned economy decreed by 5-year plans with production targets (Trung, 2006). The planning system basically followed the economic system, when government often interfered arbitrarily in the production and distribution process. LUP followed the Soviet socialist-style model of a centrally planned economy. Under this system, the resources were allocated by the state through its command directive system. The means of production belonged to public ownership. The operation of the centralized planning model was described in a simple form: the state economic units were set up in accordance with Soviet managerial style to produce a certain product (or a group of products). The production inputs and outputs were supplied and received directly by the state without analyzing the economic effectiveness and efficiency based on real demand and purchasing power from the society. In this model, the private sector was abolished and there was no opportunity for foreign capital to invest (Quang, 2003). Therefore, LUP in this period showed the bold characteristics of a bureaucratic and subsidized mechanism and met the needs of the centrally planned economy without considering the demands of land users.

By 1986, the Sixth Vietnam Party Congress officially launched the socioeconomic reform (*doi moi*) recognizing the multisectoral market economy and creating the legal framework for private sector development. In the physical planning area, despite continuing with the old method of central command planning (in the form of master and detailed plans), the physical planning system was undergoing certain changes, in which the plans were less rigid and took market factors into account. The current master plan[1] was considered as a wider spectrum of market elements, such as the plurality of development actors, the introduction of a private land-use right, the recognition of individual trade (ie, private shop-houses), and the opening to foreign capital (ie, industrial zones) (Quang, 2003). To improve the quality of LUP in recent years, it has been synchronized with overall socioeconomic development planning and detailed spatial planning for urban and rural residential areas. Notably, LUP has been tuning gradually with the market that is development in Vietnam (Vo and Nhu Trung, 2007). Thus, the efficiency of land use and demands of land users for land are emerged continuously in LUP.

Recently, LUP in Vietnam has been carried out at four levels (Fig. 7.3): national, province, district, and commune (Socialist Republic of Việt Nam, 2009).

[1] Socioeconomic development planning.

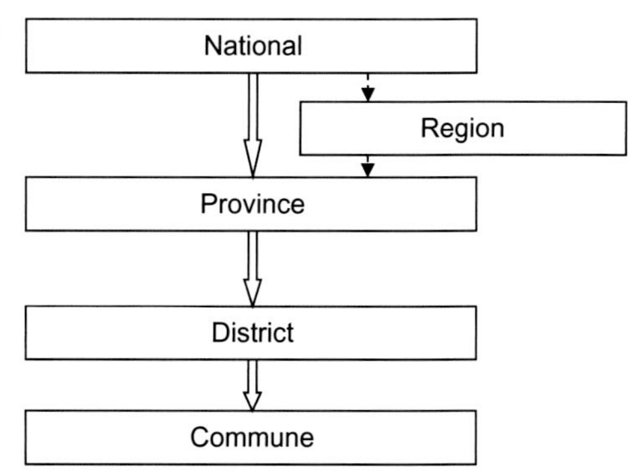

FIG. 7.3 Levels of LUP in Vietnam. *Source: Based on Socialist Republic of Việt Nam, 2009.*

Thus, it can be said that LUP has been fitted to the administrative hierarchy in Vietnam as a tool to manage and use the land resources efficiently. Besides, to concretize LUP in different scales, the Vietnam Ministry of Natural Resources and Environment (MONRE) promulgated Circular No. 19, dated February 11, 2009, that amended Circular No. 30/2004 on detailed provisions on LUP. Seven steps (see Fig. 7.4) of LUP in each level are regulated (MoNRE, 2009).

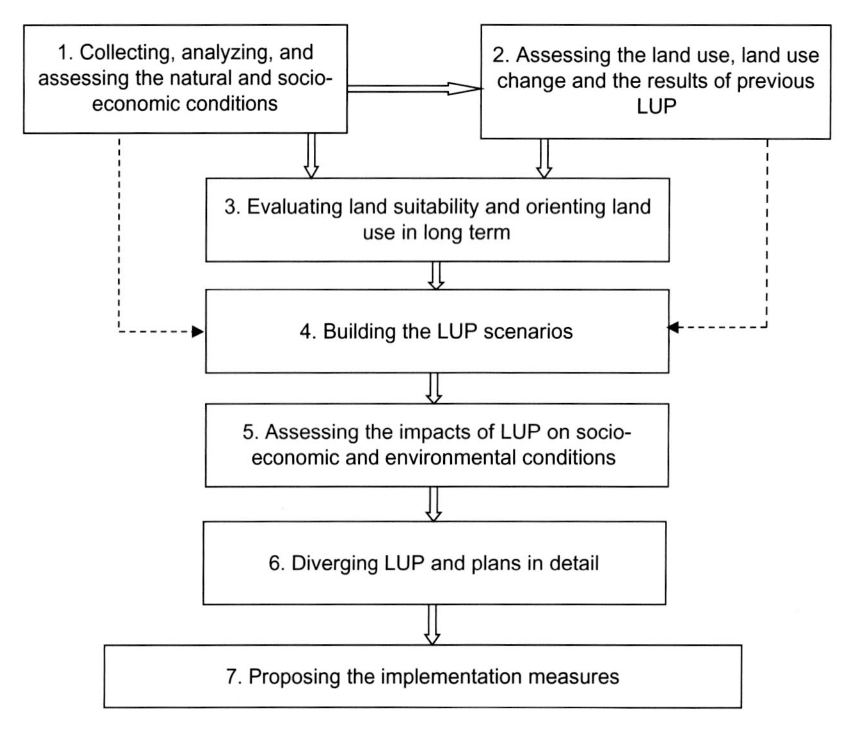

FIG. 7.4 LUP process in Vietnam. *Source: Based on MoNRE, 2009.*

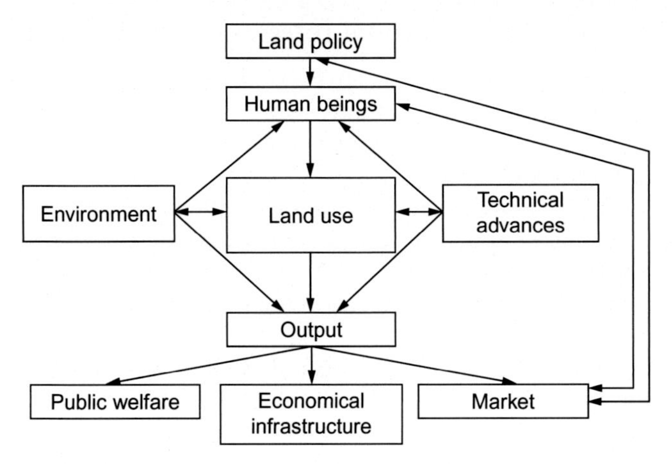

FIG. 7.5 Key factors influencing LUP in Vietnam. *Source: From Huyen, T.G., 1993.*

According to Huyen (1993), the future of LUP strategies will have to consider the population density[2] between the most populated and least populated areas, which is currently different about 17 times (GSO, 2010). Therefore, future LUP may have to seriously consider redistribution and resettlement of the population covering the entire country, not just in selected areas. Fig. 7.5 shows the key factors that influence LUP in Vietnam.

In addition to the change in the economy in Vietnam, environmental fluctuations have challenged development and LUP. To deal with these challenges, with the help of the Vietnam-Sweden program on Strengthening Environmental Management and Land Administration (SEMLA), some pilots of LUP with integrated environmental factors were carried out in some provinces. A comprehensive report was produced by SEMLA that outlined the current framework and context for LUP in Vietnam and identifies drawbacks and issues, as follows (SEMLA, 2009):

- Poorly integrated planning (with other sectors).
- Poor quality baseline data/mapping.
- Complex instructions that are difficult to follow.
- Weakness of planning expertise (lack of capacity).
- Inflexibility of plans (difficult to change).
- Lack of unified LUP strategy.
- Conflicting values and interests.
- Weak environmental planning.
- Lack of community consultation.

[2] Person per km^2.

7.3.1 Lessons Learned: Environmental Integration (SEMLA, 2009)

- Integration of environmental aspects has to been done in all the pilot programs, to a varying degree and using different methods.
- It is easy to describe current environmental problems, and hard to enforce or implement the environmental recommendations.
- Difficulties encountered includes lack of environmental data, lack of experience in predicting environmental impacts, weak cooperation among agencies, overlapping of different sector plans.
- The level of details in the environmental assessment of LUP is difficult to determine:
  - If it is too general and it is not useful.
  - If it is too detailed and it resembles an environmental impact assessment (EIA) for individual projects.
- The SEMLA integrated model has proven to be useful, especially in defining the LUP organization arrangements, environmental context, and trends analysis and public participation.
- Significant improvements can be seen when it comes to evaluating environmental issues that concern land use and also impact assessments of LUP.
- Strategic environmental assessment (SEA) as a stand-alone process is suitable for large development planning with a multitude of stakeholders and complex environmental implications, either for large areas (regions, provinces) or where major changes are planned for, and not for smaller planning and plans.
- The most effective use of SEA is when elements of SEA (public participation, alternatives development, and environmental assessment) are made an integrated part of the LUP process, and not as a parallel process resulting in duplications and weak linkages.

7.3.2 Application of the FAO Approach, PLUP, and LUPAS in Vietnam

According to the LUP process and LUP policy, the FAO approach for LUP is selected as a starting point and integrated in different LUP steps in Vietnam (Trung, 2006). Basically, 10 steps of the FAO approach and guidelines for LUP are applied specifically in 7 steps of LUP in Vietnam. Trung (2006) stated that the LUP approach is the most popular one in Vietnam. This approach and guidelines are used widely in land valuation to determine land suitability for different land-use types in the future (step 3 of the LUP process; see Fig. 7.4). Besides, the integration of biophysical land evaluation with socioeconomic and environmental appraisal is also observed. It is able to analyze the trade-offs between development targets by analyzing different scenarios. Finally, plan, implement plan, and monitor or revise the plan are applied in step 6,and step 7 of the LUP process. Therefore, the FAO approach and guidelines for LUP are useful and applied widely in different LUP levels in Vietnam.

PLUP is a bottom-up approach to apply in LUP. The domination part of this approach is the vital role of local people in making LUP achieve balanced development in society. Actually, this approach was applied in Vietnam in the early 2000s. Indeed, in 2003 PLUP was undertaken in two villages of the Mekong Delta coastal area (Huy, 2009). According to Trung (2006), PLUP is not widely used in Vietnam, but it is gaining attention from the local people in the LUP process. The advantage of this approach is that it can help to reduce the land-use conflicts by taking into account the farmers' requirements. Thus, this method is only used in the lowest level (commune level) that needs to consult with local people in Vietnam.

LUPAS is aimed at optimizing the use of resources. Through the application of LUPAS, the government's development goals are evaluated on their feasibility. Actually, LUPAS was recognized and used in Vietnam in the early 2000s. Specifically, it was applied in building LUP in Baccan Province and Cantho Province in 2002 and 2003. The results proved that it is suitable for the province level (Trung, 2006; Yen et al., 2002). However, it is not used widely in Vietnam. In particular, multiple goal linear programming is not popular for planners or authorities in Vietnam. Therefore, the application of this method in the LUP process in Vietnam is still limited.

In conclusion, in association with the changes of the economy toward a market-oriented philosophy, LUP in Vietnam also has changed gradually to supply land resource for development and sustainable development, in particular, to meet the needs of land users (households). However, LUP is obviously an open field, and needs to be approached by different methods that can help to find the best way to utilize LUP in Vietnam. The integration of environmental factors into LUP was actually germinated in Vietnam by the help of SEMLA. However, the integration was not in the process of LUP, even though it was carried out as some experiments in some provinces. Moreover, the integration was estimated as a referent document for planners to conduct in other plots in Vietnam. Furthermore, an economic analysis of the integration was not actually conducted in the plots. These limitations need to be studied in further research.

7.4 CORRELATION BETWEEN LUP AND SOCIOECONOMIC DEVELOPMENT IN MAI CHAU DISTRICT, HOA BINH PROVINCE, VIETNAM

7.4.1 Conceptual Framework

7.4.1.1 Research Area

Mai Chau District, with its complicated terrain, was conveniently selected to carry out the study. Located in a mountainous and attractive region of the province with many beautiful landscapes and traditional customs, the district is considered as one of the most beautiful districts of Hoa Binh Province in the northwest region of Vietnam. Moreover, the location of the district is also a crucial bridge between Hanoi and other provinces in the northwest region of Vietnam.

7.4.1.2 Statistical Data

To determine and analyze the correlation between LUP made in 2000 and actual socioeconomic development from 2001 to 2010 in Mai Chau District, secondary data needs to be collected, including the following:

- The results of LUP made in 2000 for the period of 10 years development from 2001 to 2010 were collected at the Department of Natural Resources Management at the district and province level.

- Based on the land-use pattern in 2010, the implementation of LUP from 2001 to 2010 was judged. Also, it was investigated by the Department of Natural Resources Management in different scales.
- Economic development in the period from 2001 to 2010, including agriculture, nonagriculture, and so forth (especially agriculture) was also collected at the different departments in the research area.

Actual social and environmental conditions from 2001 to 2010 stored regularly at the Statistical Department were used to compare with the results of LUP.

7.4.1.3 *Interview*

Interview and observation methods were used to gather vital information regarding the creation and contribution of LUP to socioeconomic development in the selected area. The interviewees were authorities and natural resources management officials at the different communes who participated in making LUP in 2000 and implemented this LUP from 2001 to 2010 in their locations. Basically, participants had to clarify the contribution of LUP to socioeconomic development of their communes.

The aim to interview the authorities at different communes in the district was to collect their judgments of economic, social, and infrastructure development in their location having connection with the LUP made in 2000. Consequently, their judgment of LUP's contributions is one of the basic assessments to address the correlation.

Furthermore, natural resources management officials' judgments reinforce the crucial information of the LUP's contributions to the natural resources management, which is integrated to determine the correlation between LUP and socioeconomic development in the research area.

The questionnaire was focused on

- Process to make LUP in the year 2000.
- Contribution of LUP to socioeconomic development.
- Effect of LUP on environmental development.

The framework is shown in the Fig. 7.6.

Accordingly, the combination between secondary data and primary data plays a vital role in determining the correlation. SPSS was used to analyze the data and linear regression indicated the correlation between LUP and socioeconomic development.

7.4.2 The Results of LUP in Mai Chau District, Hoa Binh Province, Vietnam

A summary how LUP 2000 envisioned the changes to the main land-use types is shown in Fig. 7.7. Several land use types should increase gradually. For example, agricultural land was to rise by 347ha from 2001 to 2005 and 399ha from 2006 to 2010, forest land was to increase by 3281ha from 2001 to 2005 and 4121ha from 2006 to 2010, and nonagricultural land also was to rise by 128ha from 2001 to 2005 and 100ha from 2006 to 2010. To the contrary, unused land was planned to decrease dramatically by 3757ha from 2001 to 2005 and 4621ha from 2006 to 2010.

The data in Fig. 7.8 shows the comparison between the area of land uses in LUP 2000 and the area of actual land uses in 2010. There was a difference between land-use types in LUP and actual land use in 2010.

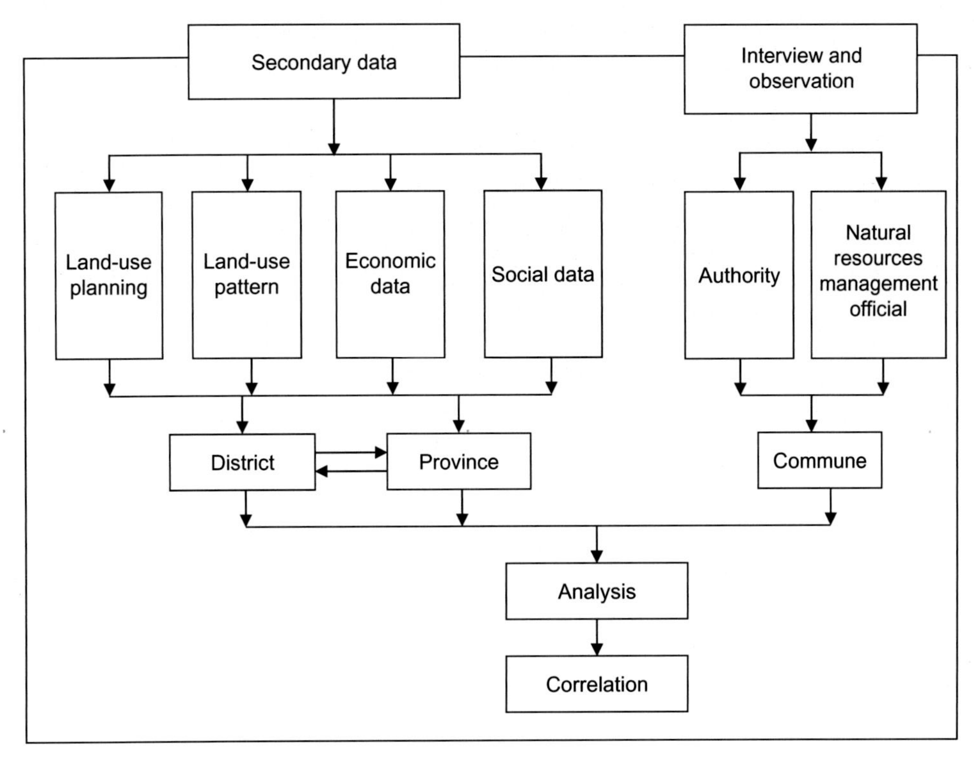

FIG. 7.6 Conceptual framework to determine the correlation.

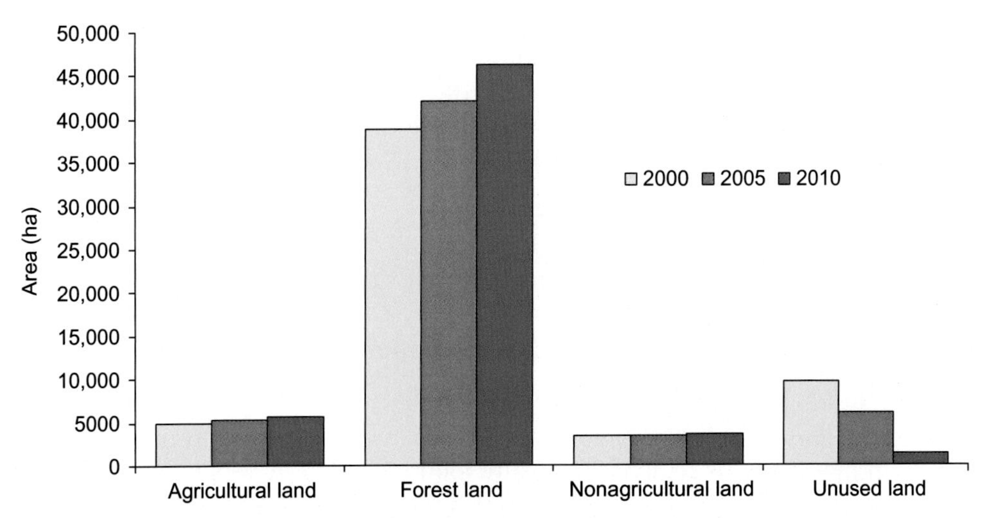

FIG. 7.7 Comparison between different land-use types in LUP in Mai Chau District. *Source: From Mai Chau People's Committee, 2001.*

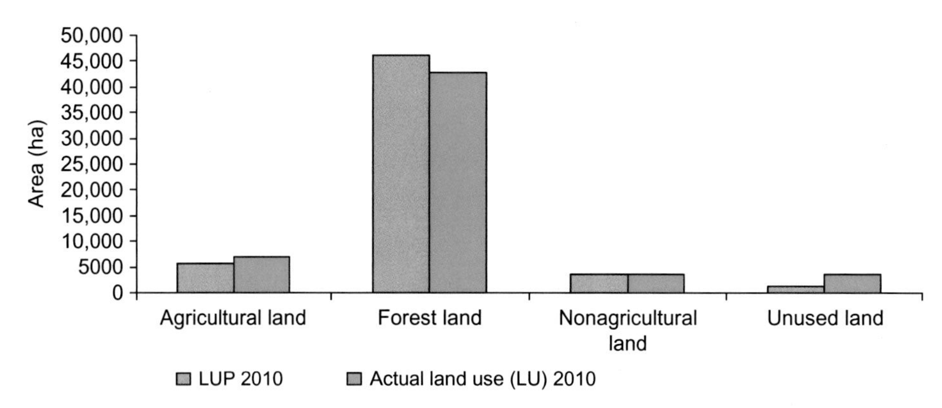

FIG. 7.8 Comparison between areas of land uses in LUP and area of actual land use in 2010 in Mai Chau District. *Source: Based on Mai Chau People's Committee, 2001d.*

The data in Table 7.1 demonstrate that actual agricultural land exceeded nearly 20% in comparison with agricultural land proposed in LUP, especially annual cropland exceeded nearly 46%. However, actual forest land only achieved 93% of the LUP plan, of which production forestland achieved nearly 52% and protection forestland reached around 183% of LUP. In the past period (2000–10), the reforestation in Mai Chau focused potentially more on protection than production forest. Moreover, actual residential land obtained nearly 105% of LUP (exceeded 5%). Land for construction of offices, public service delivery institutions, and land for national security and defense purposes only attained roughly 45% and 30%, respectively. These ratios, therefore, were very low, so it can be estimated that the purpose of using these types of land was changed in the actual implementation period. Probably, socioeconomic background assumptions of LUP 2000 as well as certain detailed planning ideas did not precisely match actual development. Land for nonagricultural production and business obtained nearly 100% of LUP, so the purposes of LUP and actual socioeconomic development were matched. Notably, unused land in LUP and actual land use were quite different. Indeed, this area in actual land use was about 3652 ha in comparison with 1351 ha in LUP, so the difference was about 270% (Fig. 7.9).

To sum up, the changes to the total stock of land-use classes tended to develop in the direction that was intended by LUP 2000.

7.4.3 Correlation Between LUP and Socioeconomic Development

7.4.3.1 Correlation Between LUP and Food Production

Commercial and industrial development in Vietnam is subjected to certain limitations, especially in mountainous regions. To ensure food for the local people has been a significant concern of farmers and authorities (FAO, 2011). Cuong (2005) demonstrated that developing agriculture and the rural economy to large-scale production would form a basis

TABLE 7.1 Results of Land-Use Planning Implementation From 2000 to 2010

Land Classification	LUP 2010 (ha)	Actual Land Use (LU) 2010 (ha)	Difference (ha)	Comparison (%)
1. Agricultural land	5749.50	6853.39	1103.89	119.20
1.1. Land for cultivation of annual crops	4393.93	6421.54	2027.61	146.15
Rice	1265.89	1244.51	−21.38	98.31
Others	3128.04	5177.03	2048.99	165.50
1.2. Land for cultivation of perennial crops	1355.57	431.85	−923.72	31.86
2. Forestland	46,176.61	42,833.77	−3342.84	92.76
2.1. Land for production forest	27,798.23	14,384.61	−13,413.62	51.75
2.2. Land for protection forest	12,857.08	23,500.97	10,643.89	182.79
2.3. Land for special-use forest	5521.30	4948.19	−573.11	89.62
3. Residential land	821.42	861.08	39.66	104.83
4. Land for construction of offices, public service delivery institutions	28.59	12.68	−15.91	44.35
5. Land for national security and defense purposes	26.00	7.82	−18.18	30.08
6. Land for nonagricultural production and business	27.98	28.07	0.09	100.32
7. Land for public use	532.76	496.00	−36.76	93.10
8. Land used for cemeteries and graveyards	215.01	183.91	−31.10	85.54
9. Land with rivers, canals, streams, and specialized water surface	1921.71	1921.71	0.00	100.00
10. Unused land	1350.80	3651.95	2301.15	270.35
Total area	56,850.38	56,850.38		

Source: Based on LUP in Mai Chau District.

for economic, political, and social stability. Thus, land users should develop and exploit effectively the natural resources in their administrative areas (Jocelyn, 2002). In the period from 2000 to 2010 in Mai Chau District, total food production increased remarkably for several reasons, such as an increase of the crop yields, and annual crop area or suitable change of the location of the annual crop with higher crop yields. The correlation between annual crop area and self-produced food is shown in Fig. 7.10.

The data indicate that total food product in Mai Chau increased gradually from roughly 13,200 tons in 2000 to 25,600 tons in 2010, while the area of annual crop also rose by nearly 53 ha in LUP and 2080 ha in actual land use throughout the same period.

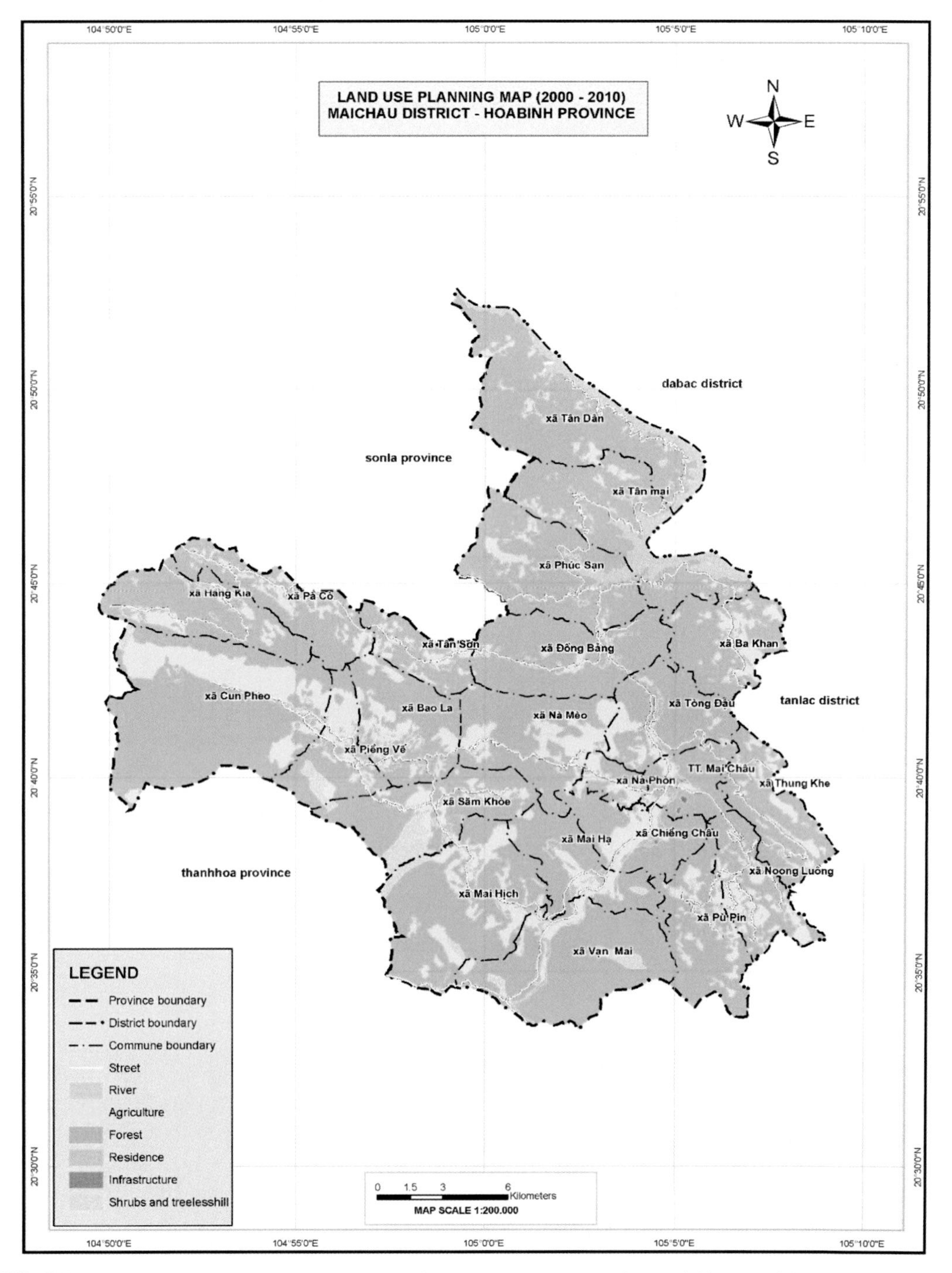

FIG. 7.9 LUP map in Mai Chau District, Hoa Binh Province. *Source: Based on Mai Chau People's Committee, 2001.*

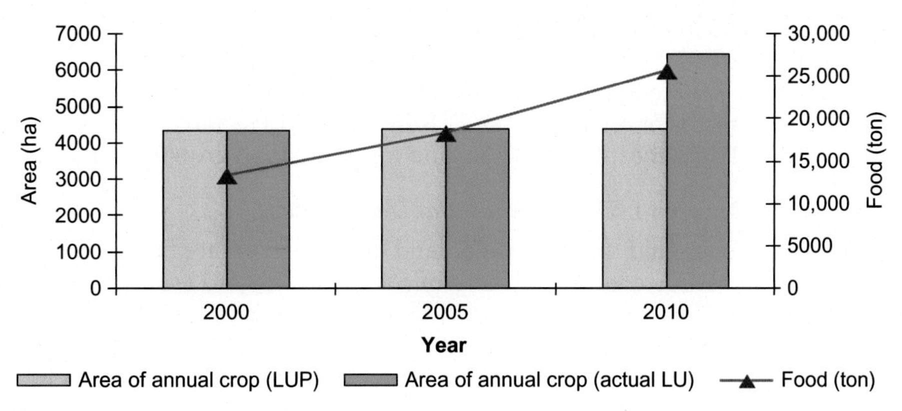

FIG. 7.10 Annual cropland and food production (2000–10).

7.4.3.2 Correlation Between LUP and Population Growth

To stably develop the society is also one of the main goals of LUP. Trends of population growth and economic development are directly related to the political stability of the government during a particular time in history (Kelly, 2004). The rate of population growth in developing countries is higher than in others, especially in Southeast Asian countries such as Vietnam and Indonesia, so the need to extend the residential area has been estimated as higher for LUP at different levels from nation to commune. Additionally, population density controls, in one form or another existing in most LUPs, can be expressed in different ways (Evans, 2004). The correlation between LUP and population growth in Mai Chau District is shown in Fig. 7.11.

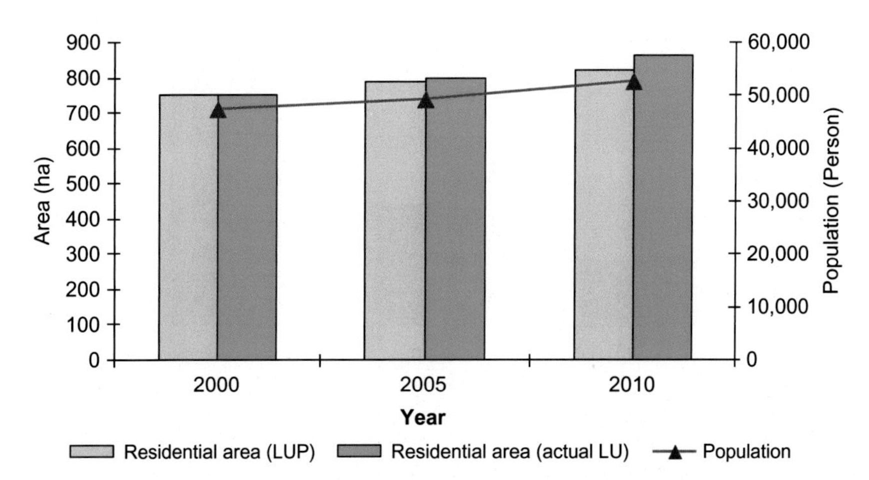

FIG. 7.11 Correlation between residential land and population growth in Mai Chau.

The figure indicates that the population of the district increased gradually from around 47,300 people in 2000 to 52,700 people in 2010, with an average population growth of 1.12% in 10 years (GSO, 2010). Residential land also rose significantly in both LUP and actual land use (LU). Indeed, the increases of roughly 70 and 110 ha were in LUP and actual LU, respectively. It is obvious that LUP was meant to provide land for population growth in such a period.

7.4.3.3 *Correlation Between LUP and Industrial Development*

Evans (2004) demonstrated that the use of land and the location of activities that operate in the LUP process possibly control the economic activities toward economic efficiency. The increase or decrease of land for economic activities is merely solved by LUP; it is a unique tool to accommodate land for different purposes throughout the specific period of development. In the first period of industrialization, land is actually significant and appeals to investors. The realization of rural industrialization and modernization demanded that industrial land rise significantly to meet the need for land and contribute to the increase of income from industry for local people (Mai Chau People's Committee, 2001).

Fig. 7.12 illustrates that land for nonagriculture and business was expanded gradually to support the demand of industrial development in Mai Chau District. Specifically, industrial land soared by around 21 ha both in LUP and actual LU from 2000 to 2010, an increase of more than three times throughout that period. The income from industry also rose dramatically from VND 5.43 billion in 2000 to VND 105.46 billion in 2010, higher by nearly 20 times. It is assumed that the increase in industrial land affected positively the industrial income of the district.

The correlation between LUP and food production, population, and industrial value is synthesized in Table 7.2. It shows that total output indicators correlate well with total assigned land use for a suitable land-use category.

Table 7.3 shows the correlation between intended change and actual change of land use in 23 communes from 2000 to 2010.

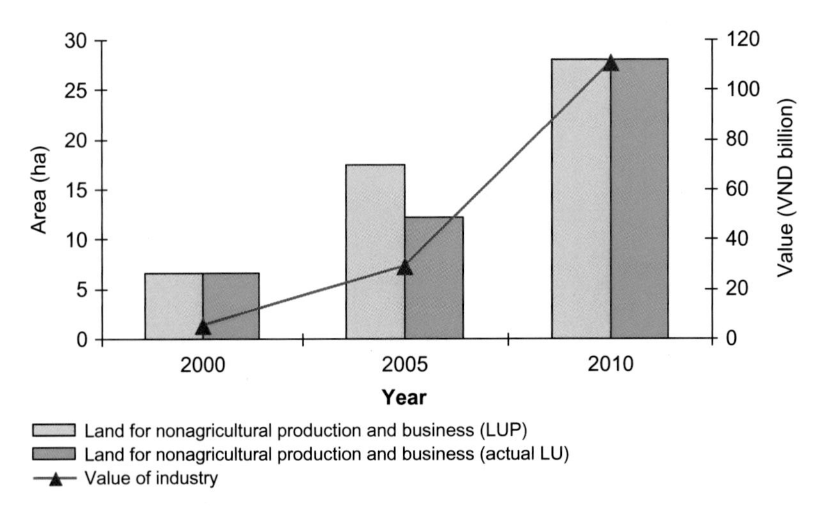

FIG. 7.12 Correlation between industrial land and the value of industry in Mai Chau.

TABLE 7.2 Correlation Between LUP and Socioeconomic Development

Dependent Variables		Independent Variables (LUP) ($n=23$)		
		Annual Cropland	Industrial and Business Land	Residential Land
1. Actual annual cropland	*R*-squared	0.579	0.086	0.069
	P-value	0.000	0.499	0.226
	Slope	1.112***	−14.140	−2.083
2. Actual industrial and business land	*R*-squared	0.043	1.000	0.064
	P-value	0.590	0.000	0.512
	Slope	−0.005	1.007***	0.051
3. Actual residential land	*R*-squared	0.000	0.008	0.400
	P-value	0.976	0.825	0.001
	Slope	0.001	−0.280	0.539***
4. Food	*R*-squared	0.579	0.068	0.069
	P-value	0.000	0.499	0.226
	Slope	4.434***	−56.388	−8.305
5. Population	*R*-squared	0.000	0.292	0.672
	P-value	0.990	0.133	0.000
	Slope	0.024	88.503	40.146***
6. Industrial value	*R*-squared	0.048	0.793	0.067
	P-value	0.573	0.001	0.502
	Slope	−0.022	3.944***	0.231

*, **, *** significant at 0.05, 0.01, 0.001, respectively.

The data in Table 7.3 proves that intended change (between actual land use in 2010 and LUP) and actual change (between actual land use in 2010 and actual land use in 2000) were significantly correlated for all land-use types. Specifically, for agriculture, 1 ha or 1% more in intended change was equivalent to 1.6 ha or 1.6% more in actual increase. For residence, 1 ha or 1% more in planned change, it increased 1.2 ha or 1.2% in actual change. In terms of industrial land, 1 ha or 1% more in intended change, the actual change increased 54 ha or 54%. For 1 ha or 1% more planned forest area, it increased 0.5 ha or 0.5% in actual change. For the unused land, the correlation was negative. In sum, a substantial impact of LUP 2000 on actual development appears at the municipality level is visible; however, as correlation coefficients (*slope*) vary and rarely approach +1.0, the actual spatial influence is limited.

Nota bene, this analysis was conducted at the municipality level, not at the level of single parcels of land to which a specific land use was assigned. The analysis indicates a high positive correlation even in potential cases where the intended changes had happened

TABLE 7.3 Correlation Between Intended Change and Actual Change of Land Use

Actual Land Use 2010—LUP (Intended Change)		Actual Land Use 2010—Actual Land Use 2000 (Actual Change) ($n = 23$)				
		Agriculture	**Residence**	**Industry**	**Forest**	**Unused**
Agriculture	*R*-squared	0.776	0.068	0.012	0.086	0.019
	P-value	0.000	0.228	0.617	0.175	0.529
	Slope	**1.619*****	0.053	−0.012	−1.924	−0.882
Residence	*R*-squared	0.082	0.789	0.008	0.162	0.035
	P-value	0.185	0.000	0.685	0.057	0.392
	Slope	3.615	**1.244*****	−0.069	18.166	8.19
Industry	*R*-squared	0.002	0.000	0.832	0.031	0.028
	P-value	0.852	0.93	0	0.419	0.446
	Slope	−40.409	−2.091	**54.055*****	619	−567.136
Forest	*R*-squared	0.163	0.093	0.024	0.416	0.308
	P-value	0.056	0.157	0.478	0.001	0.006
	Slope	−0.089	0.007	0.002	**0.51****	0.426
Unused	*R*-squared	0.039	0.054	0.009	0.114	0.589
	P-value	0.366	0.287	0.663	0.116	0.000
	Slope	0.095	−0.012	−0.003	−0.58	**−1.279*****

*, **, *** significant at 0.05, 0.01, 0.001, respectively.

somewhere else as long as these deviations balance at the municipality level. Thus, the actual spatial importance of LUP 2010 may be overestimated.

7.4.3.4 Opinion of Resource Managers and Officials

To reinforce the correlation between LUP and socioeconomic development from 2001 to 2010, the interview of natural resources management officials and authorities of 22 communes and one town in Mai Chau District was carried out with concrete questions focused on three main aspects: (1) participation in LUP; (2) contribution of LUP to socioeconomic development; and (3) effect of LUP on environment. Additionally, the area's increases and decreases of different land-use types in LUP were also extracted as independent variables.

Table 7.4 shows the influence that would be expected if LUP 2000 had worked perfectly. The table depicts three expected influences: +, positive; −, negative; and 0, no influence.

The data in Table 7.5 shows that LUP in the district was made in 2000 without local people's participation. Evans (2004) argues that a compromise with local people is very important in planning to achieve a balanced development. There was merely the participation of authorities and natural resources management officials in the making of LUP.

The contribution of LUP to economic development was claimed to be of great importance. Indeed, the contribution to socioeconomic development was rated as between 1.5 and 2.2

TABLE 7.4 Expectancy of LUP

	Independent Variables				
Dependent Variables	**Increase of Annual Cropland**	**Increase of Industrial and Business Land**	**Increase of Forestland**	**Increase of Residential Land**	**Decrease of Unused Land**
1. Contribution of LUP to economic growth	+	+	0	0	0
2 Contribution of LUP to agricultural development	+	0	−	0	0
3. Contribution of LUP to nonagricultural development	0	+	0	0	0
4. Contribution of LUP to residential development	0	0	0	+	0
5. Contribution of LUP to reforestation	−	0	+	0	+
6. Contribution of LUP to food security	+	−	−	0	+
7. Contribution of LUP to landslide prevention	−	−	+	−	+
8. Contribution of LUP to erosion prevention	−	−	+	−	+
9. Contribution of LUP to change in labor use	−	+	−	−	+

+, Positive; −, negative; 0, no influence.

TABLE 7.5 Descriptive Statistics of the Interview of Communal Officials

Variables	**Mean ($n=23$)**	**Std. Deviation**	**Min**	**Max**
Dependent variables				
1. Participation of authority in making LUP (yes=1; no=0)	1	0.0000	1	1
2. Participation of local people in making LUP (yes=1; no=0)	0	0.0000	0	0
3. Contribution of LUP to economic growth (low (<10%)=1; medium (10% to 15%)=2; high (>15%)=3)	2.0435	0.63806	1	3
4. Contribution of LUP to agricultural development (low (<10%)=1; medium (10% to 15%)=2; high (>15%)=3)	2.2174	0.73587	1	3
5. Contribution of LUP to nonagricultural development (low (<10%)=1; medium (10% to 15%)=2; high (>15%)=3)	1.4783	0.73048	1	3
6. Contribution of LUP to residential development (low=1; medium=2; high=3)	1.6957	0.55880	1	3

Continued

TABLE 7.5 Descriptive Statistics of the Interview of Communal Officials—cont'd

Variables	Mean ($n=23$)	Std. Deviation	Min	Max
7. Contribution of LUP to food security (low = 1; medium = 2; high = 3)	2.0435	0.82453	1	3
8. Contribution of LUP to landslide prevention (low = 1; medium = 2; high = 3)	1.7391	0.61919	1	3
9. Contribution of LUP to erosion prevention (low = 1; medium = 2; high = 3)	1.9130	0.59643	1	3
10. Contribution of LUP to reforestation (low = 1; medium = 2; high = 3)	1.6087	0.65638	1	3
11. Contribution of LUP to change in labor use (low = 1; medium = 2; high = 3)	1.4783	0.73048	1	3
Independent variables (LUP)				
1. Increase of annual cropland (ha)	2.2804	37.3315	−94.3100	76.2300
2. Increase of forestland (ha)	321.8461	397.9902	0.9100	1966.7500
3. Increase of residential land (ha)	3.0596	2.8041	0.3900	15.1000
4. Increase of industrial land (ha)	0.9322	2.2712	0.0000	9.5500
5. Decrease of unused land (ha)	364.2343	395.3139	55.3300	2029.8800

on a three-point scale (1: low, 2: medium, 3: high importance). The strongest influence was assumed for agricultural development (Table 7.5).

Table 7.6 shows that there is a significant correlation between the influences that municipality level interviewees *attribute* to LUP 2000 and actual socioeconomic development from 2001 to 2010. For example, the increase of annual crops and industrial land affected largely the agricultural and nonagricultural development, respectively.

7.5 CONCLUSIONS AND DISCUSSIONS

In general, LUP plays a vital role in socioeconomic development and environmental protection, in particular, the least-developed area. First of all, it is estimated to be a useful tool for the efficient use of natural resources, as well as to facilitate the institution for the use of land resource in different scales from national to commune. After that LUP contributes largely to sustainable development because it provides space for development, especially inputs for agricultural and forest development. Moreover, it can adjust the investment in long-term development, as it proposes different land-use scenarios in such a period of development. Finally, LUP is an agreed upon commitment to reconcile the benefits between the community and each individual (household) and between short-term and long-term development.

The correlation between the changes of land use in LUP and socioeconomic statistical data is evidence to prove partly the contribution of LUP to socioeconomic development throughout the period of development. Particularly, in the undeveloped area with deficient financial

TABLE 7.6 Correlation Between LUP and the Contribution of LUP to Socioeconomic Development

Variables		Independent Variables				
		Increase of Annual Cropland	**Increase of Industrial and Business Land**	**Increase of Forestland**	**Increase of Residential Land**	**Decrease of Unused Land**
1. Contribution of LUP to economic growth	*R*-squared	0.299	0.304	0.018	0.002	0.006
	P-value	0.007	0.006	0.539	0.856	0.721
	Slope	0.009**	0.155**	0.000	0.009	0.000
2. Contribution of LUP to agricultural development	*R*-squared	0.753	0.010	0.058	0.001	0.025
	P-value	0.000	0.652	0.268	0.896	0.475
	Slope	0.017***	0.032	0.000	−0.008	0.000
3. Contribution of LUP to nonagricultural development	*R*-squared	0.031	0.653	0.026	0.005	0.021
	P-value	0.420	0.000	0.464	0.752	0.510
	Slope	0.003	0.260***	0.000	−0.018	0.000
4. Contribution of LUP to residential development	*R*-squared	0.07	0.011	0.120	0.524	0.165
	P-value	0.222	0.630	0.105	0.000	0.054
	Slope	0.004	0.026	0.000	0.144***	0.001
5. Contribution of LUP to reforestation	*R*-squared	0.176	0.002	0.595	0.156	0.544
	P-value	0.046	0.838	0.000	0.055	0.000
	Slope	−0.007*	0.013	0.001***	0.095	0.001***
6. Contribution of LUP to food security	*R*-squared	0.687	0.024	0.151	0.054	0.096
	P-value	0.000	0.481	0.067	0.285	0.150
	Slope	0.018***	0.056	0.000	−0.068	0.000
7. Contribution of LUP to landslide prevention	*R*-squared	0.134	0.000	0.528	0.208	0.506
	P-value	0.086	0.963	0.000	0.029	0.000
	Slope	−0.006	−0.003	0.001***	0.101*	0.001***
8. Contribution of LUP to erosion prevention	*R*-square	0.149	0.018	0.441	0.144	0.403
	P-value	0.069	0.537	0.001	0.074	0.001
	Slope	−0.006	−0.036	0.001***	0.081	0.001***
9. Contribution of LUP to change in labor use	*R*-squared	0.096	0.611	0.017	0.004	0.012
	P-value	0.150	0.000	0.549	0.769	0.622
	Slope	0.006	0.251***	0.000	−0.017	0.000

*, **, *** significant at 0.05, 0.01, 0.001, respectively.

support, land resources and land allocation become the vital keys to change the economic structure toward nonagriculture and increase the income of local people. Besides, the correlation also demonstrates the effect of LUP implementation on actual development, which is one of the backgrounds to propose different land-use scenarios in LUP for the next period of development.

The changes of land use in LUP (independent variables) and actual socioeconomic development adjustment (dependent variables) of local authorities and natural resources management officials, who made and realized LUP, demonstrate the correlation as well. Indeed, LUP played a vital role in food security and contributed to socioeconomic development, such as economic growth, residential development, change of the labor force, and environmental protection (landslide and erosion prevention).

According to such evidence, we can restate that LUP contributes remarkably to socioeconomic development and environmental protection in the least-developed area where local people's low income and complicated terrain are major issues. Thus, improving the quality of LUP is estimated as one of the vital ways to realize the different development scenarios with separate conditions, including socioeconomic, physical, environmental, and so on. However, the creation and realization of LUP depends on the development's purposes in specific periods as well.

References

Albecht, D., Eller, E., Fledderman, A., Janz, K., Kunzel, W., Riethmuller, R., 1996. Experience of land use planning in Asian projects: selected insights. In: The Working Group on Land Use Planning for the Asian-Pacific Region.

Amler, B., Betke, D., Eger, H., Ehrich, C., Kohler, A., Kutter, A., et al., 1999. Land Use Planning: Methods, Strategies and Tools. GTZ, Berlin.

Bristow, M.R., 1981. Planning by demand: a possible hypothesis about town planning in Hong Kong. Asian J. Public Administration, 199–223.

Crowley, J.R., Hall, J.L., Bruce MacDuogall, E., Passarello, J., Thomson, F.J., 1975. Land use planning. Supporting Paper 3.

Cuong, T.H., 2005. Market Access and Agricultural Production in Vietnam. University Hohenheim, Germany.

Evans, A.W., 2004. Economics and Land Use Planning. Blackwell Publishing, Oxford, UK.

Fagerström, M.H.H., Messing, I., Wen, Z.M., 2003. A participatory approach for intergrated conservation planning in a small catchment in Loess plateau, China: Part I. Approach and methods. Antena 54, 255–269.

FAO, 1990. Participation in Practice: Lessons from the FAO People's Participation Programme. Food and Agriculture Organization, Rome.

FAO, 1993. Guidelines for Land-Use Planning. Food and Agriculture Organization of United Nations. Food and Agriculture Organization of United Nations, Rome, Italy.

FAO, 2005. Land Cover Classification System: Classification Concepts and User Manual, Software Version 2.0. Food and Agriculture Organization of the United Nation, Rome.

FAO, 2007. Land Evaluation: Towards a Revised Framework. Land and Water Discussion Paper 6.

FAO, 2011. Vietnam and FAO: Achievements and Success Stories. FAO Representation in Vietnam.

FAO, UNEP, 1999. The Future of Our Land: Facing the Challenge. Guidelines for Integrated Planning for Sustainable Management of Land Resources. Food and Agriculture Organization, Rome, Italy.

FAO/UNEP, 1997. Negotiating a Sustainable Future for Land: Structural and Institutional Guidelines for Land Resources Management in the 21st Century. Food and Agriculture Organization/United Nations Environment Programme, Rome.

GSO, 2010. National Statistical Data in 2010. National Statistical Office, Vietnam.

Hessel, R., Van den Berg, J., Kabore, O., Van Kekem, A., Verzandvoort, S., Dipama, J.-M., et al., 2009. Linking participatory and GIS-based land use planning methods: a case study from Burkina Faso. Land Use Policy 26, 1162–1172.

Huy, M.Q., 2009. Building a Decision Support System for Agricultural Land Use Planning and Sustainable Management at the District Level in Vietnam. Georg-August-Universität, Göttingen.

Huyen, T.G., 1993. Country profile: land use in Vietnam: facts and figures. Sustain. Dev. 1 (3), 4–7. MCB University Press.

Jocelyn, A.S., 2002. Do Rural Infrastructure Investments Benefit the Poor? Evaluating Linkages: A Global View, A Focus on Vietnam. School of International and Public Affairs, Columbia University and the World Bank, Vietnam.

Kelly, S.B., 2004. Community Planning: How to Solve Urban and Environmental Problems. Rowman & Littlefield Publishers, USA.

Laborte, A.G., Roetter, R.P., Hoanh, C.T., 2002. The land use planning and analysis system of the systems research network in Asia. In: Proceedings of the 2nd International Conference on Multiple Objective Decision Support Systems for Land, Water and Environmental Management, Australia.

Loi, N.V., 2008. Use of GIS Modelling in Assessment of Forestry Land's Potential in Thua Thien Hue Province of Central Vietnam. Georg-August-Universität, Göttingen.

Mai Chau People's Committee, 2001. The report on land use planning of Maichau District, Hoa Binh, Vietnam.

MoNRE, 2009. Circular No. 19, 2009. Detail provisions on land use planning. Hanoi, Vietnam.

Puginier, O., 2002. Hill Tribes Struggling for a Land Deal: Participatory Land Use Planning in Northern Thailand Amid Controversial Policies. Humboldt University, Berlin.

Quang, N., 2003. Review of Existing Planning System: Obstacles and Strategies Moving Toward Innovative Planning Approaches. Case Study of Ha Tinh Planning System. GTZ, PDP-HaTinh, Vietnam.

Roetter, R.P., Hoanh, C.T., Laborte, A.G., Van Keulen, H., Van Ittersum, M.K., Dreiser, C., et al., 2005. Integration of systems network (SysNet) tools for regional land use scenario analysis in Asia. Environmental Modelling Software 20, 291–307.

SEMLA, 2009. Integrated Land Use Planning: Results and Lessions Learnt. Strengthening Environmental Management and Land Administration, Sweden Cooperation Programme, Hanoi, Vietnam.

Socialist Republic of Việt Nam, 2009. In: Decree 69/2009/CP Addendum on Land Use Planning, Land Price, Land Recall, Compensation, Support and Relocation. The Government of Socialist Republic of Vietnam, Hanoi. August 13.

The National Land Use Planning Commission, 1998. Guidelines for Participatory Village Land Management in Tanzania. Peramiho Printing Press, Peramiho, Tanzania.

The Socialist Republic of Vietnam, 2003. Landlaw in 2003. The Government of Socialist Republic of Vietnam, Hanoi.

Trung, N.H., 2006. Comparing Land Use Planning Approaches in the Mekong Delta, Vietnam. Wageningen University, Wageningen, Holland.

Trung, N.H., Quang Tri, L., Van Mensvoort, M.E.F., Bregt, A., 2004. GIS for participatory land use planning in the Mekong delta, Vietnam. In: The 4th International Conference of Asian Federation of Information Technology in Agriculture and Natural Resources, Bangkok, Thailand.

UN, 2009. Case Studies on Community Participation. United Nations.

Vo, D.H., Nhu Trung, T., 2007. Land administration for poverty reduction in Vietnam. "Good land administration – it's role in the economic development". In: International Workshop, Ulaanbaatar, Mongolia, June 27–29.

Weng, Q., 2010. Remote Sensing and GIS integration: Theories, Methods, and Applications. McGraw Hill, USA.

Yen, B.T., Pheng Kam, S., Quang Ha, P., Thai Hoanh, C., Huy Hien, B., Castella, J.-C., et al., 2002. Exploring land use options for agricultural development in Bac Kan province, Vietnam. In: Paper Symposium 30, 17th World Congress of Soil Science.

SECTION V

ADAPTIVE LIVELIHOOD IN RESPONSE TO CHANGE

CHAPTER

8

Coping Mechanisms of the Ethnic Minorities in Vietnam's Uplands as Responses to Food Shortages

V.X. Tinh

Institute of Anthropology, Hanoi, Vietnam

8.1 INTRODUCTION

Before the global food crisis, over 800 million people had daily food shortages throughout the world (World Food Summit, 1996). Until now in Vietnam, together with the food price, which increased twofold in 2008 compared with previous years, food security is problematic, especially for the ethnic minority uplanders. Although Vietnam, as a multiethnic country with 50 of the country's total of 53 ethnic minority groups living in the upland, it is the second largest rice exporter in the world. However, food shortages are relatively widespread in ethnic minority groups (MARD, 2001; Tinh, 2002a; Bui, 2003). In the northern mountainous regions, some Hmong households had food shortages between 5 and 6 months per year (Tinh, 2002b). Though Vietnam has achievements in the implementation of governmental policies to assist ethnic minorities, ethnic groups in the country's uplands still face many difficulties and challenges with food security. The purpose of this chapter is to identify the factors that caused the food shortages of ethnic minority groups in Vietnam's uplands and present the coping mechanisms they have used to deal with food shortages.

8.2 FOOD SECURITY AROUND THE WORLD

8.2.1 Definition of Food Security

Food security is defined in many different ways. In 1986, the World Bank defined food security as "access by all people at all times to enough food for an active, healthy life," while in 1992, the Food and Agriculture Organization of the United Nations (FAO) and the World Health Organization (WHO) defined food security as "the access of all people to the food

http://dx.doi.org/10.1016/B978-0-12-805453-6.00008-5

needed for a healthy life at all times" (cited in Von Braun, 1999, p. 41). According to the Action Plan of the World Food Summit held in Rome in November 1996: "food security exists when all people, at all times, have physical and economic access to sufficient, safe and nutritious food to meet their dietary needs and food preferences for an active and healthy life."

The antonym of food security is food insecurity. There are two main types of food insecurity: chronic and transitory. Chronic food insecurity is a long-term problem caused at the household level by lack of income or assets to produce or buy adequate food for the household, while transitory food insecurity is a short-term food security problem caused by a shock to the food production or economic system, where income or resources necessary to adjust to the shock are not available (Gladwin et al., 2001; Von Braun, 1999).

8.2.2 Approaches to Food Security

The definition of "food security" can be used at different levels: household, regional, national, and global. At the household level, food security is understood as "...sustainable ability to supply nutritious food to all individuals of the household, regardless of female or male, adults or little ones, health or ill-health" (cited in Action Aid, 2000, pp. 21–22); at the regional level, it means the ability to obtain sufficient food to meet the needs of all households in the region; at the national level, it refers as the ability of a country to obtain sufficient food to meet the needs of all of the households in the country; and at the global level, it implies a worldwide ability to obtain sufficient food to meet the needs of all households throughout the world (Action Aid, 2000).

Over the last four decades, the awareness of food security has shifted through three paradigms: (1) from the international and national levels to the household and individual levels; (2) from a focus on food to a focus on livelihood, or from food security to sustainable livelihood; and (3) from outsiders' indicators to insiders' conceptions (Maxwell, 1996). Lofgren and Richards (2003) focus on two levels of food security: macro and micro. At the macro level, if food production in the region as a whole falls far short of food requirements, then it is necessary for most countries to turn to imports for a large share of domestic food consumption, while at the micro level, food security depends on the ability of individual households to meet their food requirements (Lofgren and Richards (2003)). In addition to national and household levels, Von Braun (1999) argues that food security should study at three different levels: national, household, and socioeconomic and demographic. At the national level, food security can, to some extent, be monitored in terms of supply and demand indicators; that is, the quantities of available food versus needs. At the household level, it may be measured by direct surveys of dietary intake in comparison with appropriate nutritional norms. However, this measurement method is costly due to the time required for data collection and processing. At the socioeconomic and demographic level, food security can be properly analyzed by using proxy variables such as real wage rates, employment, price ratios, and migration.

During the period between the 1960s and 1970s, food security mainly focused on the international and national levels. The focus has recently shifted to the household and individual levels (Rigg, 2001), as there is a paradox that some countries, Brazil for example, export a great amount of food; however, this country still has many people suffering from food shortages (Hollist et al., 1987). In many cases, a country has sufficient food to meet its demand, but the food distribution is unequal and irregular, then it leads to a situation in which one person has a surplus of food, while another lacks food. This phenomenon may also occur at the household level, meaning that if there is sufficient food for all household members, but the food distribution is irregular, some members of the household will be food insecure. Therefore, to ensure food security at the household level, both obtaining sufficient food for the family and equal distribution of food among family members should be taken into account.

8.2.3 Poverty Reduction and Food Security in Vietnam

In Vietnam, the concept of food security and its indicators are closely related to the concept and indicators of poverty. There are two different poverty lines: one was set by the National Program on Hunger Eradication and Poverty Reduction, and the other was based on the results of the Vietnam Household Living Standard Surveys in 1999. The poverty indicator for households set by the National Program in the period of 1996–2000 was a household whose average income per capita was less than 13kg of rice per month. For mountainous areas and islands, this indicator was less than 15kg (equal to US$3). During the period 2000–5, the poverty indicator was 80,000 VND (US$4.40) for mountainous areas and islands; and 100,000 VND (US$5.50) for rural areas in the plains. From the year 2005 onward, the indicator was 200,000 VND (US$11) and 260,000 VND (US$14.40) for mountainous areas and islands and rural areas in the plains, respectively. In addition, the Vietnam General Statistics Office's definition of "poverty" was households having an income lower than the minimum income needed to cover essential needs of food, clothing, housing, health care, and education. This income bracket is defined as "hunger" (or food shortage) to differentiate between the "poor" and "very poor" populations.

Regarding policies and poverty reduction strategies in Vietnam, two important programs were launched recently emphasizing poor communes. The first one was the National Program on Hunger Eradication and Poverty Reduction. This program was implemented in the period of 1998–2000. The second one was the socioeconomic development program, also known as Program 135. This program has been implemented since 1999, targeting 2374 of the most difficult communes in the country with a special focus on mountainous and remote communes. The program emphasized building infrastructure such as electricity, schools, clinics, and irrigation systems. The program also supports forestry and agricultural extension activities, and capacity building for the local people. To integrally resolve all matters concerning sustainable development and poverty reduction, the government has also created guidelines to integrate the poverty reduction goals with other socioeconomic development programs, including the 10-year strategy on socioeconomic development (2000–10), the strategy on economic growth and poverty reduction, and the 5-year socioeconomic development plan.

According to international poverty standards, the rate of poor households in Vietnam has dramatically decreased from 51.8% in 1993 to 24.1% in 2004. Therefore, the number of poor households in the country has been reduced by nearly 60% (Table 8.1).

The poverty level in Vietnam differs with respect to each region and ethnicity. The rate of poor households decreased most dramatically in the northeast region, from 86.1% in 1993 to 31.7% in 2004, and decreased most slowly in the northwest region, from 86.1% in 1993 to 54.4% in 2004 (Statistical Publishing House, 2005). Most of the poor live in the countryside and mountainous areas (FAO, 2005) where the poverty rate is still quite high.

TABLE 8.1 Rate of Poverty in Vietnam 1993–2004 (in percentage)

	1993	1998	2002	2004
Poverty rate	*58.1*	*37.4*	*28.9*	*24.1*
City	25.1	9.2	6.6	10.8
Countryside	66.4	45.5	35.6	27.5
Food poverty	*24.9*	*13.3*	*9.9*	*7.8*
City	7.9	4.6	3.9	3.5
Countryside	29.1	15.9	11.9	8.9

Source: From General Statistics Office, 2005. Household Living Standard Survey in 1998, 2002 and 2004 (www.gso.org.vn).

8.3 METHODOLOGY

This study applied the sustainable livelihood approach (SLA), which presents the main factors that affect people's livelihoods, and typical relationships between these (DFID, 1999). At the household level, the study looks into five types of household assets (or capitals) where food shortage, directly and indirectly, affect household assets and vice versa. The details of these assets and their measurement indicators are discussed below.

- *Natural capital*: The natural capital asset includes land and forest resources that are used by a household and how these resources are connected to their food status. In the case of food shortages, householders have to sell or lease a portion or all of their land. Householders may have to change the way they use their land or change the type of cultivation to reduce investment in their agricultural production. They may have to go to the forest and exploit timber and nontimber products to replace the family's food supply. Some householders even have to eat food collected from nature. Changes in capital use may lead to other impacts on their household living standard. Selling land means they do not have land to cultivate in the future. It will seriously endanger their ability to earn a living.
- *Social capital*: This capital source comes from relationships within the family and among other relatives, neighbors, friends, colleagues, and others, and determines how these relationships contribute to a guaranteed food supply in the household. When a family is short of food, they may need help from one of these relationships. Help can mean direct food assistance or cash. With support from relatives and acquaintances, a household can move to a new location or another city to earn a living. This may bring about unexpected results, such as a household being unable to pay debts, or being placed into a worse living situation.
- *Human capital*: Human capital is assessed according to the following factors: professional skills, employment ability, education, and health conditions in connection with a household's food sources. In the case of food shortages, household members have to use their knowledge to earn a living by doing different jobs such as woodworking, embroidery, knitting, and so on. A household may have to work for others in their community or in other places. Some households have to cut down on the number of

meals in a day or reduce the nutritional value of each meal. This action may bring about serious side-effects. Work performance of each household member is weakened by overwork. They may also be contaminated by social evils in their working environments. The reduction in the number of meals in a day and in the nutritional value in each meal may lead to a degradation of the family's labor force and the shortage of Vitamin A, iodine, and iron among household members. This deficiency may cause serious developmental problems in young children, such as low weight, Down syndrome, handicaps, or other developmental problems. Malnutrition may also cause unpredictable diseases.

- *Financial capital*: Financial capital concerns the spending and income of a household in relation to its food sources. When people lack food, they may have to find a substitute income source. They must accept any job, despite potential health risks. They have to cut back on spending and this affects some aspects of their lives. For example, owing to a modest income, they may tighten investment in their children's study, or even ask their children to give up their studies. When a family member is sick, they cannot afford the medical fees.
- *Physical capital*: Physical capital refers to different kinds of valuable assets such as a house, furniture, means of transportation, and so on. In the case that the household lacks food, they must mortgage their house, production tools, or furniture to get more money. This action will bring about negative effects. The household will have a less suitable residential location. Lack of production means will reduce the household's capability for work.

Several methods were employed for field data collection. First, focus group discussions were arranged with the rich and poor groups in the study sites. Second, in-depth interviews were conducted with local government officials and knowledgeable villagers. This step also involved transect walks throughout the study areas in different directions. Finally, a structured questionnaire was used to conduct a survey of 98 households in two villages (34 households in Pieng Pho and 64 in Binh Son 1). The secondary data was also obtained during field data collection, including reports, statistical figures, and national surveys.

8.4 OVERVIEW OF THE STUDY AREA

This 3-year study (2005–8) was conducted in two upland villages of Ky Son District, Nghe An Province; namely, Pieng Pho (Pha Danh Commune) and Binh Son 1 (Ta Ca Commune). Pieng Pho Village is located in the south of Pha Danh Commune, it is 15 km away from the center of the commune, and 2 km from the center of the district. The village had 34 Thai households with 187 members. The main local livelihood includes wet-rice, swidden, livestock, and forest. The local people also earn extra income from off-farm activities including stonecutters, woodworkers, and handicraft weaving. Binh Son 1 (village) is located nearby National Highway No. 7B, 2 km away from the center of Ky Son District. There were 64 Kho Mu households, with 412 members in the village. The main locals' livelihoods include cultivating swidden rice, maize, and cassava, and exploiting the natural resources: wet-rice, swidden, and forest. The local people also earn extra income from off-farm activities.

8.5 FINDINGS AND DISCUSSION

8.5.1 Food Poverty and Shortage in Pieng Pho and Binh Son 1

In the study villages, poor and food-poor (hungry) households constitute a major part of the population (54.54%), of which 13.13% are food-poor and 41.41% are poor (see Fig. 8.1). These rates vary significantly across the sites. In Pieng Pho Village, 6% of the local households are food-poor and 26% are poor, while in Binh Son 1, these figures are 17% and 50%, respectively.

Table 8.2 presents income and expenditures by group in the study sites. Generally, the income per head in Pieng Pho Village was higher than that of Binh Son 1. The income per capita within the poor household group in Pieng Pho was about 294%, above the current poverty line, which was set at US$6.80/per capita/month for the countryside, and was 278% of the income per capita of the group of wealthy households in Binh Son 1 Village.

In Binh Son 1, poor households have an average income that is approximately equal to 60% of the current food poverty line. Both the poor and wealthy groups have average incomes that are slightly higher than the food poverty line (15.3% higher for the wealthy group and 29.8% higher for the poor group, respectively). In Binh Son 1, the poor household group spent nearly double the amount of the poor group in Pieng Pho. However, the spending of those in

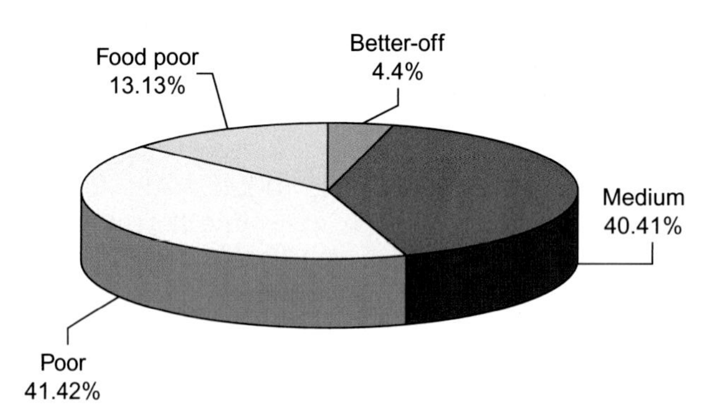

FIG. 8.1 Classification of household economy.

TABLE 8.2 Income and Expenditures by Group in the Study Villages (Unit: US$/person/month)

	Pieng Pho		BinhSơn 1		Average	
Socioeconomic Groups	Income	Expenditure	Income	Expenditure	Income	Expenditure
Better-off	37.8	17.5	7.9	4.7	15.4	7.9
Medium	21.5	14.7	10.2	9.0	15.5	11.7
Poor	16.3	12.1	8.9	5.5	10.2	6.6
Food-poor	15.0	9.2	4.2	4.7	5.1	5.1
Average	20.7	14.0	8.3	6.3	11.9	8.6

the poor group in Binh Son 1 was about 61.5% of their average income, while in Pieng Pho the monthly spending of those in the poor group was 111.8% of their average monthly income.

In the food-poor households, the main proportion of their expenditures was dedicated to food. In Pieng Pho Village, this group spent over 85% of total expenditures for food. The respective figure in Binh Son 1 was 78%. In contrast, spending for education accounted for only 3% of the total expenditures (in both villages), no more than the expenditures they spent on rituals in the same year. Taxes and fees accounted for around 2%.

Food shortages are relatively common in the study villages. Sixty-eight percent of the population ran into food shortages, lasting at least a month (see Table 8.3). The situation varied across sites. In Binh Son 1, the food shortages were more serious, affecting around 92% of the population, including households in the better-off group, while in Pieng Pho only 24% of the inhabitants encountered this situation. In general, food shortages in both villages had a statistically significant relationship with household socioeconomic class, with households of higher socioeconomic class less likely to face food shortage problems.[1]

The duration of food shortages that local households faced was between 1 and 8 months (see Fig. 8.2). Sixty percent of food-poor families experienced food shortages between 1 and 3 months. Thirty-three percent experienced food shortages from 4 to 6 months, 3% for 7–8 months, and the remaining 4% for over 8 months. In general, villagers in Pieng Pho suffered from food shortages for less than six months, while some households in Binh Son 1 suffered from food shortages for more than 8 months per year.

TABLE 8.3 Food Shortages in Study Villages

	Pieng Pho		Binh Son 1		Total	
Socioeconomic Groups	**Number**	**%**	**Number**	**%**	**Number**	**%**
Food sufficiency	*26*	*76*	*5*	*8*	*31*	*32*
Better-off	2	6	0	0	2	2
Medium	19	56	3	5	22	22
Poor	5	15	2	3	7	7
Food-poor	0	0	0	0	0	0
Food insufficiency	*8*	*24*	*59*	*92*	*67*	*68*
Better-off	0	0	2	3	2	2
Medium	2	6	16	25	18	18
Poor	4	12	30	47	34	35
Food-poor	2	2	11	17	13	13
Total	34	100	64	100	98	100

[1] The correlation coefficient between household socioeconomic class and the likelihood to face food shortages is 0.4314 ($p < 0.0001$) and that between household socioeconomic class and duration of food shortages is 0.5151 ($p < 0.0001$).

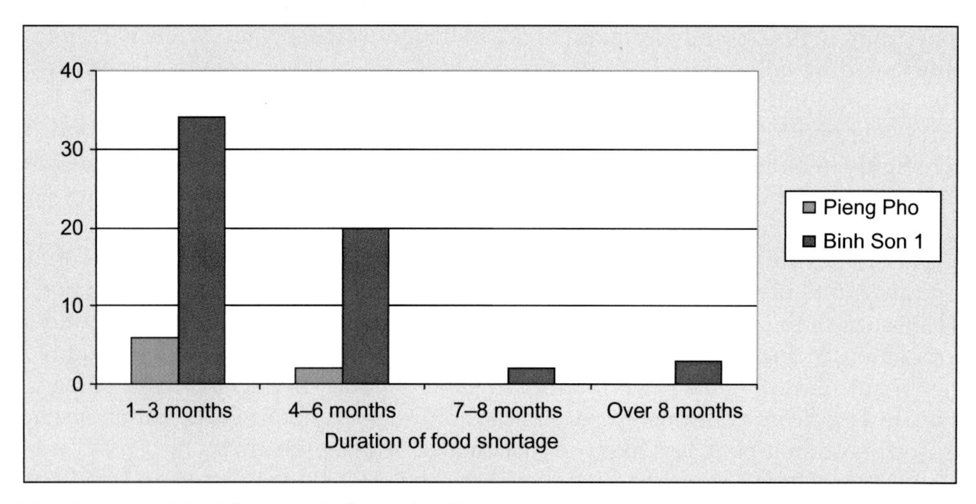

FIG. 8.2 Duration of food shortages in the study villages.

TABLE 8.4 Rice Production Per Capita in the Study Villages (Unit: kg/person/year)

Socioeconomic Groups	Pieng Pho	Binh Son 1	Total sample
Better-off	225	177	189
Medium	177	238	210
Poor	211	206	207
Food-poor	140	145	145
Total sample	186	203	198

In terms of rice production, there is a significant variation between socioeconomic groups. Generally, the better-off, medium, and poor households have higher rice productivity than previous years (see Table 8.4). Interestingly, while food shortages were common in Binh Son 1, rice production outputs in this village in general, and among the village's food-poor households in particular, were higher than the corresponding figures in Pieng Pho Village. The reason for this situation is that income sources in Pieng Pho Village were more diversified than that of Binh Son 1; therefore, although rice productivity in Pieng Pho Village is low, the number of food-poor people in this village was smaller than that of Binh Son 1.

8.5.2 The Impacts of Five Assets on Food Security

This section discusses the impacts of the five capital assets on food security in the study villages.

- *Natural capital*

In general, local households have limited agricultural land (Table 8.5). Only 2 out of 98 households (2%) in both villages have paddy fields. Almost all households (95%) have

TABLE 8.5 Home Gardens and Fishponds Per Capita by Socioeconomic Class in the Study Sites (Unit: m^2/person)

Socioeconomic Groups	Upland Field	Home Garden	Fishpond
Pieng Pho	*854*	*31*	*19*
Better-off	138	13	9
Medium	861	39	24
Poor	917	16	9
Food-poor	1125	25	19
Binh Son 1	*1768*	*3.14*	*1.39*
Better-off	1450	–	3.00
Medium	1947	5.91	3.70
Poor	1694	2.40	–
Food-poor	1745	1.23	0.44
Average	1505	11	7

swidden fields. Poor and food-poor households are more likely to use swidden fields for rice cultivation than households of other socioeconomic groups.

Swidden fields are mostly located on relatively sloping hills. In Pieng Pho, households do not have the right to field ownership, but in fact, the community respects each family's exploitation efforts, and regulations ensure that households do not use the land for purposes outside of growing paddy. In Binh Son 1, swiddens are allocated to households, but not given "red book." Therefore, people use the land out of habit and sometimes hand over the land to others. There are also differences in the plant growing conventions in the two villages. The Thai in Pieng Pho more sensibly organize and shift crops to adapt to the location's natural and economic conditions. Dry rice is grown on the upper land for 1 or 2 years, then the cultivation of that rice will stop for 2 to 3 years. When the "bop bop" trees cover the fields, the people burn the trees and grow rice again.

In swidden cultivation, the fallow period plays a significant role in the recovery of soil nutrients. However, in the study sites, a 2- to 3-year fallow period is not long enough to fertilize soil, because biomass from "bop bop" leaves does not contain enough nutrition to restore soil fertility. Almost all villagers mentioned that upland rice productivity was much lower than previous years due to soil nutrient depletion as a result of shortening the period for which their land can lie fallow. In Binh Son 1, almost all households cultivated paddy rice. In the cultivated land, most ground-cover plants are one-seed layer weeds and reeds, and there are not "bop bop" trees as in Pieng Pho Village. It may be simply be a difference between two different types of land. In general, "bop bop" trees create thicker and better ground-cover in the fields, compared with covered weeds and reeds. According to collected data, swidden rice production in Binh Son 1 is much slower than in Pieng Pho.

Per capita area of home gardens and fishponds in both villages is limited and not dependent on socioeconomic classes. On average, each person gets only 11 m^2 of home garden land and 7 m^2 of fishpond land.

In relation to forest resources, in both villages forests are used by the community. Following regulation here, households have rights to exploit wood, forest vegetables, and other forest products. However, they cannot undergo shifting cultivation in the forest.

- *Human capital*

In general, there is considerable variation in the number of people among households belonging to the different economic groups (see Table 8.6). However, there is no relation between the number of members of a household and its food shortages.[2] Yet, the number of laborers per household tends to be lower in the poor and hunger households. While the households belonging to two medium and better-off groups all have an average number of laborers per household higher than that of the surveyed samples, and the rate of households belonging to the poor and hunger group is lower than the general average level. The result of correlation analysis confirms a statistical relation between the labor force and the economic category of the surveyed households.[3] Similarly, the rate of laborers per capita also has a statistical link with the household economy. The households of the hunger group have a labor rate per capita lower than that of households in the other groups.[4]

As Table 8.7 shows, the education level of the household head has an important relationship with household socioeconomic group. The results show that education level of

TABLE 8.6 Human Assets Across Socioeconomic Groups

Socioeconomic Groups	Household Size	Number of Labor Force Per Household	Labor Force Per Capita	Household With Political Power
Better-off	8.00	4.75	0.59	0.50
Medium	6.30	4.05	0.64	0.48
Poor	5.83	3.41	0.59	0.15
Food-poor	6.85	3.23	0.47	0.08
Average	6.24	3.70	0.59	0.29

TABLE 8.7 Education Level and Age of the Household Head by Socioeconomic Groups

	Education Level	Age	Position[a]
Better-off	7	41	0.50
Medium	7	43	0.38
Poor	5	38	0.10
Food-poor	3	45	0.08
Average	6	41	0.22

[a] *The number in this column shows the rate of households whose heads have political positions in the local area.*

[2] Correlation coefficient between household size and its food shortage is 0.0439 ($p < 0.6677$).

[3] Correlation coefficient between number of laborers and its food shortage is 0.1993 ($p < 0.0492$).

[4] Correlation coefficient between the rate of laborer per capita and its food shortage is 0.2231 ($p < 0.0273$).

household heads in poor and food-poor groups were much lower than average and they were also lower than those of the medium and better-off groups. Similarly, households with political position (mostly the head of the household) are often in the medium and better-off groups. It is very rare that households with political position fall into the poor or food-poor groups. Correlation analysis results confirm a statistical relationship between education level, household political position, and socioeconomic class.[5] However, the age of the household head does not have a statistical relationship with socioeconomic class.[6]

The education level reflects clearly the quality of human resources. While in Pieng Pho Village, there are 28 people who are civil servants (including 17 teachers and 6 health staff), in Binh Son 1 Village there is no civil servant. In Pieng Pho, the civil servants have made remarkable contributions to their households' income. In this village, the education level of the household heads and household members is higher than that of those in Binh Son 1. Regarding general education, up to 2006 Pieng Pho Village had 53 students completing 12th grade; Binh Son 1 Village had only 3 students. At the college and university level, Pieng Pho had 23 students, while Binh Son 1 did not have any students. The good education gives the people of Pieng Pho Village an advantage in adopting advanced cultivation techniques and in market access.

Health condition is also an important factor of human capital that has impact on food security. Through the survey, it can be seen that there is only one indicator of both the Thai and the Khơ Mu peoples, which are quite equivalent: the rates of asthma, mental disorder, and rheumatism of the Thai and Khơ Mu, which are 17.6% and 15.9%, respectively. The other indicators of both ethnic peoples are very different. This shows that the health of Thai people is much better by far than that of the Kho Mu. Particularly, while only 2.9% of households of the Thai people have patients suffering from respiratory diseases and malaria, this rate for the Khơ Mu is 12.7%; the rates of women with symptoms of sickly complexion, especially during pregnancy, of the Thai and Khơ Mu are 0% and 22.2%, respectively; the rates of undersized children are 2.9% and 19%, respectively; and the rates of sickly people who can't work are 2.9% and 22%, respectively (see Table 8.8).

Data obtained from the in-depth interviews clearly show that the level of a household's income is strongly influenced by a number of unwell family members. In Pieng Pho, almost all

TABLE 8.8 Situation of Some Diseases in the Two Research Sites (Unit: %)

No.	Index	Thai	Kho Mu
1.	Household members (especially children) get respiratory diseases, malaria	2.9	12.7
2.	Asthma, mental disorders, rheumatism	17.6	15.9
3.	Women with pale complexion (especially during pregnancy)	0	22.2
4.	Undersized children	2.9	19.0
5.	Weak health, cannot do hard work (according to interviewees)	2.9	22.0

[5]Correlation coefficient between socioeconomic class and education of household head is 0.3966 ($p<0.0002$) and with political position of household head is 0.3267 ($p<0.0010$).

[6]Correlation coefficient between socioeconomic class and age of household head is 0.0339 ($p<0.7402$).

poor households suffer from diseases caused by the food shortages. As a result, they spend a great deal of money for medical treatment. Table 8.8 shows the general health status in study sites. The results show that most women in the study areas have headaches and dizziness, which is a symptom of anemia and malnutrition. Although villagers do not ingest enough nutrients due to food shortages, and most male villagers often drink wine, leading to physical health weakness. In Binh Son 1, ritual ceremonies were quite popular as villagers believe that it can treat diseases while these ceremonies do not exist in Pieng Pho.

- *Social capital*

The central premise of social capital is the social networks (family relationship, relatives, community), and local regulations (traditions, accessing mechanism to the program, projects and types of services). Both the Thai and the Kho Mu have traditions of closed family attachments, family surnames, and communities. For the Thais in Pieng Pho, most people share two main surnames, Lo and Luong, so most residents in the village are relatives. The close attachment by ancestral heritage enables them to support each other in production and in important events such as building houses, weddings, and funerals, and in difficulties such as food shortages.

In Pieng Pho, assistance in the community can be seen in three forms: exchanging labor, giving hands (often staying to have a meal), and material assistance. Each form of assistance applies in line with particular events and circumstances. The demand for assistance varies from household to household; however, villagers can call for additional support in some circumstances. For example, normally each household needs around 12–30 laborers for field cultivation, but a family could not supply these labors, especially during peak times, such as land preparation, transplanting, or harvesting. In this case, the family calls for assistance from the community. For a house building event, it may need around 200 laborers to cut and deliver timbers from the jungle. Material assistance occurs when a family celebrates their new house. In this case, each household contributes two bottles of wine and three kg of rice. For a funeral event, each family contributes a bunch of firewood, two bottles of wine, and three kg of rice. In the Kho Mu community, the assistance can be different based on the nature of the relationships.

The impact of assistance is also different. In the villages in better condition, such as Pieng Pho Village, the assistance from families, relatives, and the community bring about certain effects to reduce and eradicate poverty. To the contrary, in Binh Son 1, the assistance only helps people to survive; almost all households are poor. The side-effects of such assistance can create a poverty trap, as a person who receives assistance is a result of growing dependence on relatives and community. Traditional cultivation is also considered as a special type of capital. The two surveyed communities have different traditional cultivation: Thai ethnic people have rich experience in wet-rice cultivation and gardening, while Kho Mu people have rich experience in swidden cultivation. After moving from their native places in Que Phong District (Nghe An Province) to here 20 years ago, Thai people have shifted from wet-rice cultivation to swidden cultivation. They still have used a number of techniques of wet-rice cultivation for swidden and garden cultivation, such as strictly obeying the seasonal schedule, ensuring the perfect cultivation cycle, especially taking care for rice-intensive farming of the house and jungle gardens, and so on. Thus, their productivity of swidden rice is higher than that of Khơ Mu people. They get more income from other agricultural products as well.

TABLE 8.9 Access to Training and External Supports by Socioeconomic Groups

	Training (2003–6)		External Support (2006)	
Socioeconomic Groups	**Number**	**%**	**Number**	**%**
Pieng Pho	*32*	*94*	*32*	*94*
Better-off	2	100	2	100
Medium	20	95	20	95
Poor	8	89	8	89
Food-poor	2	100	2	100
Binh Son 1	*36*	*56*	*60*	*94*
Better-off	1	50	2	100
Medium	10	53	16	84
Poor	20	63	32	100
Food-poor	5	5	10	91
Average	68	69	92	94

Access to development projects is different between villages. In terms of technical training, 69% of households received training during 2003–6 (see Table 8.9). While there was no significant difference in the participation of households from different socioeconomic classes, variation existed between the two villages. Almost all households (94%) in Pieng Pho were involved in at least one training event during this period, while this figure in Binh Son 1 was 56%. However, only 38% thought this training was useful to them. Around 21% found the training of little use; 15% thought the training was not useful at all; and 26% never tried to apply what they learned from the training. Forty-seven percent of the households who did not participate in training mentioned that they lack information about the training, 18% mentioned the difficult language used in the training, and 18% considered the long distance to the training location. For some others, they were ineligible for the training.

In terms of access to the market, all surveyed households generally assessed that the improvement of local infrastructure over the last several years has contributed to improved market access for local people. At present, most of the farm products are sold in the village. In terms of access to external supports from development projects and programs, there was a presence of state-funded development programs in both villages, especially Programs 134 and 135, which primarily targeted the poor and food-poor. Besides, infrastructure works like power grids, irrigation systems, and village health posts, there was also support from these programs to individual households. In 2006, 94% of households received support such as credit, food, subsidized seedlings/seeds, materials for house construction, home materials, and other consumer products (see Table 8.9). However, not only poor and food-poor households were eligible for such supports, but also for other classes including the better-off. Unusually, not all the poor and food-poor households were able to receive support.

- *Financial capital*

The financial capability of households plays an important role in ensuring food for households, particularly when food production does not meet food demands.

The cash income in two survey sites in 2006 had a remarkable difference in terms of its value and proportion of the total income of households (see Table 8.10). On average, the cash income of a household in the two villages was approximately 80.7 thousand VND (about US$4)/per person/per month, making up about 37.5% of its total income. However, while in Binh Son 1, on average, a person had only 43.6 thousand VND (about US$2)/per month (21.9% of household income in total), and in Pieng Pho, each person got 198.6 thousand VND (over US$9)/per month (53.2% of household income in total). The differences in cash income of households belonging to two economic groups in two villages were quite big. While most of the income of the better-off households in Pieng Pho was cash (making up 75% of total income), in Binh Son 1, the cash income of the better-off households only was 30.5% of total income. Similarly, the proportion of cash income of hunger households in Pieng Pho was 30.5% of total income while it was only 6.2% of household income in total in Binh Son 1. In general, the cash income of the people in Pieng Pho Village was higher than the total income of people in Binh Son 1 Village, on the average about 131.2% in 2006.

In both villages, the value of cash income among hunger households is very low, compared with the other households. In general, the cash income of hunger households is only 14.5% of the general average income level and 7.2% of better-off households' income level. In Pieng Pho, the cash income of poor households in 2006 was 82.3 thousand VND (about US$4.50)/per capita/per month, equal to 41.4% of the average income of the whole village, and 15.9% of the average income of better-off households. In Binh Son 1, the cash income of hunger households was only 4.7 thousand VND/per person/per month (about US$2), equal to 10.7% of the average income of the whole village and 14.2% of better-off households' average income.

TABLE 8.10 Cash Income Per Capita in 2006 by Socioeconomic Group (Unit: US$/person/month)

Socioeconomic Groups	Crops	Husbandry	Off-Farm	Total
Pieng Pho	*200.001*	*759,064.2*	*1,227,276.8*	*19,865,411.0*
Better-off	–	648,963.6	45,229,225.1	51,718,828.7
Medium	200.001	840,124.7	1,396,587.7	22,369,012.4
Poor	260.001	682,753.8	246,121.4	929,145.1
Food-poor	180.001	93,750.5	729,174.0	823,094.6
Binh Son 1	*30.00001*	*152,860.8*	*178,371.0*	*331,261.8*
Better-off	50.00002	375,692.1	60,760.3	436,512.4
Medium	40.00002	161,510.9	432,442.4	593,993.3
Poor	30.00001	174,661.0	80,180.4	254,871.4
Food-poor	10.00001	19,650.1	27,370.1	47,030.3
Average	**80.0004**	**327,191.8**	**480,012.7**	**807,284.9**

Although cash income of food-poor households in Pieng Pho was relatively high, it remained unstable. In 2006, around 88.5% of the cash income (27% of total income) of food-poor households came from off-farm activities. In fact, working as hired laborers was the only source of off-farm income that those households could generate, compared to at least two different sources for households from other socioeconomic groups (see Fig. 8.3). Food-poor households did not have access to stable monthly cash income sources, such as jobs in local state organizations, minor trading, or handicrafts, which somewhat influenced the financial viability of these households.

Timing in harvesting and spending plays an important role in the financial capability of local households, particularly those in the poor and food-poor groups. For surveyed households, particularly those without a stable monthly cash income (poor and food-poor households), income is highly seasonal (see Fig. 8.4). Harvest times occur primarily at the end of the year, while spending occurs throughout the year. In addition, major expenditures—particularly the traditional New Year, various spring rituals (January-March), farm inputs (seeds, varieties, fertilizers), and education costs—rarely coincide with harvesting time.

In terms of access to official loans, 52 out of 98 surveyed households (53% of the sample) borrowed money from official sources (eg, local banks or through development projects) during 2003–6. By the end of 2006, the total lending to both villages was US$19,800 (see Table 8.11), or US$400 per borrower. However, variation existed between the two villages. While 79.4% of households in Pieng Pho borrowed money from official sources, only 39.1% of households in Binh Son 1 did so. Moreover, average borrowing per household in Pieng Pho was over 200% as much as that in Binh Son 1.

Households who did not borrow money gave various reasons (see Fig. 8.5). Lack of collateral was mentioned by most people (24%). After that there was the fear of not being able to repay the loan, inability to pay off current debts, and ineligibility for the loan. Some

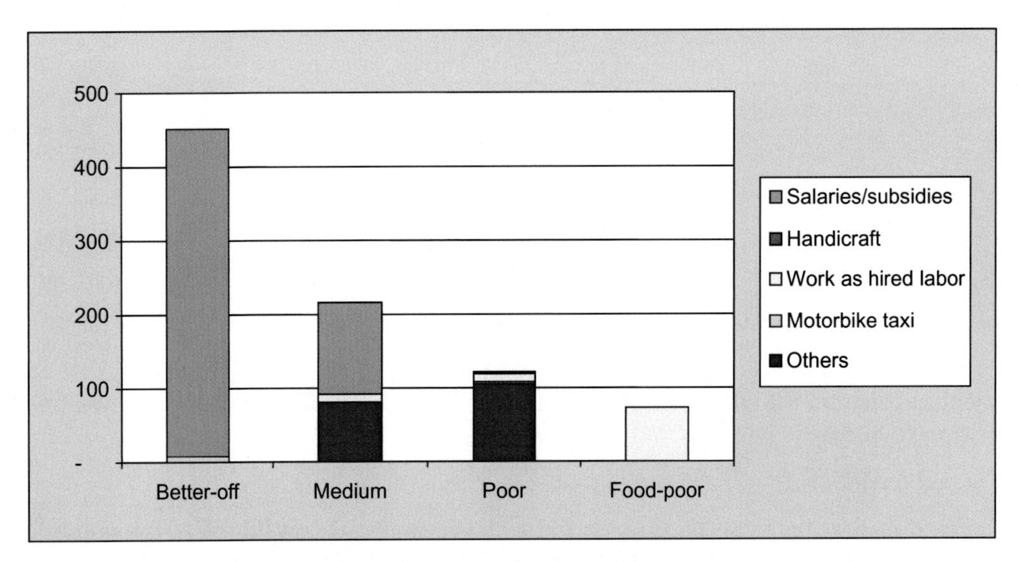

FIG. 8.3 Income from off-farm activities of Pieng Pho Village in 2006.

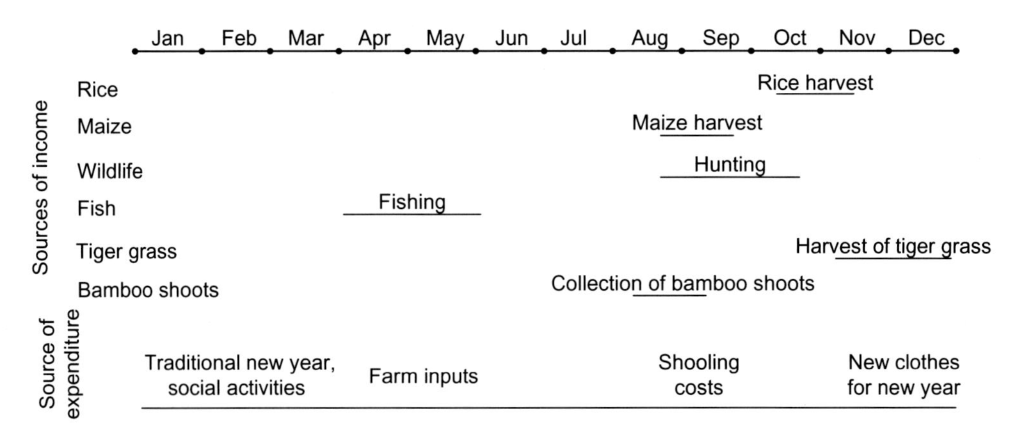

FIG. 8.4 Timing for major incomes and expenditures of a food-poor household. *Source: Village survey in Binh Son 1.*

TABLE 8.11 Borrowing From Official Sources Between 2003 and 2006, by Socioeconomic Group

Socioeconomic Groups	**Households Borrowing Money Over the Group Total**		**Total Borrowing (US$)**	**Average (US$)**
	Number	**%**		
Pieng Pho	*27*	*79.4*	*24,350,013,500*	*9,019,500*
Better-off	2	100	220,001,200	11,000,600
Medium	16	76.2	1,695,009,400	10,594,560
Poor	7	77.8	410,002,300	5,857,300
Food-poor	2	100	11,000,600	5,500,300
Binh Son 1	*25*	*39.1*	*1,125,006,200*	*4,500,200*
Better-off	0	0	–	–
Medium	9	47.4	425,002,400	4,722,300
Poor	12	37.5	520,002,900	4,333,200
Food-poor	4	36.4	180,001,000	4,500,200
Average	**52**	**53.1**	**35,600,019,800**	**6,846,400**

Number in parentheses refers to percentage of households borrowing money over the group total.

households also mentioned reasons like having no demand for a loan or still waiting for their turn to borrow.

- *Physical capital*

Physical capital here can be understood as basic material conditions of the community, such as electricity, roads, schools, stations, and the material condition of households, such as the house and furniture. In terms of electricity, only Binh Son 1 is connected with national

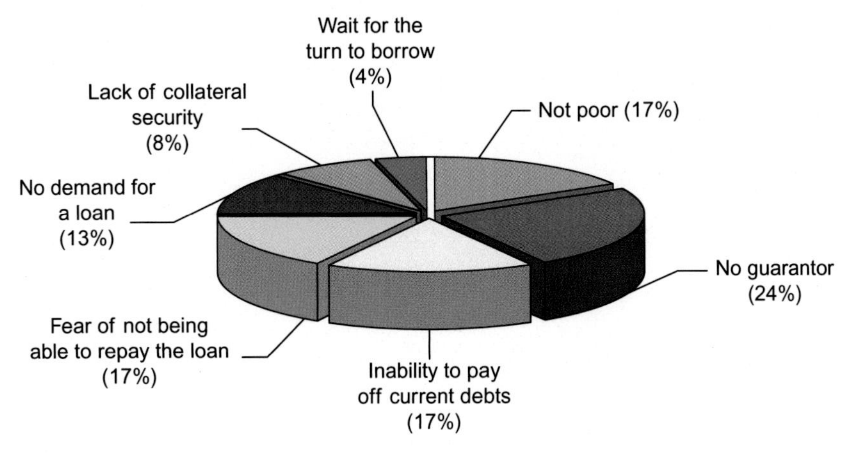

FIG. 8.5 Reasons for limiting the borrowing of money during the period 2003–6.

network electricity, while Pieng Pho is still using small hydropower (with 22 generators). Although they have access to network electricity, the Kho Mu only use power for lighting and rarely use it for other purposes. There are 11 televisions in 11 households (16.8% of the total households), one electric cooker, one electric fan, and one grinding machine. We can say that although they have access to network electricity, people here are too poor to improve their lives. In Pieng Pho, even with small hydropower, the village has 12 televisions in 12 households (33.3% of the total households).

The two surveyed villages enjoy quite favorable transportation conditions leading to the center of the district and Nam Can frontier pass. Pieng Pho is located near Highway 7A, and Binh Son 1 is located near Highway 7B. However, in Pieng Pho, when people want to sell products, the trucks can only stop at the top of the slope, several hundred meters from the village, whereas in Binh Son 1, trucks can go inside the village. Regarding school conditions, in Binh Son 1, the schools are located near the commune's secondary and high schools, and in Pieng Pho the schools are far from the commune's schools, but nearer to the center of the district. Therefore, pupils often join the primary school in Ta Ca Commune, and a joint secondary and high school in the district. Basic health care facilities in the two villages are similar to education facilities.

Among various types of physical capital, the house is the most important because of its great value. In the people's lives, besides eating expenses, they spend much of their money to build houses, more spending than for any other event. Between the two ethnic groups, the Thais have more solid houses than the Kho Mu: 29.4% compared with 14.3% and similarly, the Thais have 61.8% medium houses compared with 38.1% of the Kho Mu. On the contrary, 44.4% of the Kho Mu houses are temporary houses compared with 5.9% of Thai houses (see Table 8.12).Through interviews, it is known that some Kho Mu households use the privileged credit fund (reserved for production development) to build houses.

Beside houses, people use a large part of their savings to buy motorbikes, televisions, and radios, because these things are both needs and also the manifestation of the family's prosperity. Results from the survey in early 2006 show that the Thai ethnic group has 22 motorbikes (of which six families own two) while the Kho Mu have only eight. There are respective figures for the following indices: video player—11 and 4, respectively; radio—7 and 3, respectively; table, chairs and wardrobe—11 and 6, respectively. From interviews, we

TABLE 8.12 Types of House Construction

	Types of House				
Ethnic Groups	Good	Medium	Temporary	Tent	Without House
Thai	10	21	2	0	1
	29.4%	61.8%	5.9%	0%	2.9%
Kho Mu	9	24	28	2	0
	14.3%	38.1%	44.4%	3.2%	0%
Kinh	1	0	0	0	0
	100%	0%	0%	0%	0%
Total	20	45	30	2	1
	20.4%	45.9%	30.6%	2.0%	1.0%

found that most expenses for purchasing comforts, especially motorbikes, are not normally completely affordable, and people have to borrow from siblings or relatives. Therefore, each time they spend money for more than basic demands they often incur debt, which somewhat affects the usage of food in the family.

8.5.3 Measures to Deal With Food Shortages

8.5.3.1 Main Causes of Food Shortages Among the Study Households

Research results showed that there were five major types of causes for food shortages in 2006: (1) lack of labor, (2) lack of land, (3) natural calamities, (4) schooling costs, and (5) too many children (Fig. 8.6). Other minor causes include the lack of off-farm jobs, inadequate capital, and insufficient hybrid varieties. In this section, we focus on three major causes: land, labor, and natural calamities.

Inadequate labor was regarded as the main cause of food shortages by most households. Surveyed results showed that 35 out of 67 households (52%) who ran into food shortage problems in 2006 were because of this issue. The average number of laborers per households is 3.37 and the rate of laborers per capita is 0.53, relatively less than the sample average of 3.7 and 0.59, respectively (see Fig. 8.6).

Lack of land was the second reason leading to food shortages. Of the 67 households with food shortage problems in 2006, 33 (49%) thought their food insufficiency was related to the amount of production land. In reality, however, per capita land size of these households was 1607 m^2 of upland field, 6 m^2 of home gardens, and 8 m^2 of fishponds. Except for home gardens, land holdings of these households were even higher than the sample average. Consequently, their land problem was more about the quality of the land or the suitability of the land to the current crops than about quantity.

The third most prevalent cause for food shortages was calamities. Twenty-three households (34%) considered this as a major cause for their food shortages in 2006. Such calamities as a disease (for livestock), pests (for crops), and bad weather (floods and droughts) were most often mentioned.

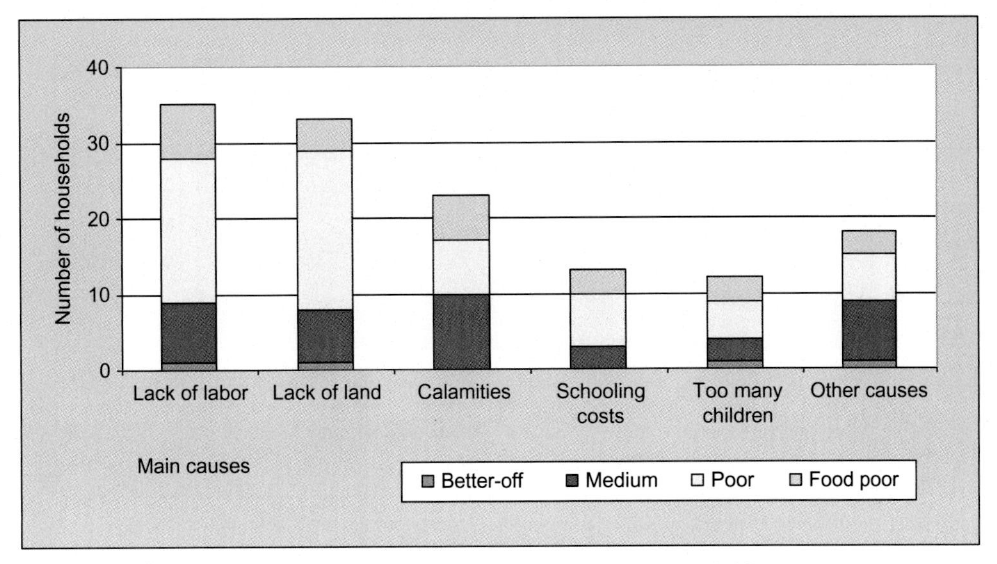

Note: There may be more than one cause for food shortage for each household

FIG. 8.6 Main causes of food shortages.

However, in the researcher's view, the main reason for food shortages in poor households, especially in the Kho Mu community, was due to quality and labor capability. A question is raised: Why do both villages have the same natural capital (land cultivation on mountainous fields), the same accessibility to markets, service, and social policies, but 92% of Kho Mu households are poor whereas only 24% of Thai households are poor? The answer is not due to the lack of cultivating land, lack of a labor force, or natural disasters. The labor qualities of the Thai in Pieng Pho, as analyzed above, are the development of education, health conditions, and labor capabilities reflected in production skills (including the application of wet-rice cultivation practices in mountain fields), the ability to access markets, the ability to shift crop structure, and the ability to get income from other professions, diversifying plants, cattle, and other work. In fact, if we make a comparison, the labor quality and ability among the Thais in Pieng Pho is more competitive than the Kho Mu in Binh Son 1. In conclusion, we can say that the five types of capital (natural, human, social, financial, and physical) are all connected to food security, among which, human and social capital play very important roles.

8.5.3.2 Measures to Deal With Food Shortages

To deal with food shortage problems, people in the study villages have applied various measures, including working as hired labor, exploiting forest products, borrowing money with high interest rates, selling household physical assets (including animals), and others (see Fig. 8.7). For many households, working as hired laborers in return for cash or food in the case of need was the most widespread tactic to deal with food shortages. Of the 67 households with food problems in 2006, 49 (73%) applied this measure. In addition, 35.5% of households who did not have problems with food shortages in 2006 also thought they would have done the same.

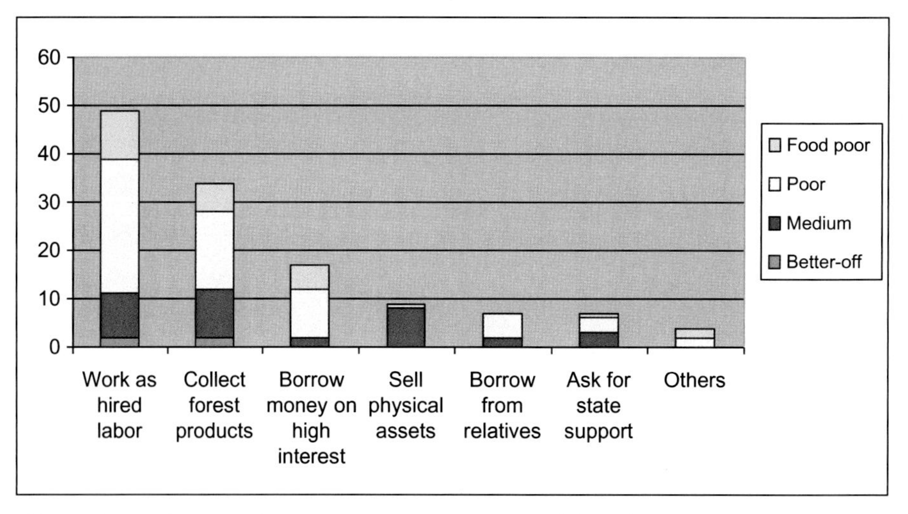

Note: One household may employ more than one measure to cope with food shortage

FIG. 8.7 Measures to cope with food shortages. *Source: Household census.*

Collecting forest products was also considered an important measure to cope with food shortages. Thirty-four out of the 67 households (51%) with food problems in 2006 applied this measure. In addition, 32% of households who did not have food shortages in 2006 also thought they would have relied on the forest to overcome their food problems. Products collected from the forest were used for two purposes: for home consumption and for sale in the market for money to buy food. Ms Lo Thu Loan in Pieng Pho village is an example. In 2006, crop failure due to drought led to four months of food shortages for her household. To meet food demands, she had to collect banana flowers and leaves from the forest and sell them in the district market. The money was used to purchase rice for the whole family. On average, she was able to earn US$1.10 per visit to the forest and market.

Borrowing money to buy food in case of shortages was also mentioned by local people as one solution. Fortunate households may be able to borrow from their relatives. These were households with infrequent food shortage problems. Many others had to borrow money with high interest rates to have cash for food. This measure is highly temporary as households with chronic food problems (mostly the food-poor) who had to borrow money with high interest rates would end up spending most of their crop harvest to pay back the loan and the interest. As a consequence, the vicious cycle begins again in the next period.

In addition, some households sold their assets (including cattle, buffalo, and pigs) to meet food demands in case of shortages. However, this measure is not applicable to the food-poor, as they are often the ones without any valuable assets.

Besides immediate measures, local households also applied long-term strategies. Two important mechanisms are to borrow money from official sources to invest in farm production and to clear more land for cropping.

It is important to note the mutual support for the households with food shortages from within the community. In both villages, the community plays an active role in helping its members in case of need. In Binh Son 1, for example, local people contributed rice to help those with food problems. Friends and relatives are also willing to help in case of need.

8.6 CONCLUSIONS AND RECOMMENDATIONS

In the difficult situation of the Vietnam upland region at present, the food security of ethnic minority peoples remains an important issue. Through the investigation among Thai and Kho Mu people in Ky Son District, Nghe An Province, it can be seen that the food security of the ethnic peoples in this locality sticks closely to the five capital resources. Regarding *resource capital*, the ethnic peoples are facing the challenge of farming land shortage because of the growth of population and exhausting of swidden land. Regarding *social capital*, the relationship of families, lineages, and village communities of the two ethnic peoples plays the important role in mutual help for production and food. Regarding *human capital*, the personal ability, health condition, and living practices of Thai people are better than those of Kho Mu people. Regarding *financial capital*, although the condition to access to the credits and developmental resources of the state of the two communities is similar; the property resource of the Thai families is better than that of the Kho Mu ones. Regarding *material capital*, the transportation distance and condition from the two communities to the district center are favorable but while Kho Mu people have the power grid, the Thai people still have used self-sufficient hydroelectric power.

In response to the food shortages, Kho Mu people mainly borrow at interest and exploit natural products while Thai people develop production and occupations to increase their income. By the investigation, it can be seen that among the above five capital resources, only human capital and property accumulation in the financial capital have a big difference between Thai and Kho Mu people. However, ensuring food security of Kho Mu people is much weaker than that of Thai people. This shows that for the food security of the ethnic peoples in this area, the human capital is the most important.

To increase the food security of Thai and Kho Mu people, together with pushing up the reclamation of more wet-rice fields and increase the fertility for farming land; bringing into play community assistance, setting up the community food fund; rationalizing the payments in households; and constructing the electric grid for the Thai village as soon as possible, it is important to increase the productive capacity for people and integrate their economy into the market economy. To increase this capacity, we need to develop education, increase knowledge of agricultural cultivation, and push up the health care for people.

Acknowledgments

This chapter is based on outcomes of the cross border project on "Coping Mechanisms of the Ethnic Minorities in Upland Areas of Vietnam and the Lao-PDR as Responses to the Food Shortage: Strengthening Capacity and Collaboration in Studying between the Institute of Anthropology, Vietnam and the Institute for Cultural Research, Lao-PDR".

Author would like to thank the Rockefeller Foundation for financial support to carry out the project. A special thank you to researchers who actively involved in the project, including Dr. Pham Quang Hoan, Dr. Tran Hong Hanh, Dr. Tran Van Ha, Dr. Dang Thi Hoa, Dr. Nguyen Quang Tan, Dr. Mai Van Thanh, Dr. Nguyen Danh Thin and Dr. Dao Quang Vinh. Without valued support from these researchers, the project would not have been accomplished.

References

Action Aid, 2000. International Trade and Food Security (Reference Book). National Political Publishing House, Hanoi.

Bui, M.D. (Ed.), 2003. Some Issues Regarding Poverty Reduction of Ethnic Minorities in Vietnam. Social Sciences Publishing House, Hanoi.

DFID, 1999. Sustainable Livelihoods Guidance Sheets. DFID, London.

FAO, 2005. Agricultural trade reform offers benefits, but the poor could be left behind. Rome, Italy.

Gladwin, C.H., Thomson, A.M., Peterson, J.S., Anderson, A.S., 2001. Addressing food security in Africa via multiple livelihood strategies of women farmers. Food Policy 26 (2), 179–180.

Hollist, W., Laldand, F., Tullis, L.M., 1987. Pursing Food Security. Lynne Rienner Publisher, Boulder and London.

Lofgren, H., Richards, A., 2003. Food security, poverty, and economic policy in the Middle East and North Africa. In: Lofgren, H. (Ed.), Food, Agriculture, and Economic Policy in the Middle East and North Africa. Research in Middle East Economics, vol. 5. JAI Press/Elsevier, Amsterdam.

MARD, 2001. The state of food security in Vietnam: progress since the 1996 world food summit: key indicator. Hanoi.

Maxwell, S., 1996. Food security: a post-modern perspective. Food Policy 21 (2), 155–170.

Rigg, J., 2001. Food security, vulnerability and risk: linking food, poverty and livelihoods. In: Paper Presented at the Workshop on Sustainable Livelihoods in Southeast Asia, Hanoi, Vietnam, April.

Statistical Publishing House, 2005. Vietnam Implements Millennium Development Goals. Statistical Publishing House, Hanoi.

Tinh, V.X., 2002a. An toanluongthuccuacacdantocthieu so o vungcao Viet Nam duoitac dong cuayeu to xa hoi va van hoa (Impact of Social and Cultural Factors on Food Security of Ethnic Minorities in Upland, Vietnam), trong: Trung tam nghiencuu tai nguyenvamoitruong, Phattrien ben vung o mien nui Vietnam: 10 namnhinlaivacac van de datra (Center for Natural Resources and Environment Studies, Sustainable Development in Upland of Vietnam: Looking Back 10 Years and Development Issues). Agricultural Publishing House, Hanoi.

Tinh, V.X., 2002b. Looking for Food: The Difficult Journey of the Hmong in Vietnam. Land Tenure Center, Wisconsin-Madison University, USA.

Von Braun, J., 1999. Food security - a concept basis. In: Kracht, U., Schulz, M. (Eds.), Food Security and Nutrition: The Global Challenge. LIT VERLAG, Münster.

World Food Summit, 1996. http://www.fao.org/wfs/index_en.htm.

CHAPTER

9

Home Gardens in the Composite Swiddening Farming System of the Da Bac Tay Ethnic Minority in Vietnam's Northern Mountain Region

N.T. Lam, N.H. Duong†*

*Vietnam National University of Agriculture, Hanoi, Vietnam †Center for Agricultural Research and Ecological Studies, Hanoi, Vietnam

9.1 INTRODUCTION

Home gardens can be defined as the land surrounding a house on which a mixture of annual and perennial plants are grown, together with or without animals, and largely managed by the household members for their own use or commercial purposes (Millat-e-Mustafa, 1998). The home gardens are a source of edible, medicinal, and other useful plants (Godbole, 1998), food security, and biodiversity conservation (Valere et al., 2014; Mariel et al., 2009; Galhena et al., 2013; Amilcar et al., 2008). For instance, Godbole (1998) found 120 plant species in the Konyak home gardens in northern Nagaland, India. Hung et al. (2001) found 84 plant species in the mixed garden of the Da Bac Tay ethnic minority in Vietnam's northern mountain region. Home garden systems, as well as the composite swiddening of the Da Bac Tay ethnic minority, have survived throughout the centuries as the result of long-term adaptation of cultivated plants and cultural techniques to local ecological conditions (Rambo, 1998). Through the development stages, the garden economy has gradually gained an important role and position in the social economy, although it only covers 5–10% of the cultivated surface of a farm household's total area. Beside economic gains, the garden also contributes to solving local ecological and human cultural problems and to improve cultural life within the community. Garden development also helps to reduce the pressure on local forest resources.

However, home gardens are often ignored as an important part of traditional upland farming systems by scientists and extension workers due to their small size and the low

Redefining Diversity and Dynamics of Natural Resources Management in Asia Volume 2
http://dx.doi.org/10.1016/B978-0-12-805453-6.00009-7

economic value of species (Bishwajit et al., 2013). Thus, the study of home gardens is very important for enhancing local knowledge of natural resources management as well as for community development. In this chapter, we investigate the structure and functions of home gardens within the local livelihood system of the Da Bac Tay ethnic minority. Furthermore, we also analyze the interaction between the home garden and the other components of the composite swiddening system to maintain its sustainability. Finally, this study will help promote this form of horticultural practice in Vietnam's northern mountain region.

9.2 MATERIALS AND METHODS

9.2.1 The Study Site

This study was conducted in Tat Hamlet of Tan Minh Village, Da Bac District, Hoa Binh Province in Vietnam's northwestern mountains (latitude 20°N and longitude 105°E; Fig. 9.1). This hamlet is inhabited by the Da Bac Tay ethnic minority group who have practiced composite swiddening for at least a century (Rambo, 1998).

The location of the hamlet is about 300 m above sea level at the valley floor, with the peaks of surrounding hills and mountains reaching an elevation of 800–950 m. The climate is monsoon tropical with a dry season from October to January, followed by a period of little rainfall from February to April and a rainy season from late April to September. The average air temperature over 20 years (1978–98) is 22.9°C, and the average annual rainfall over the same period is 1824 mm (Dung et al., 2002). The hamlet has a total natural land area of 743 ha, of which only 20% has a slope of less than 25 degrees, with a narrow valley floor being flat enough for permanent settlement, roads, and paddy fields (Rambo and Vien, 2001). The surrounding mountains have slopes of 30–60 degrees, and are occupied with active swidden fields, fallowed swiddens, grasses, bushes, and secondary forests. Previous studies have found that a variety of species are planted in the home gardens for multiple purposes such as consumption, medicine, timber, and firewood (Hung et al., 2001). However, the home garden diversity in space and time, and its interactions with the other parts of the composite swidden agroecosystem, have not yet been studied in detail.

9.2.2 Methods

This study used a wide range of data collection methods. First, direct observation and description were carried out to understand the structure and function of each species in the home garden. Semistructured interviews were also carried out to investigate the variation of home garden species in time. Then, 30 households were randomly selected for structured interviews. Finally, data analysis was carried out using descriptive and comparison methods. The plant names were cross-checked with a checklist of plants available in the Da watershed (Loc and Cuc, 1997).

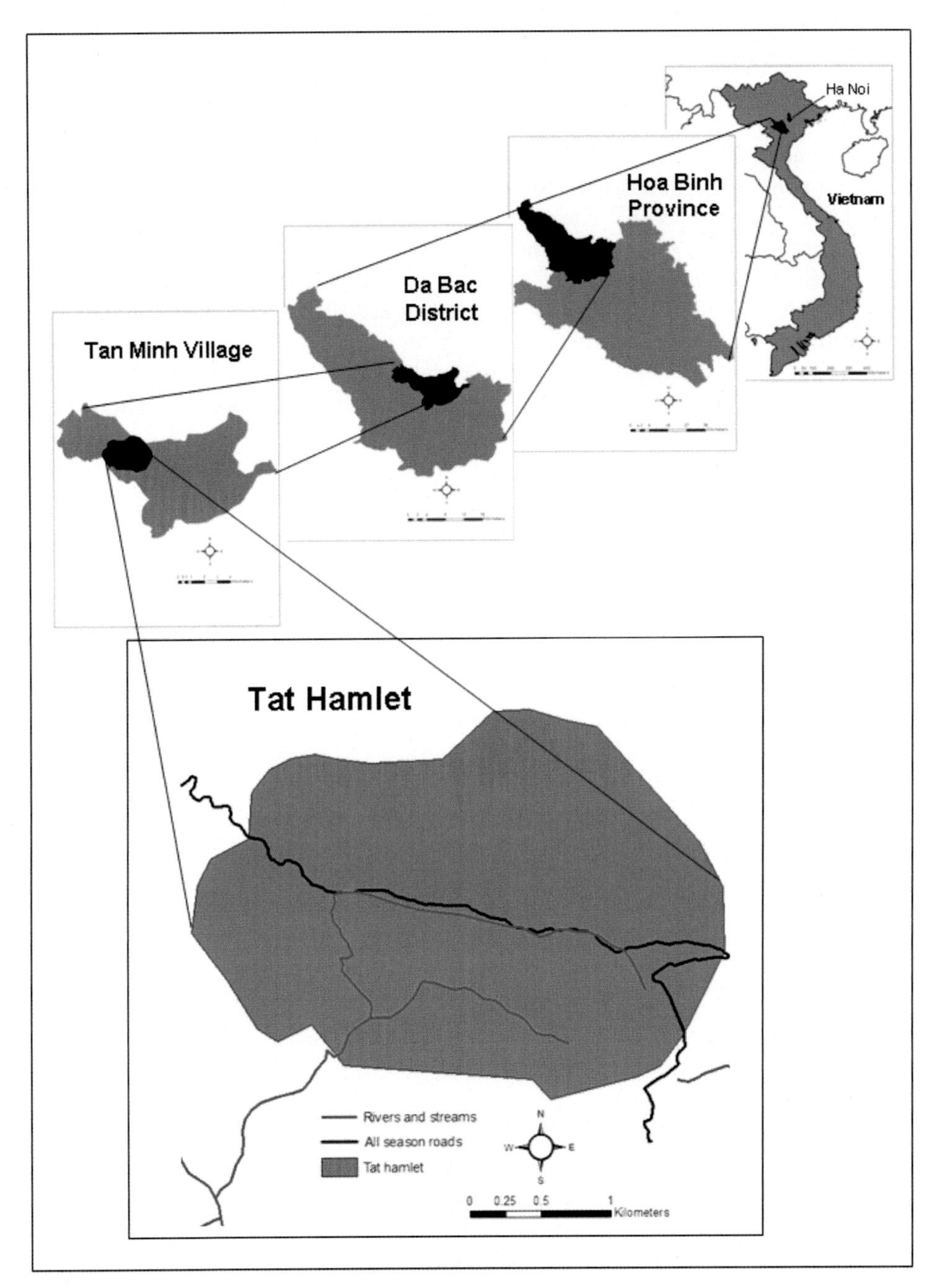

FIG. 9.1 Location of Tat Hamlet in northern Vietnam. *From Vien, T.D., Rambo, T., Lam, N.T., 2009. Farming With Fire and Water: The Human Ecology of a Composite Swiddening Community in Vietnam's Northern Mountains, Kyoto Area Studies on Asia. Kyoto University Press and Trans Pacific Press, Kyoto.*

9.3 RESULTS AND DISCUSSIONS

9.3.1 Home Gardens Characterization

Home gardens in Tat vary widely in their makeup around the houses of the hamlet. The average area of home gardens is 347 m^2, and most of the households have gardens of 50–300 m^2, the largest being 1500 m^2. The standard deviation of the size distribution is not high. Besides the home garden, 80% of the individual households cultivates vegetables in about 20–40 m^2 (Table 9.1), which are enclosed by bamboo fences to protect them from free-ranging livestock (Rambo, 1998). Many types of vegetables are planted including small colza, cabbage, kohlrabi, and *Coriandrum sativum*. while green onions are sometimes planted in raised wooden troughs. Marigolds and other ornamental flowers are sometimes also grown in the home garden. Several households have clumps of tea in the corners of their home gardens.

9.3.2 Species Diversification in the Mixed Gardens of a Da Bac Tay Community

The home gardens in Tat Hamlet have a great variety of interactions taking place vertically, horizontally, and temporally. Table 9.2 shows that there is a wide range of plant species in the home gardens of Tat Hamlet. According to the research results, 69 species were planted in the mixed gardens of the sampled households in Tat Hamlet. Most of the plants in the gardens were planted by the farmers. There are some newly cultivated plants in the gardens such as Japanese persimmon, orange trees, and some kinds of vegetables. The most common plants in the home gardens are banana, small colza, Japanese persimmon, chinaberry tree, jack tree, sauropods, mango, pineapple, guava, papaya, taro, and medicinal plants. Many households

TABLE 9.1 Household Distribution by Home Garden Area ($n = 30$).

Area (m^2)	Frequency ($n = 30$)	%
≤20	1	3.33
21–50	0	0.00
51–200	13	43.33
201–300	9	30.00
301–500	2	6.67
501–1500	5	16.67
Total	30	100.00
Average = 347.87		
Min = 16		
Max = 1500		

Based on Tat hamlet household survey, 2004.

TABLE 9.2 List of Species Found in Mixed Home Gardens of a Da Bac Tay Community

No.	Plant	Local Name	Scientific Name	English Name	Number of Household	%
Fruit tree						
1	Chuối	Cỏ quấy	*Musa paradisiaca* L.	Banana	27	90.00
2	Hồng	Hồng	*Diospyros kaki*	Japanese persimmon	22	73.34
3	Mít	Cỏ Mí	*Artocarpus heterophyllus* Lamk.	Jacktree	19	63.33
4	Dứa	Cỏ hôi nác	*Pandanus amaryllifolius* Roxb.	Fragrant screw pine	15	50.00
5	Xoài	Xoài	*Mangifera indica* L.	Mango	15	50.00
6	Ổi	Mạc ối	*Psidium guajava*	Guava	14	46.67
7	Bưởi	Cỏ Búc	*Citrus grandis* (L.) Osb.	Pomelo	12	40.00
8	Đu đủ	Cỏ tành cọ	*Carica papaya* L.	Papaya	12	40.00
9	Chanh	Mạc lưu	*Citrus limon* (L.) *Burm. f.*	Lemon	9	30.00
10	Mơ	Mạc phụng	*Prunus armeniaca* L.	Apricot	8	26.67
11	Nhãn	Nhãn	*Dimocarpus longan* Lour.	Longan	8	26.67
12	Trứng gà	Trứng gà	*Pouteria Lucuma* (Jacq.) *Moore & Stearn Lucuma.*	Egg fruit	8	26.67
13	Mận	Mạc mắn	*Prunus domestica* L.	Plum	7	23.33
14	Vải	Mạc cái	*Litchi sinensis* Radlk.	Litchi	7	23.33
15	Mía	Ọi	*Saccharum officinarum* L.	Sugarcane	5	16.67
16	Cam	Cam	*Citrus sinensis* (L.) Osb.	Orange	4	13.33
17	Táo	Táo	*Ziziphus mauritiana* Lamk.	Indian Jujube	4	13.33
18	Roi	Roi	*Syzygium samarangense*	Water Apple	2	6.67
19	Sấu	–	*Dracontomelon duperreanum* Pierre.	Dracontomelum	2	6.67
20	Phật thủ	Dứt mư	*Citrus medica* var. *sarcodactylis* (Sieb.)	Fingered citron	1	3.33
21	Gấc	Mạc khau	*Momordica cochinchinensis* (Lour.) Spreng.		2	6.67
22	Đào	Cỏ cai	*Prunus persica* (L.)	Batsch	1	3.33
23	Sổ	Sổ	*Dillenia indica* L.	–	1	3.33
24	Trám	–	*Canarium album* (Lour.)	Raeusch	1	3.33
25	Khế	Mạc phương	*Averrhoa carambola* L.	Carambola	4	13.33
26	Cau	Cỏ lang	*Areca catechu* L. Betel-Nut	Areca tree	2	6.67
27	Hồng bì	Vòng	*Clausena lansium* (Lour.)	Skeels	1	3.33

Continued

TABLE 9.2 List of Species Found in Mixed Home Gardens of a Da Bac Tay Community—cont'd

No.	Plant	Local Name	Scientific Name	English Name	Number of Household	%
Vegetable						
28	Rau cải canh	Phác cạc	*Brassica juncea* (L.)	Small colza	24	80.00
29	Rau ngót	Phác ọt	*Sauropus androgynus* (L.) Merr.	Sauropus	18	60.00
30	Giềng	Cỏ khà	*Alpinia gagnepainii* K.	Gagnepain	9	30.00
31	Su hào	Su hào	*Brassica oleracea L. gemmifera* Zinh.	Kohlrabi	8	26.67
32	Gừng	Khỉnh	*Zingiber officinale* Rossoe.	Ginger	7	23.33
33	Hành	Họm tiêu	*Allium fistulosum*	Onion	7	23.33
34	Xả	Khỉnh trơ	*Cymbopogon citratus* (DC.) Stapt.	Lemon grass	7	23.33
35	Lá lốt	Khắc ơ lật	*Piper lolot* C. DC.	Piper lolot	4	13.33
36	Ớt	Ượt	*Capsicum frutescens* L.	Chili pepper	4	13.33
37	Cà tím	Mạc Khựa	*Solanum melongena* L.	Egg plant	3	10.00
38	Cải bắp	Cải bắp	*Brassica oleracea* L.	Cabbage	3	10.00
39	Mướp	Mạc buộc	*Luffa cylindrica* (L.) M. J. Roem.	Egyptian cucumber	2	6.67
40	Rau muống	Rau muống	*Ipomoea aquatica* Forssk.	Water spinach	2	6.67
41	Tỏi	–	*Allium porrum* L.	Garlic	3	10.00
42	Đậu ván	Mạc bùa bẹp	*Dolichos lablab*	Broad beans	3	10.00
43	Cà pháo	Mạc khựa đòn	*Solanum undatum* Poir.	Egg plant	1	3.33
44	Kinh giới	Họm kinh giới	*Elsholtzia ciliata* (Thunb.) Hyland.	Marjoram	1	3.33
45	Mồng tơi	–	*Basella rubra* L.	Ceylon spinach	1	3.33
46	Rau chua	Phác xốm	*Fagopyrum cymosum*		1	3.33
47	Rau chuôi	Phác mược	*Dichrocephala auriculata*		1	3.33
48	Tía tô	Tứn đảnh	*Perilla frutescens* var. *crispa* (Thunb) Hand-Mazz.	Perilla	1	3.33
49	Rau mùi	Hỏm	*Coriandrum sativum*	*Coriandrum sativum*		
50	Rau muống chua	Rau muống chua		Sour water spinach	1	3.33

TABLE 9.2 List of Species Found in Mixed Home Gardens of a Da Bac Tay Community—cont'd

No.	Plant	Local Name	Scientific Name	English Name	Number of Household	%
Root plant						
51	Dọc mùng	Pặc ọc	*Colocasia gigantea*	Taro	11	36.67
52	Khoai lang	Ón	*Ipomoea batatas*	Sweet potato	3	10.00
53	Khoai sọ	Mạc Phược	*Colocasia esculenta* (L) Schott.	Taro	2	6.67
54	Sắn	Cỏ mên cỏ	*Manihot esculenta* Crantz	Cassava	1	3.33
55	Dong riềng	Đót	*Canna orientalis*	Edible canna	1	3.33
Medicinal plants						
56	Nghệ	Hạn	*Curcuma domestica* Val.	Turmeric	6	20.00
57	Rẻ quạt	Rẻ quạt	*Belamcanda chinensis* (L.) DC.	Belamcanda	5	16.67
58	Sâm đại hành	Sâmđại hành	*Eleutherine bulbosa* (Mill.)	Urban	3	10.00
59	Cảng hạo	–			1	3.33
60	Hương nhu	–	*Ocimum gratissimum* L.	Holy basil	1	3.33
61	Lá bỏng	Bợ bục	*Kalanchoe pinnata* (Lamk.) Oken.	Life plant	1	3.33
62	Láng	–	*Crinum giganteum* Andr.		1	3.33
Timbers and other trees						
63	Xoan	Cỏ hiên	*Melia azedarach*	Chinaberry tree	20	66.67
64	Lát hoa	Lát	*Chukrasia tabularis* A. Juss	Seeds winged	4	13.33
65	Sau sau	–	*Liquidambar formosana* Hance.	Oriental sweetgum	1	3.33
66	Chè	Cỏ che	*Thea sinensis*	Tea	3	10.00
67	Cọ	Có	*Livistona tonkinensis* Magalon	Palm	1	3.33
68	Hoa bóng nước	–	*Impatiens balsamina* L.	Garden Balsam (Touch-me-not)	1	3.33
69	Hoa mào gà	–	*Celosia argentea* L.	Cox comb	1	3.33

Based on Tat hamlet household survey, 2004.

make bamboo frames for climbing plants such as cucumber, Egyptian cucumber, *Luffa acutangula*, and *Dolichos lablab* (broad bean).

Vertical diversification was often observed as multiple layers of vegetation occur in home gardens of the Da Bac Tay. The top layer is made up of timber and fruit trees including chinaberry trees, textured wood, jack tree, pomelo, Japanese persimmon, longan, sapodilla, and plum. The next lower layer is made up of trees of orange, lemon, and banana plants. The lowest layer, under the shade of these trees, are vegetables, tubers, medicinal plants, aromatic spices, pineapple, ginger, saffron, chili, edible canna, small colza, taro, medicinal plants, and grasses.

This type of mixed planting in the gardens takes full advantage of the soil and contributes to the production of multiple products. The type of intercropping found are fruit trees intercropped with vegetables, perennial fruit trees intercropped with annual crops, and timber trees intercropped with fruit trees and vegetables. In all cases, the trees are planted in such a way as to not compete with the lower story plants and vegetables. Most households do not construct fences, or they are constructed in a cursory way, made out of bamboo and easily destroyed by cattle and easily entered by chickens.

Most of the seeds that farmers plant in the gardens are saved by the farmers from their last crops. In some cases, households buy the seeds from hawkers who visit the hamlet or from Tu Ly market. Sometimes the seeds are not very good, they do not sprout, and do not grow successfully. Recently, Japanese persimmon was provided to some households by the district extension service.

Diversification was also observed and quantified by analyzing how often the different plant species occur at the different home gardens. The cropping calendar illustrates timings for different plants to be grown in the home gardens. From this, it is observed that between January and March, there were fewer plants cultivated in the gardens. During that time, if any farmer wanted home garden products, such as vegetables, they needed to purchase them from hawkers or in the market. From April to the end of December more plants are grown, including small colza, sauropus, kohlrabi, squash, onion, garlic, salad, *C. sativum*, and luffa (Table 9.3). Farmers try to space these plants temporally so that garden products are always available. For example, farmers grow small colza year-round, except during the three hottest

TABLE 9.3 Cropping Calendar of Some Kinds of Vegetable in Home Gardens

Plant	1	2	3	4	5	6	7	8	9	10	11	12
Small colza									●	—	●	
Sauropus				●	●							
Kohlrabi										●	—	●
Ginger		●	—	—	—	—	●					
Squash	●	—	—	—	—	—	—	●				●—●
Fiber melon				●	—	—	—	—	—	●		
Garlic, onion										●	—	●
Salad										●	—	●
Coriandrum sativum									●	—	●	
Luffa				●	—	—	—	—	—	●		

Note: According to lunar calendar.

months (June, July, and August). They also grow potatoes during September and October and start harvesting them from November to March. The harvesting continues till April and May of the next year. They can also grow lettuce from August to the following April. In addition, they vary the planting spatially. For example, they grow sauropus around the fence, under the shadow of a big tree. From February to August, farmers often gather bamboo shoots in the forest for selling and eating, along with forest "vegetables" such as phac hac, phac khau tom, and mac khoanh.

9.3.3 Factors Affecting Garden Diversity

A number of factors affecting garden diversity were uncovered during the semistructured interview. First, personal choice is the main factor influencing the diversity of what is grown. Home gardens are mainly cultivated by women and they often prefer that their households consume vegetables. Second, the age of the household has a considerable affect on the diversity found in the gardens. For instance, most older farmers have a nice garden with numerous plant species, while younger households have poorer gardens with limited species (see Appendix 9.1).

Seasonal variation also impacts the variety of plants species in the home garden. The number of plant species is higher in the summer season than during the other seasons. For instance, Hung et al. (2001) found that 84 species occur in 15 sampled households in the summer, while our survey found only 70 species in 30 households in the winter (see also Appendix 9.1).

Access to information is also important, and this is much improved since national electrification came to the hamlet in 2000 (Gia et al., 2004). Thanks to the extension programs, through training and capital support, many new plant species have been introduced to the home gardens. Some of the examples are citrus, Japanese persimmon, orange, and Thieu litchi. Some households also go to the nearest market in Cao Son or the district town to buy vegetable seeds and fruit tree seedlings for their garden, further diversifying what is found in their gardens. Vien (2004) also found land-use changes in Tat Hamlet responding to market fluctuations, perhaps leading to further diversification.

9.3.4 Home Garden Production

Although home gardens occupy only a small area compared to other components of the composite swiddening system under the individual household management system (Fig. 9.2), they are considered to be a highly efficient form of land use (Millat-e-Mustafa, 1998). Tat residents reported that only women took care of the gardens. Chemical fertilizers, ash, manure, and pesticides are always applied to the garden. Because of intense human management and large material inputs, the gardens maintain productivity and the plant species can be easily changed to respond to internal and external factors such as home consumption, food security, raising livestock, and market fluctuations. It is difficult to assess the actual yield released from the mixed gardens of the Da Bac Tay community. This is because local people harvest the vegetables, fruit, firewood, timber, and medicinal plants any time they want to and it is not possible to record every instance of something being harvested. The survey conducted in 1998–99 found that 32% of home gardens provide some cash income, ranging from 5000 to

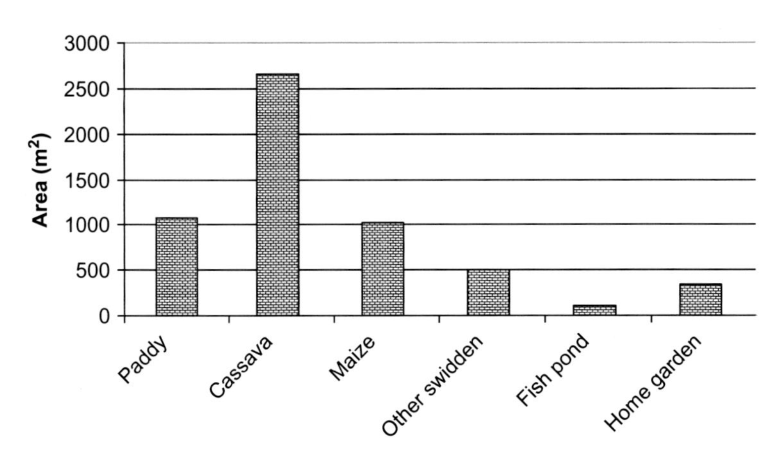

FIG. 9.2 Overall land-use pattern (average area per household) 2004.

200,000 VND per year, mainly through the sale of bananas and vegetables. However, these figures do not clearly reflect the importance of the gardens. Fig. 9.2 illustrates the importance of the home garden in Tay society due to its contribution to the household economy during both the dry and rainy seasons. Some households had also sold their vegetables to their neighbors.

Some gardens were observed to be infested by weeds and unproductive trees. These gardens were generally found in younger, or newer, households. This suggests that young households in Tat Hamlet need more time to develop their home gardens (see Appendix 9.1).

9.3.5 Expanding Home Garden Components to Other Parts of the Composite Swiddening

Home gardens interact with other components of the composite swiddening system. Rambo (1998) showed these interactions between Tay household gardens and other components of the composite swiddening system through a diagram. The earlier studies reported that many kinds of vegetables are planted in the swidden during the cropping season (the rainy season) such as taro, mustard, chili, pumpkin, and squash (Vien, 1998; Rambo, 1998; Cuc and Terry, 2001). During the cropping season, local farmers have to stay on their swidden fields to take care of their crops. Consequently, a part of the garden element shifts to the swidden component. The household members also go to their swiddens to bring vegetables to their homes for daily meals in the main settlement area in Tat Hamlet.

Wild vegetables were still abundant in the forests in Tat Hamlet. Many types of forest vegetables were collected as nontimber forest products for home consumption. Hung et al. (2001) found the number of common species shown in Table 9.4. Hung et al. (2001) reported that *Masus japonicus*, *Fagopyrum cymosum*, *Forest litchi*, *Ficus*, *Garcinia multiflora*, and *Dioscorea permilis* are the preferred species.

In the past, local people only collected vegetables from their own garden and swidden. Compared to previous years, Tat Hamlet has changed in where vegetables are sourced from. Now vegetables are obtained from the home garden, fallow field, and forest, or purchased from neighbors, hawkers, and markets. In the dry season, the two main sources for vegetables

TABLE 9.4 Edible Wild Vegetables, Fruits, and Tubers Used in Tat Hamlet

Vietnamese Name	Tay Name	Latin Name	Use
Rau chua	Xồm lồm	*Fagopyrum cymosum*	Cooking sour soup
Rau tàu bay	Pạc tàu bịn	*Gynura crepidioides*	As vegetable
Cà dại	Mạc khoáng	*Solanum torvum*	Fruit eaten as vegetable
Hoa chuối	Mặc Pị	*Musa uranoscopos*	Use flower as vegetable
Vải rừng		*Nephelium cuspidatum*	Fruit eaten
Diếp dại		*Lactuca indica*	As vegetable
Ngoã	Mặc ngoã	*Ficus fulva*	As vegetable
Quả nóng	Mặc nau	*Saurauia* sp.	Fruit eaten
Sung	Mặc not	*Ficus* sp.	Fruit eaten
Quả dọc	Mặc lục	*Garcinia multiflora*	Cooking sour soup
Quéo	Mặc quéo	*Artocarpus* sp.	Fruit eaten
Hạt tiêu rừng	Mặc khèn	*Zanthoxylum nitidum*	Used as spice like black pepper
Củ mài	Mền Pà	*Dioscorea persimilis*	Tubers eaten
Củ nâu	Mặc Pậu	*Dioscorea cirrhosa*	Tubers eaten/source of brown dye for cloth
Khoai mại	Mặc bầu	*Dioscorea opposita*	Edible starch extracted. Tubers are toxic so must be carefully processed
Hà thủ ô	Hoọc chạc	*Fallopia multiflora*	
Ngấy		*Rubus alceifolius*	Boil with drinking water
Hồng bì	Mặc khẹn	*Clausena lansium*	Fruit eaten
Sim	Mặc nin	*Rhodomyrtus tomentosa*	Fruit eaten
Rau đắng	Pặc khau tốm		Use as vegetable

From Hung, D.T., Trung, T.C., Cuc, L.T., 2001. Agroecology. In Bright Peaks, Dark Valleys: A Comparative Analysis of Environmental and Social Conditions and Development Trends in Five Communities in Vietnams' Northern Mountain Region. National Political Publishing House, Hanoi.

are gardens and hawkers, while in the rainy season, vegetables are collected from the home garden and forest. The hawkers said that they do not prefer to bring vegetables to Tat Hamlet during the rainy season because the farmers can eat bamboo shoots for the whole rainy season in place of vegetables. The rainy season in Tat Hamlet runs from June or July to October and the dry season runs from November to May of the next year. During the time from May to October, most of the bamboo grows rapidly such as "nứa đại, nứa tép, măng đắng, măng ngọt, and măng luồng." Collecting bamboo shoots for sale is profitable for the farmers, so most households go into the forest and gather bamboo shoots; this activity has brought significant cash income to households, and each household can get 3–5 million VND in 1 year. Bamboo shoots are also consumed very quickly in the hamlet. Bamboo shoots are collected from the community forest and from a part of Phu Tho Province.

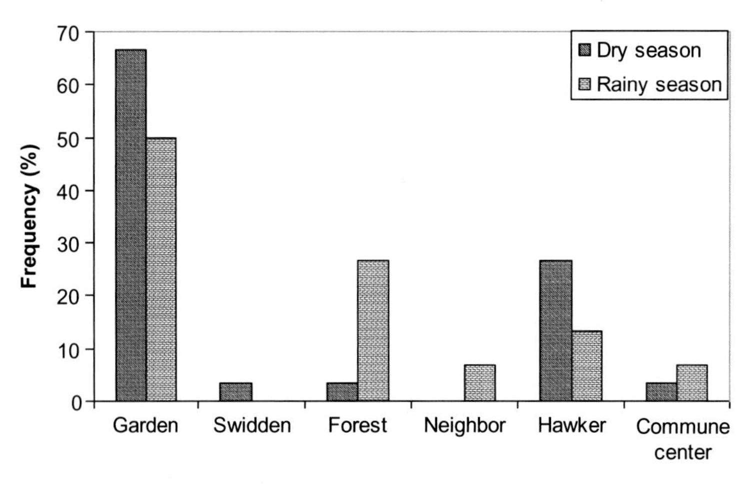

FIG. 9.3 Frequency of households to obtain different sources of vegetables in the dry and rainy seasons ($n=30$). *Based on Tat hamlet household survey, 2004.*

The dry season is the appropriate time to plant vegetables and is the reason why home gardens are the second most important activity to households. During the dry season, there are less forest vegetables, and there are few bamboo shoots in the forest. In former times, they used to gather many kinds of vegetables from the forest but nowadays, the farmer only gather Indian taro, banana stems for animal feed, and bamboo shoots. Table 9.5 shows the seasonal calendar for forest vegetables in Tat Hamlet.

Elements of the home gardens are also connected to the paddy fields. During the winter season, the paddy fields form a suitable land for growing potatoes, kohlrabi, and mustard. The harvested products are used for home consumption, livestock feed, and are sold in the market. Some households want to extend their gardens to a larger area, where products can be oriented to the market economy.

9.3.6 Some Difficulties in Gardening

Most farmers lack technical and capital support. In the past the extension services supported some households through providing new varieties of trees, such as the Japanese persimmon, orange, and Thieu litchi. However, this support was limited in scope (7–10 plants for each household). These efforts also did not work as there was no demand (yet) in the market for these trees' products. Many households do not actively manage their gardens and let the plants grow in a "natural" way. In general, extension activities in support of home gardens are lacking.

Available labor for home garden work is also limited. Women spend a lot of time in the swidden field, and men spend a lot of time collecting timber and hunting, while older people collect firewood and "forest" vegetables and children go to school. Thus, there is a labor shortage with regard to home gardens.

Another problem associated with home gardens is soil quality. In general the soil is poor, and appropriate (flat) areas are limited. In the dry season these areas also lack water. Market demand for many of the crops that are found in home gardens is also weak, so farmers are

TABLE 9.5 Seasonal Calendar for Collecting Forest Vegetables in Tat Hamlet

Forest vegetable	1	2	3	4	5	6	7	8	9	10	11	12
Bambooshoot (nứa đại)												
Bambooshoot (nứa tép)												
Bitter bambooshoot (Mang dang)												
Bambooshoot (luồng)												
Sinarundinaria sp. (lánh anh)												
Sweet bambooshoot												
Măng lau												
Dioscorea persimilis (củ mái)												
Phắc tin chắp												
Phắc hác												
Phắc khau tồm												
Mặc khoánh												
Pothos grandis S. Buchet												
Papaya												
Bananas flower												
Lá kìa												

Based on Tat hamlet household survey, 2004.

hesitant to invest in the inputs needed to increase the gardens' productivity. Finally, Tat Hamlet is distant from the city center and from its central market, making it difficult to sell products that are produced.

9.4 CONCLUSIONS AND RECOMMENDATIONS

Home gardens are a part of the composite swiddening farming system in Tat Hamlet, which have an important role in the system. They mostly exist in relation to other components of the cultivation systems and are integrated into the cultural life of the Da Bac Tay ethnic minority in Vietnam's northern mountain region. There are a wide variety of plant species comprising more than 69 common plants, in the home gardens. The number varies per garden and is related to plant spacing, structure, and seasonal fluctuation. Home gardens contribute to Da Bac Tay households not only through the production of vegetables and fruit, but also as a stable food supply during most of the seasons. These home gardens incorporate medicinal plants and improve the environment surrounding the homestead. Moreover, the home gardens in Tat Hamlet are integrated with other components in the composite swiddening system. Home gardens are usually managed by the women in the household. One of the difficulties in further developing the gardens is that local people lack capital investment, areas for garden expansion, and technical know-how.

To further develop home gardens in Tat Hamlet, the area devoted to them should be expanded as much as possible. Trees that are not useful should be cut down, and plants that are suitable to local conditions should be cultivated. Plants that can be consumed and/or have high economic value should be grown together with medicinal plants and plants that can be used as fodder. Finally, fertilizer (manure or chemical) should be applied to encourage plant growth. Extension workers, key farmers/hamlet leaders, and scientists who work in the area should help the hamlet households in all of these areas as well as introduce new gardening techniques, help with finding markets, encourage the farmers to participate in horticulture, and build demonstration models in the hamlet. All of these will help hamlets like Tat to further develop and expand their home gardens and ultimately improve the lives of the hamlet population.

APPENDIX 9.1 List of Plant Species in the Home Garden and Their Function in Tat Hamlet ($n = 15$ Sampled Households)

No.	Vietnamese Name	Tay Name	Scientific Name	Frequency	Uses
1	Hồng đất	Cọ Hồng đất	*Rosa* sp.	13.3	1
2	Su hào	Mặc Su Hào	*Brassica gongylodes*	13.3	3
3	Đậu tương	Mặc Đậu tương	*Glycine soja*	13.3	3
4	Trứng gà	Pặc xay cày	*Pouteria Lucuma*	13.3	1
5	Hoa loa kèn	Cọ hoa kèn	*Lilium longiflorum*	13.3	6
6	Khế	Mặc pương	*Averrhoa carambola*	13.3	1,3
7	Gấc	Mặc Gấc	*Momordica cochinchinensis*	13.3	3,4
8	Bạc Hà		*Mentha piperita*	13.3	3,4
9	Sâm đại hành		*Eleutherine subaphylla*	13.3	4
10	Dâu tằm	Cọ Mòn	*Morus alba*	13.3	4
11	Lưỡi hổ	Cọ Lín sữa	*Sansevieria trifasciata*	13.3	6
12	Xoan	Cọ Hiên	*Melia azedarach*	26.6	5
13	Táo	Cọ Táo	*Ziziphus mauritiana*	26.6	1
14	Đu đủ	Cọ Mặc Tách	*Carica papaya*	26.6	1,3
15	Ngô	Klé	*Zea mays*	26.6	2
16	Dền gai	Pặc hồm	*Amaranthus spinosus*	26.6	3
17	Cà Chua	Mặc cà chua	*Lycopersicon esculentum*	26.6	3

APPENDIX 9.1 List of Plant Species in the Home Garden and Their Function in Tat Hamlet (n = 15 Sampled Households)—cont'd

No.	Vietnamese Name	Tay Name	Scientific Name	Frequency	Uses
18	Xạ can	Cọ han	*Belamcanda chinensis*	26.6	4
19	Trầu không	Bợ Pu	*Piper betle*	26.6	8
20	Xương Rồng	Cọ họt ngước	*Cereus peruvianus*	26.6	6
21	Đậu ván	Mặc Đậu ván	*Dolichos lablab*	26.6	3
22	Hoa Cúc	Cọ hoa cúc	*Tagetes erecta*	26.6	6
23	Cau	Mặc lang	*Areca catechu*	26.6	8
24	Na	Mặc Na	*Annona squamosa*	26.6	1
25	Xà lách	Pặc Xà lách	*Lactuca sativa*	26.6	3
26	Mướp đắng	Mặc Cà noi	*Momordica charantia*	26.6	3,4
27	Hẹ		*Allium ramosum*	26.6	3,4
28	Hoàng Tinh	Mặc Tọ	*Maranta arundinacea*	26.6	2
29	Hoa mào gà		*Aerva sanguinolenta*	26.6	6
30	Thuốc lá,	Cọ thuốc lá	*Nicotiana tabacum*	26.6	8
31	Hoa Hè		*Sophora japonica*	26.6	8,4
32	Mã đề	Pặc Lá mã đề	*Plantago major*	26.6	4
33	Lá mơ	Bợ Mơ	*Paederia foetida*	26.6	3,4
34	Bông Bụt	Cọ ải huôn	*Hibiscus rosa-sinensis*	26.6	4,6
35	Vông đồng	Cọ dầu	*Hura crepitans*	26.6	6
36	Hoa 10 giờ		*Portulaca pilosa*	26.6	6
37	Nhãn	Cọ nhàn	*Dimocarpus longan*	33.3	1
38	Ổi	Cọ Ổi	*Psidium guajava*	33.3	1
39	Cọ	Cọ Có	*Livistona cochinchinensis*	33.3	7
40	Tỏi	Họm Bùa	*Allium sativum*	33.3	3,4
41	Sả chanh	Pịnh chơ	*Cymbopogon citratus*	33.3	3,4
42	Mướp	Mặc cái	*Luffa cylindrica*	33.3	3
43	Khoai lang	Mặc on	*Ipomoea batatas*	33.3	2,3
44	Ớt chỉ thiên	Mặc ớt chỉ thiên	*Capsicum annuum*	33.3	3
45	Lạc	Mặc Họ	*Arachis hypogaea*	33.3	3
46	Cây doi		*Syzygium jambos*	33.3	1
47	Bưởi	Cọ Púc	*Citrus grandis*	40	1,4
48	Dứa	Cọ Hoài na	*Ananas comosus*	40	1,3

Continued

APPENDIX 9.1 List of Plant Species in the Home Garden and Their Function in Tat Hamlet (n = 15 Sampled Households)—cont'd

No.	Vietnamese Name	Tay Name	Scientific Name	Frequency	Uses
49	Đào	Cọ mặc Cai	*Prunus persica*	40	1,6
50	Bí xanh	Mặc pác	*Benincasa hispida*	40	3
51	Rau ngót	Pặc ót	*Sauropus androgynus*	40	3
52	Mùi tàu	Hợm Plu Lào	*Eryngium foetidum*	40	3
53	Cà trắng	Mặc Cựa	*Solanum album*	40	3
54	Kinh giới	Hợm Kinh giới	*Elsholtzia ciliata*	40	3,4
55	Cải trắng	Pặc cạt	*Brassica rapa*	40	3
56	Cải Dưa	Pặc cạt xôm	*Brassica juncea*	40	3
57	Đậu đen	Mặc đậu đen	*Vigna unguiculata*	40	3
58	Vải	Cọ Pặc vải	*Litchi chinensis*	40	1
59	Dong Riềng	Mặc đót	*Canna orientalis*	40	2,3
60	Nghệ	Cọ Han Lửa	*Curcuma longa*	40	3,4
61	Ngải cứu	Pặc Lá ngái	*Artemisia vulgaris*	40	3,4
62	Bắp cải	Mặc bắp cải	*Brassica oleracea*	40	3
63	Cải củ	Mặc cà cộ	*Raphanus sativus*	40	3
64	Sắn dây	Mịn Cọ chừa	*Pueraria thomsonii*	40	3,4
65	Dọc mùng	Pặc ọc	*Colocasia gigantea*	40	3,4
66	rau đay	Pặc rau đay	*Corchorus capsularis*	40	3
67	Sả chanh	Cọ Chu Khau	*Cymbopogon citratus*	40	3,4
68	Mít	Cọ mí	*Artocarpus heterophyllus*	46.6	1,2
69	Mía	Cọ Oi	*Saccharum officinarum*	46.6	3
70	Sắn	Cọ mên cọ	*Manihot esculenta*	46.6	2
71	Chanh	Cọ Lựu	*Citrus limonia*	47	1,3
72	Mận	Cọ mặc phục	*Prunus salicina*	53.3	1,6
73	Bí đỏ	Cọ mặc úc	*Cucurbita pepo*	53.3	3
74	Hành	Hợm Tiêu	*Allium fistulosum*	53.3	3,4
75	Khoai Sọ	Mặc Pược	*Colocasia antiquorum*	53.3	3
76	Mơ	Cọ Mặc Phục	*Prunus mume*	60	1,6
77	Rau ngò		*Coriandrum sativum*	60	3,4
78	Ớt nhà	Mặc ượt	*Capsicum frutescens*	60	3

APPENDIX 9.1 List of Plant Species in the Home Garden and Their Function in Tat Hamlet (n = 15 Sampled Households)—cont'd

No.	Vietnamese Name	Tay Name	Scientific Name	Frequency	Uses
79	Lá lốt	Pắc ơ lật	*Piper lolot*	60	3
80	Gừng	Cọ khịnh	*Zingiber officinale*	60	3,4
81	Riềng	Cọ khả	*Alpinia officina*	60	3,4
82	Chuối	Cọ cuôi	*Musa paradisiaca*	73	1,3
83	Chè	Cọ chè	*Camellia sinensis*	80	8

Notes: 1: Fruit trees; 2: Food crops; 3: Vegetables; 4: Medicinal plants; 5: Timber trees; 6: Ornamental plants; 7: Palm trees; 8: Tea and betel plants.
From Hung, D.T., Trung, T.C., Cuc, L.T., 2001. Agroecology. In Bright Peaks, Dark Valleys: A Comparative Analysis of Environmental and Social Conditions and Development Trends in Five Communities in Vietnams' Northern Mountain Region. National Political Publishing House, Hanoi.

References

Amilcar, R., Márquez, C., Schwartz, N.B., 2008. Traditional home gardens of Petén, Guatemala: resource management, food security, and conservation. J. Ethnobiol. 28 (2), 305–317.

Bishwajit, R., Habibur, R., Jannatul, F., 2013. Status, diversity, and traditional uses of homestead gardens in northern Bangladesh: a means of sustainable biodiversity conservation. SRN Biodivers. 2013, http://dx.doi.org/10.1155/2013/124103. Article ID 124103.

Cuc, L.T., Terry, R.A. (Eds.), 2001. Bright Peaks, Dark Valleys; A Comparative Analysis of Environmental and Social Conditions and Development Trends in Five Communities in Vietnams' Northern Mountain Region. National Political Publishing House, Hanoi.

Dung, V.D., Vien, T.D., Lam, N.T., 2002. Soil erosion, leaching, and soil nutrient status in composite swiddening system, Hoa Binh Province. In: Paper Presented at the Agroecosystem Analysis Workshop in Khon Kaen University, Khon Kaen, Thailand, 12–14 July 2002.

Galhena, D.H., Russell, F., Maredia, K.M., 2013. Home gardens: a promising approach to enhance household food security and wellbeing. Agric. Food Sec. 2, http://dx.doi.org/10.1186/2048-7010-2-8.

Gia, B.T., Dung, P.T., Quang, D.V., Calkins, P., 2004. Production and marketing of agro-forest products in northwest upland of Vietnam; a case study in Tat Village, Tan Minh, Da Bac, Hoa binh province. In: Paper Presented at the Workshop on Marketing of Non-Timber Forest Products Funded by the Ford Foundation in Hoa Binh Province, April 2004.

Godbole, A., 1998. Homegardens: traditional systems for maintenance of biodiversity. In: Applied Ethnobotany in Natural Resource Management Traditional Homegardens; Highlights of a Training Workshop Held at Kohima, Nagaland, India, 18–23 June 1997.

Hung, D.T., Trung, T.C., Cuc, L.T., 2001. Agroecology. In: Bright Peaks, Dark Valleys: A Comparative Analysis of Environmental and Social Conditions and Development Trends in Five Communities in Vietnams' Northern Mountain Region. National Political Publishing House, Hanoi.

Loc, P.K., Cuc, L.T., 1997. Da River Watershed (Checklist of Plants) [Luu vuc song Da (Danh luc thuc vat)]. Agricultural Publishing House, Hanoi.

Mariel, A.S., Stain, R.M., Sara, L.C.R., 2009. Home gardens sustain crop diversity and improve farm resilience in Candelaria Loxicha, Oaxaca, Mexico. Hum. Ecol. 37 (1), 55–77.

Millat-e-Mustafa, M., 1998. Overview of research in homegarden systems. In: Rastogi, A., Godbole, A., Shengji, P. (Eds.), Applied Ethnobotany in Natural Resource Management Traditional Homegardens; Highlights of a Training Workshop Held at Kohima, Nagaland, India, 18–23 June 1997. International Centre for Integrated Mountain Development, Kathmandu, Nepal, pp. 13–38.

Rambo, A.T., 1998. The composite swiddenning agroecosystem of the Tay ethnic minority of the northwestern mountains of Vietnam. In: Land Degradation and Agricultural Sustainability: Case Studies from Southeast and East Asia. Regional Secretariat, the Southeast Asian Universities Agroecosystem Network (SUAN), Khon Kaen University, Khon Kaen, Thailand, pp. 43–64.

Rambo, A.T., Vien, T.D., 2001. Social organization and the management of natural resources: a case study of Tat Hamlet, a Da Bac Tay ethnic minority settlement in Vietnam's northwestern mountains. Southeast Asian Stud. 39 (3), 299–324.

Valere, K.S., Belarmain, F., Barthelemy, K., Achille, E.A., Alix, F.R.I., Rodrigue, C.G., Sebastian, C., Mohammad, E.D., Romain, G.K., 2014. Home gardens: an assessment of their biodiversity and potential contribution to conservation of threatened species and crop wild relatives in Benin. Genet. Resour. Crop. Evol. 61 (2), 313–330.

Vien, T.D., 1998. Soil erosion and nutrient balance in swidden fields of the composite swiddenning agroecosystem in the Northwestern Mountains of Vietnam. In: Land Degradation and Agricultural Sustainability: Case studies from Southeast and East Asia. SUAN Regional Secretariat, Khon Kaen University, Khon Kaen, Thailand, pp. 65–87.

Vien, T.D., 2004. Changes in the composite swiddening system in Tat Hamlet in Vietnam's northern mountains in response to integration into the market system. In: Hisao, F., et al. (Eds.), Ecological Destruction, Health, and Development: Advancing Asian Paradigms. Kyoto University Press, Kyoto, Japan.

CHAPTER

10

How Agricultural Research for Development Can Make a Change: Assessing Livelihood Impacts in the Northwest Highlands of Vietnam

N.H. Nhuan[*,†], *E. van de Fliert*[*], *O. Nicetic*[‡]

[*]The University of Queensland, Brisbane, QLD, Australia [†]Vietnam National University of Agriculture, Hanoi, Vietnam [‡]The University of Queensland, Brisbane, QLD, Australia

10.1 INTRODUCTION

Assessing the impacts of agricultural research for development (AR4D) projects is crucial for sustainable development. The selection of an appropriate impact assessment method for a particular project helps to obtain good impact assessment indicators at different levels of contributions. The findings on impact assessment are an important resource, not only for learning about the impacts of AR4D but also for informing effective development policies and strategies and guiding long-term sustainable development in the target areas (Cramb et al., 2003; Krall et al., 2003, p. 329). A review of the impact assessment approaches used in developing countries highlights the weaknesses in existing approaches regarding their objectives and the methods used to achieve these objectives (Nguyen et al., 2015). The major limitations of conventional impact assessment approaches, as identified in a review of the literature, are an overemphasis on direct research outputs; a short-term, quantitative, and economic focus; a one-way communication approach; donors' prioritization of cost-effectiveness; researchers' lack of understanding of local culture and languages; and poor feedback mechanisms (Nguyen et al., 2015). These limitations lead not only to unconvincing evidence in the impact assessment findings but also to the limited utilization of the findings for local development and for the effective implementation of future agricultural research interventions.

This chapter discusses the characteristics and limitations of impact assessment approaches in agricultural research projects with a particular focus on recent AR4D projects in the Northwest Highlands of Vietnam. The chapter presents a holistic impact assessment

Redefining Diversity and Dynamics of Natural Resources Management in Asia Volume 2
http://dx.doi.org/10.1016/B978-0-12-805453-6.00010-3

framework that utilizes the sustainable livelihoods framework developed by the UK Department for International Development (DFID, 1999) as a lens for impact analysis and adopts the participatory impact assessment (PIA) approach to gather reliable data and empower local stakeholders in the impact assessment process. The study reported in this chapter tested the holistic impact assessment framework in three agricultural research projects in the Northwest Highlands to validate the appropriateness of the proposed framework and gain insights into how AR4D projects that adopted participatory processes can make better changes in people's lives.

10.2 RESEARCH METHODOLOGY

To collect data, this study employed participatory techniques in combination with documentary research. The documentary research included a review of the literature on existing development theories and practices such as the sustainable livelihoods framework and PIA approach and their fit with AR4D. It also examined secondary sources on the impact assessment approaches to agricultural research project, including AR4D initiatives implemented by the research institutions and development agencies active in the Northwest Highlands. These include the Northern Mountainous Agriculture and Forestry Science Institute (NOMAFSI), Plant Protection Research Institute (PPRI), Center for Agricultural System Research and Development (CASRAD), Vietnam National University of Agriculture (VNUA), Tay Bac University (TBU), Australian Centre for International Agricultural Research (ACIAR), and French Agricultural Research Centre for International Development (CIRAD).

The study used the participatory techniques of focus group discussions (FGDs) with farmers, semistructured interviews with farmers, and in-depth interviews with key informants including local leaders, agricultural extension staff, and agricultural researchers who were actively involved in AR4D initiatives in the Northwest Highlands. The research team conducted fieldwork in Moc Chau and Yen Chau districts of Son La Province, covering the implementation sites of three agricultural research projects: the ACIAR Northwest Project,[1] the CIRAD ADAM Project,[2] and the NOMAFSI Project.[3] The study selected four communes in two districts, namely, Phieng Luong and Muong Sang communes in Moc Chau District (ACIAR Northwest Project), Chieng Hac Commune in Moc Chau District (CIRAD ADAM Project), and Chieng Dong Commune in Yen Chau District (NOMAFSI Project).

The study adopted the purposive sampling method for the selection of participants for both the FGDs and in-depth interviews to include the major participating researchers, extension staff, and farmers in the three selected projects that formed the focus of the study. The study involved both farmer researchers (ie, farmers actively involved as coresearchers in

[1] "Improved market engagement for sustainable upland production systems in the Northwest Highlands of Vietnam."

[2] "Support for agroecology extension in mountainous areas of Vietnam" (in French: "*Appui au Développement de l'Agro-écologie en zone de montagne du Vietnam*," hence the abbreviation "ADAM").

[3] "Integrated measures for sustainable maize production on sloping lands of the northern mountainous regions of Vietnam."

maize research components of the selected research projects) and other farmers (farmers not involved in the selected research projects) in the research location. The research team facilitated eight FGDs (with four to five farmers in each group) and eight in-depth interviews with village and commune leaders. The research team conducted in-depth interviews in Son La Province and in Hanoi with 6 local agricultural extension staff and 15 agricultural researchers from NOMAFSI, CIRAD, VNUA, and TBU who had been actively involved in agricultural research projects in the Northwest Highlands in recent years. The research team also conducted 29 semistructured interviews with individual farmers. To collect the data, the research team made three field trips to Son La Province (in Dec. 2012, Sep. 2013, and Jul. 2014).

Initial findings were reported to the participants to elicit their feedback and validate the data. All the gathered primary data and information were recorded, reviewed, and translated into English. Thematic analysis was used for data analysis and interpretation of the research findings. This involved cleaning the quantitative and qualitative data, coding the data based on its themes or patterns, and carrying out the analysis with the assistance of the latest Statistical Package for the Social Sciences (SPSS) software.

10.3 FINDINGS AND DISCUSSION

10.3.1 Shift Toward AR4D Projects in the Northwest Highlands

The Northwest Highlands is characterized by high ethnic diversity and steep mountain ranges and sloping highlands. The topography is high and rugged in the northwest and decreases toward the southeast area along the border with China and the plateau region. The highlands region includes the six provinces of Son La, Lai Chau, Dien Bien, Hoa Binh, Yen Bai, and Lao Cai. It has a total natural area of 5.07 million ha, which accounts for 15.32% of the whole country (NOMAFSI, 2012). The population of the region is 5.06 million people (GSO, 2014). The highland is identified as one of the poorest regions of the country. According to the latest general census, the rate of poor and marginally poor households in the region stood at about 36% compared to about 14% for the whole country (MoLISA, 2014). Although rapid growth of the market economy has pushed the socioeconomic development of the Northwest Highlands, it has also generated social, economic, and environmental challenges including poverty and widening economic gaps within the region and between different ethnic groups; barriers to market integration; and emerging environmental problems such as soil erosion and degradation, biodiversity loss, and deforestation (Castella, 2012). In addition, agricultural research and extension programs have not paid adequate attention to the engagement of local communities and their knowledge in agricultural development activities (Thai et al., 2011). These factors have led to the unsustainable development in the Northwest Highlands.

Recognizing the problems facing the Northwest Highlands, since the late 1990s there has been a great deal of investment made by Vietnamese government research institutes and international development agencies in the form of various socioeconomic development policies and research initiatives. Most of these agricultural research projects have targeted economic development through increasing agricultural production and improving market engagement. However, a shift toward a research for development (R4D) approach, targeting the immediate use of research outputs for development purposes, became apparent in the late

2000s. Some explicit AR4D initiatives were implemented by internationally funded projects in the Northwest Highlands. These AR4D projects have adopted participatory approaches in an attempt to better link the research with development, but with varying approaches to farmer engagement, ranging from merely using farmers as field laborers to the involvement of farmers as coresearchers.

Looking at examples of the government-funded projects first, NOMAFSI is a leading organization carrying out agricultural research in the Northwest Highlands. In the last 5 years, the institute has carried out more than 140 agricultural research projects, most of which have been conducted in the Northwest Highlands. Other national research organizations and universities such as PPRI, CASRAD, and VNUA have also conducted a large number of agricultural research projects in the highlands. The major objectives of these agricultural research projects are the selection and development of high-yielding crops, mobilization of indigenous knowledge and experiences of minority ethnic communities on plant gene resources, plant protection and biodiversity conservation, agricultural value chain development, and market engagement for small agricultural producers. Participatory techniques such as participatory rural appraisal (PRA) tools were adopted in the implementation and evaluation processes of several of these agricultural research projects for data collection (Ha et al., 2010; Harrison, 2002).

An example of the internationally funded development activities in the region is the Sustainable Land Use and Rural Development in Mountainous Region of South East Asia Program (the Upland Program) initially implemented in 2000 by University of Hohenheim in collaboration with VNUA, Thai Nguyen University, and the National Institute of Animal Husbandry. This multidisciplinary research program aims to make a scientific contribution to the improvement of natural resource conservation and the livelihoods of the rural people in the northern uplands of Vietnam through a wide range of activities in agriculture and food science (Neef and Neubert, 2011, p. 181). The program sees participatory processes as a key crosscutting component in the implementation of its activities (Neef and Neubert, 2011).

Another example of an active international development agency in the region is ACIAR, which, since late 2007, has focused its work on the northwest and south central coast regions of Vietnam. Multidisciplinary teams in the Northwest Highlands have carried out agricultural research projects funded by ACIAR. These agricultural research projects in the Highlands have focused on three key areas: (i) the production and marketing of local agricultural and forestry products (high-value temperate fruits, maize, and vegetables), (ii) the production and marketing of beef cattle, and (iii) sustainable agroforestry management to improve market integration and sustain livelihoods for local communities (ACIAR, 2009; Van de Fliert, 2008). Unlike conventional agricultural research projects, some recent ACIAR-funded AR4D projects in the Northwest Highlands have modified their approaches to include participatory processes and appropriate communication strategies for planning, experimentation, monitoring, and evaluation through collaborative research mechanisms. The aim of these modified approaches is to develop the decision-making capacity of local people.

10.3.2 Weaknesses in Existing Impact Assessment Approaches to Agricultural Research

The review of documentation on the agricultural research projects conducted by active research institutions and development agencies in the Northwest Highlands, the FGDs with farmers, and the in-depth interviews with agricultural researchers and extension staff led to the identification of five major weaknesses in existing impact assessment approaches in the region. First, most research projects only undertake an end-evaluation at a single point in time or they only implement a short-term impact assessment rather than a long-term impact assessment. Second, the impact assessment of these research projects pays more attention to the objectives of researchers and donor agencies than to the interests of local communities. Third, despite the fact that most agricultural research projects claim to apply a participatory approach, top-down planning, implementation, and monitoring and evaluation approaches are still dominant, especially in most government-funded agricultural research projects. These top-down approaches lead to low levels of empowerment of local communities and in most cases miss capturing what local communities perceive to be their needs. Fourth, there is a gap in researchers' understanding of the diverse local cultures and languages. This gap sometimes leads to limited communication and unreliable impact assessment findings. Fifth, the indicators currently used for impact assessment are aimed at measuring the return on investment or cost-effectiveness for the donor organization, rather than fostering the sustainability of the local community. Moreover, the mechanisms used to report the findings and obtain feedback from the local community are not clear and often simply do not exist.

Compared to the Vietnamese government-funded projects, international donor-funded agricultural research projects in the Northwest Highlands have had a stronger participatory orientation and a broader scope of impact considerations. However, there is still no clear strategy for assessing the long-term social, economic, human, physical, and natural impacts on the sustainable livelihoods of local communities. The impact assessments of both domestic-funded and international-funded research projects are very weak in terms of sharing the impact assessment findings and obtaining feedback from the key stakeholders, especially local beneficiaries. Table 10.1 summarizes the major dimensions of the impact assessments conducted by domestic- and international-funded research projects in the highlands.

From the summary it is concluded that most projects in the Northwest Highlands—both the Vietnamese government-funded projects and the international agency-funded projects—use a top-down approach with limited attention paid to the cultural diversity and complexity of the region. The impact assessments of most agricultural research projects in the region have had a short-term and economic focus. Projects have directed their efforts toward measuring direct research outputs, reporting scientific findings, and analyzing cost-effectiveness to report to donors and funding agencies rather than targeting the sustainable livelihoods of local communities. A lack of mechanisms for sharing the impact assessment findings with and getting feedback from stakeholders, especially local communities, has resulted in limited opportunities for local people to understand and reflect on the projects' desired impacts. Because of these weaknesses, the impact assessment findings provide limited evidence on how the agricultural research projects have contributed to sustainable development in the Northwest Highlands.

10.3.3 Development of a Holistic Framework for the Impact Assessment of AR4D

The weaknesses in the impact assessment objectives and methodologies in agricultural research projects in the Northwest Highlands clearly demonstrate the need for a holistic

TABLE 10.1 Comparison of Impact Assessment Approaches in Agricultural Research Projects in the Northwest Highlands

Dimension	Vietnamese Government-Funded Projects	International Development Agency-Funded Projects
Impact assessment approaches and methods	o Top-down approach o No impact assessment, or sometimes the impact assessment is conducted at the end of a project o Use of mainly quantitative methods for data collection and analysis	o Top-down approach in most projects but bottom-up approach in some recent AR4D projects o Impact assessment often implemented at the end of a project o Use of both quantitative and qualitative methods for data collection and analysis
Impact assessment indicators	o Mainly short-term and economic-focused indicators (eg, changes in production outputs and income) o Aimed at direct scientific outputs (technology development and publications) and project performance rather than local sustainability	o Mainly short-term and economic-focused indicators (eg, changes in production outputs and income) o Aimed at direct scientific outputs (capacity building and publications) and cost-effectiveness for donors rather than local sustainability
Stakeholders' participation in impact assessment processes	o Project implementers are evaluators who define impact assessment indicators o Local communities, extension staff, and government staff are information givers o No participation of private sector (private companies and traders) or NGOs in impact assessment process	o External specialists or researchers are evaluators who define impact assessment indicators o Local communities, extension staff, and government staff are information givers o Limited participation of private sector (private companies and traders) and NGOs in impact assessment process
Dissemination of impact assessment findings	o No mechanism for sharing impact findings with and getting feedback from local communities o The sharing of impact assessment findings among research partners (research institution, development agencies, and local governments) is very weak	o Limited or no mechanisms for sharing impact assessment findings with and getting feedback from local communities o Efforts made to share impact assessment findings among research partners (research institutions, development agencies, and local governments), mostly through publications and media products

impact assessment framework based on the sustainable livelihoods framework as a lens for impact analysis. Evaluation processes should be participatory using qualitative and quantitative methods to measure the livelihoods outcomes and impacts of an AR4D. The use of this type of framework would support sustainable social change and development in this remote and culturally diverse region. This study attempted to develop such a framework by answering the following five interrelated questions.

(1) *Why should AR4D projects carry out impact assessments?*

Although the assessment of AR4D impact could play an important role in the development of the Northwest Highlands, scholars have made few attempts to assess existing impact assessment approaches in AR4D projects in the region. Local government, development agencies, and research institutes have limited understanding about the contribution of impact assessment findings to development. Therefore, this study's analysis of current impact assessment strategies and testing of alternative impact assessment approaches—in the course of developing a holistic impact assessment framework for AR4D in the highlands—represents a significant step forward. An appropriate impact assessment framework will help to increase understanding of AR4D projects in the Northwest Highlands. It will also contribute to the formulation of suitable development strategies for the highlands in the longer term. The proposed impact assessment framework can also be utilized in other regions with similar levels of socioeconomic disadvantage and cultural diversity so that the impacts of projects can be maximized toward the goal of sustainable development.

(2) *Who are the relevant users of impact assessment findings?*

AR4D aims at contributing to long-term development through the provision of innovations that support sustainable farming systems. Because different development stakeholders have different roles in different phases of an agricultural research project, they have different interests in the expected impacts of the project (Lilja et al., 2001). Identification of the relevant users of impact assessment findings will help to better utilize the findings to achieve sustainable development. In a 2014 Food and Agriculture Organization of the United Nations (FAO) e-mail conference on the ex-post impact assessment of agricultural research, researchers agreed that the two main objectives of the assessment of agricultural research impact are accountability and learning (Ruane, 2014). Accountability refers to the effective analysis of the resources used for agricultural research, and learning means drawing lessons for more effective implementation of future agricultural research.

In addition, the conventional impact assessment approach often pays attention to accountability and cost-effectiveness for donors but ignores the fact that farmers are important end users of the impact assessment findings, because if the farmers do not understand fully the impacts of the new technology, they will have no incentives to scale-out this technology to sustain the identified impacts. From a sustainable livelihoods perspective, the proposed holistic impact assessment framework identifies five key user groups of impact assessment findings: farmers as end users; agricultural extension staff; researchers; policy makers; and private sector actors such as traders and companies. These can be further categorized into two main groups of users of impact assessment findings: local users (local farmers, extension staff, traders, and local policy makers) and external users (agricultural researchers from national and international research institutions, universities, international development agencies, and high-level policy makers). Depending on the objectives of the particular AR4D project, each user group has certain roles in the utilization of the impact assessment findings for the scale-out of the research outputs.

(3) *What types of impacts should AR4D assess?*

Because of the complexity of AR4D projects, their research interventions can generate a wide range of impacts. According to Cramb and Purcell (2001), the impacts of a participatory technology development project include intermediate and long-term livelihood impacts and the research innovation process itself. Marasas et al. (2001) developed a comprehensive impact assessment framework in which the impacts of agricultural research are divided into three major groups: the direct research products, the intermediate impacts, and the people-level impacts. Several scholars have identified that analysis of AR4D processes and impacts is compatible with the principles of the sustainable livelihoods framework because of the mutual interactions between AR4D and livelihood assets, development policies and institutions, and the shared context in which livelihood strategies are combined for better outcomes and impacts (Adato and Meinzen-Dick, 2002; Carpenter and McGillivray, 2012). In the present study, the sustainable livelihoods framework was adapted and utilized as a lens to identify four key groups of impact assessment indicators: (i) direct research outputs, (ii) livelihood impacts, (iii) institutional impacts, and (iv) impacts in the vulnerability context.

The direct research outputs of AR4D include new or improved plant varieties or agricultural technologies, sets of recommendations, and publications (Marasas et al., 2001). Because the pathway from research outputs to impacts is often affected by multiple internal and external factors (eg, livelihood assets base, vulnerability context, and institutional environments), assessing the direct outputs of AR4D not only helps to analyze the effectiveness of an AR4D project but also to understand how livelihood outcomes and impacts could be achieved. Funding agencies, research institutions, and local government also need to know the research outputs that have resulted from an intervention so that they can make better decisions. The direct outputs of AR4D include a trained workforce of local institutions capable of delivering outreach activities and outreach material to support these activities and new technologies and capacity of the local institutions' workforce to facilitate innovation processes. In many cases, research institutions, local extension agents, and target communities could immediately apply the direct research outputs, leading to better outcomes and impacts.

The sustainable livelihoods framework provides the parameters for a comprehensive conceptual analysis of *how* AR4D projects achieve impacts and *what* impacts they achieve. As stated above, this study utilized the sustainable livelihoods framework as a lens to identify four key groups of impact assessment indicators for AR4D projects: direct research outputs, livelihood impacts, institutional impacts, and impacts in the vulnerability context. The diagram in Fig. 10.1 shows the four key groups of impact assessment indicators, and Table 10.2 presents a detailed description of these groups. Depending on the research objectives, the specific social context of the research locations, and the length of time in which the assessment takes place after a project is completed, the impact assessment set out in the framework could focus more on either short-term or long-term impacts.

(4) *What methods and resources should AR4D projects use to measure impacts?*

PIA, which was initially practiced in South Asia and East Africa by international development agencies and nongovernmental organizations (NGOs) as an extension of PRA, is an alternative approach to impact assessment (Robinson, 2002). Unlike conventional top-down impact assessment approaches, PIA aims to measure the real impacts created

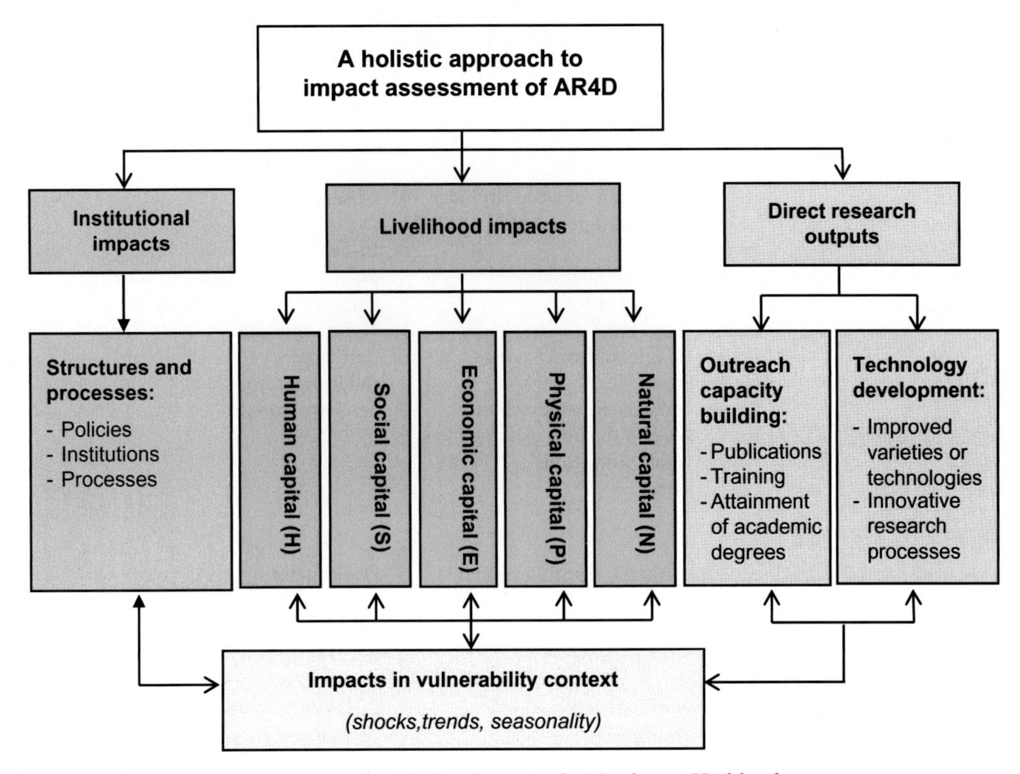

FIG. 10.1 Holistic impact assessment for AR4D projects in the Northwest Highlands.

TABLE 10.2 Key Groups of Indicators and Methods Used in the Assessment of AR4D Impact in the Northwest Highlands

Impact Type	Key Indicators	Key Methods and Tools
I. Direct research outputs		
Technology development	Achievement of scientific products compared to expected outputs: *improved agricultural technology and innovative research process*	o Documentary research o In-depth interviews with key researchers from research institutions and with local agricultural extension staff o Observation
Capacity building	Changes in capacity of research organizations: *publications, training, and attainment of academic degrees*	o Documentary research o In-depth interviews with researchers and local agricultural extension staff

Continued

TABLE 10.2 Key Groups of Indicators and Methods Used in the Assessment of AR4D Impact in the Northwest Highlands—cont'd

Impact Type	Key Indicators	Key Methods and Tools
II. Livelihood impacts		
Livelihoods capital: – Human – Social – Economic – Physical – Natural	Positive changes in livelihood capital: – Human (knowledge and skill, health) – Social (trust, membership, informal safety net, communication) – Economic (income and savings, credit) – Physical (roads, transportation, sanitation, healthcare) – Natural (soil protection, biodiversity)	o Observations o Documentary research o In-depth interviews with key informants (local leaders, extension staff and agricultural researchers) o FGDs with farmers using visual participatory tools (Venn diagram, ranking, radar diagram, 10-seed techniques) o Semistructured interviews with farmers
III. Institutional impacts		
Policies, institutions, and processes	Changes in policies, institutions and processes (formal and informal): *development policies and development strategies, culture, scaling-up opportunities, research organizational capacity, research collaboration, and research for development strategies*	o Observations o Documentary research o In-depth interviews with key informants (local leaders, extension staff, and researchers) o FGDs with farmers using visual participatory tools (Venn diagram, seasonal calendar, resource mapping) o Semistructured interviews with farmers
IV. Impacts in the vulnerability context		
Vulnerability context: – Shocks – Trends – Seasonality	Changes in: – Shocks (human or animal health, natural disasters, and sudden economic changes) – Trends (migration, resource use, and other indicators such as prices, governance, and technologies) – Seasonality (production, price, employment, and health)	o Observations o Documentary research o In-depth interviews with key informants (local leaders, extension staff and researchers) o FGDs with farmers using visual participatory tools (seasonal calendar, resource mapping) o Semistructured interviews with farmers

by a development project rather than accounting for aspects of its implementation such as inputs and service delivery, structure construction, and training (Catley et al., 2008). PIA not only helps to generate information and statistical data on the extent of the change created by development activities but also helps to empower local communities. It therefore focuses on enhancing the equal and equitable participation of stakeholders and beneficiaries in defining and assessing the impact indicators. True participation in impact evaluation is the goal itself, and PIA empowers people by equipping them with the capability to change their own lives (Pretty, 1995a; Van de Fliert, 2010). However, the types and levels of participation also depend on who the main users of the PIA findings will be.

Cromwell et al. (2013) discern five key steps in participatory approaches to impact assessment: identifying the interested stakeholders; establishing the stakeholders' expectations; identifying the priority evaluation criteria and defining the impact assessment indicators; agreeing on methods with the stakeholders; and collecting and analyzing data in collaboration with the relevant stakeholders.

An appropriate impact assessment framework is one that aims to both fully measure the impacts of an AR4D project and empower local stakeholders in the impact assessment processes. The PIA approach, with a wide range of participatory data collection methods and tools such as FGDs, visual data collection techniques, in-depth interviews, direct observation, and semistructured interviews, is an effective way to collect qualitative and quantitative data about the impacts of AR4D projects. Because of limited education levels, poor economic conditions, language barriers, and high levels of cultural and natural diversity in the Northwest Highlands, the adoption of participatory and visual techniques can engage the most disadvantaged groups in the assessment of AR4D impact and in local livelihood development processes. The collaboration among stakeholders (eg, farmers, extension staff, and researchers) is strengthened.

As just discussed, the PIA approach has many advantages over conventional top-down impact assessment approaches. However, scholars raise three concerns about the use of participatory methods that need to be considered. First, project evaluators might co-opt "participation" and "participatory" as merely fashionable terms. A rapid and uncritical adoption of participatory methods will result in weak evidence if the assessment process has not maintained a clear focus on sustaining the impacts of the project (Pretty, 1995b; Robinson, 2002). Such communication practices do not enhance the engagement of local people in impact assessment processes nor facilitate these people's utilization of impact assessment findings to achieve livelihood impacts. Second, participatory processes are often time-consuming. An insufficient allocation of time and human resources to the participatory sessions leads to weak commitment among the stakeholders to the assessment processes (Robinson, 2002). Third, the outcomes of participatory evaluation are driven by group dynamics, which can cause a distorted view of reality to unaware evaluators (Campbell, 2001).

(5) *How should AR4D projects communicate the impact assessment findings to support the sustainability of impacts?*

Sharing the impact assessment findings helps the key stakeholders to understand the contribution made by AR4D to the development of target communities and regions. In the design of appropriate communication strategies for communicating the findings on AR4D impact in such a diverse setting context as the Northwest Highlands, project implementers should pay careful attention to how and for what purposes the findings can be utilized for development. As discussed earlier in relation to the second question posed in the development of the proposed framework, there are two main groups of users of impact assessment findings: local users (such as local farmers, extension staff, traders, and local policy makers) and external users (such as agricultural research institutions, universities, international development agencies, and high-level policy makers). Appropriate strategies are required to communicate the impact assessment findings to each of these stakeholder groups.

For the local stakeholders and beneficiaries, the project's evaluation facilitators can communicate the impact assessment findings immediately or soon after the completion of the assessment process. The project can communicate the findings in both the impact assessment process and the impact sharing process. Evaluation facilitators should base their communication of the findings with local stakeholders and beneficiaries on the particular social context and available time and financial resources. The communication of findings can take place in the field or at places that are most convenient for the participation of different stakeholders. Facilitators should use simple and understandable language when disseminating impact assessment findings, due to the different levels of education and skills of local stakeholders.

For the research institutions, donors, and policy makers, after sharing the findings and getting feedback from the key stakeholders and beneficiaries, the project can produce conventional written reports and publications shared with research partner organizations. The dissemination of visual products such as video, photo stories, and posters at international conferences, university seminars, and agricultural extension training can be an effective way to sustain the impacts. The promotion of research products and innovations through websites and electronic forums such as e-mail conferences also helps to share the impact assessment findings with a wide audience at low cost.

In conclusion, the application of a holistic impact assessment framework for AR4D is achievable by mixing different data collection and analysis methods and by integrating more than one conceptual framework. Answering the five questions discussed above in the development of the proposed framework highlighted the common steps in the process of utilizing and integrating development theories and practice into a holistic impact assessment framework for AR4D. The first step is designing the AR4D project based on the identification of its possible impacts. Depending on the specific social context of the research site and the expected lapse of time between the completion of the project and its evaluation, the project can set out to focus on short-term or long-term impacts. The second step focuses on identifying the key groups of impact indicators. The third and fourth steps are selecting the appropriate assessment strategies and the specific methodology and techniques for data collection and analysis to measure the relevant impact assessment indicators. The last step is conducting participatory workshops or meetings with local communities to share the impact assessment findings and get feedback from local stakeholders to help verify the findings and facilitate the use of the research outputs on a large scale. The findings can also be reported to funding and implementation agencies, and can be shared among research partners and development agencies through publications, media, and other electronic means for better implementation and assessment of future AR4D interventions (Fig. 10.2).

The application of a comprehensive impact assessment framework such as the one briefly outlined in this study can be expected to face a number of challenges. First, the holistic framework may not work well if the research teams lack good facilitation skills or lack a deep understanding of the local culture and an appreciation of the complexity of the research context. Second, because social, human, economic, and environmental impacts are more likely to be achieved if AR4D projects are designed to deliver measurable impacts, the impact pathway and the causal links between outputs and impacts should be well integrated into the impact assessment process. Third, even an impact evaluation that adopts a holistic framework may be insufficient to measure the full contribution of an AR4D project to local changes. For example, the effective application of participatory approaches in the Northwest Highlands is hampered by the dominance of conventional research approaches and top-down political power structures. Fourth, no standard set of participatory communication techniques can fit different communities and locations.

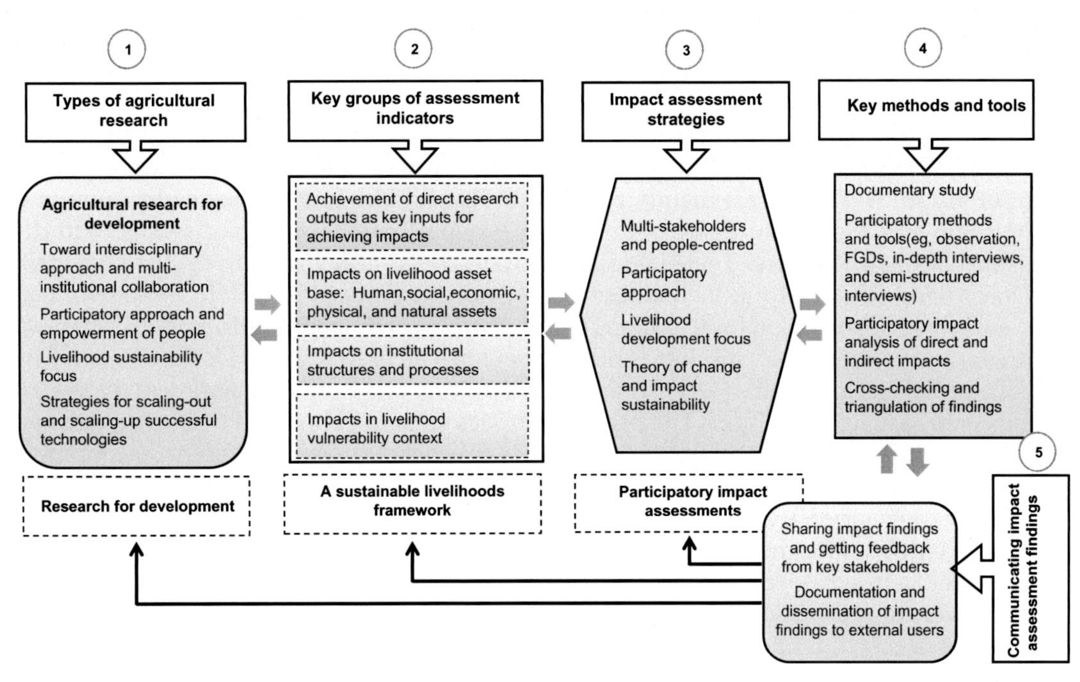

FIG. 10.2 Holistic impact assessment framework for AR4D.

The types and expected levels of participation of stakeholders in the impact assessment process will depend on the relevant users of the assessment findings for the particular project. In each case, evaluators should consider carefully the appropriate time allocation and the suitable location for each participatory activity. The effective sharing of impact assessment findings among the different groups of relevant stakeholders and at different levels is necessary to help sustain the identified impacts.

10.3.4 Assessing the Impacts of AR4D Projects in the Northwest Highlands

10.3.4.1 Overview of the Three Selected Agricultural Research Projects

The study selected three recently completed agricultural research projects in the Northwest Highlands for testing the proposed holistic impact assessment framework; namely, the ACIAR Northwest Project, the CIRAD ADAM Project, and the NOMAFSI Project. These projects are self-proclaimed as AR4D projects, and shared the aim of achieving significant economic impacts and improved livelihoods for local communities; however, they varied in the research approaches and communication strategies used for research design, implementation, monitoring, and evaluation. One project attempted to make the research process an adaptive and innovative process that facilitated both the farmers' adoption and the institutionalization of the research outputs. The other two projects focused on achieving direct research outputs as scientific impacts that could be disseminated on a wider scale through conventional communication channels.

The ACIAR Northwest Project funded by ACIAR was a complex project that ran from May 2009 to Dec. 2013. The project focused on two major systems in the Northwest Highlands region; namely, the maize-based and temperate fruit-based farming system, both of which had been driven in recent years by a rapid transition to a market economy. The overall objective of the project was "to increase smallholder engagement in competitive value chains associated with maize-based and temperate fruits-based farming systems and to use this engagement to improve land and crop management in these rapidly transforming sectors" (Stur et al., 2013). The project mainly carried out its activities in a total of eight communes in Mai Son and Moc Chau districts in Son La Province and Tam Duong and Sin Ho districts in Lai Chau Province. The Australian-commissioned organization was the University of Queensland (UQ) and the Vietnamese-led organization was NOMAFSI. These two main implementing organizations collaborated with other research institutions, universities, local partners, and international development agencies such as PPRI, CASRAD, VNUA, and TBU and the provincial and district Department of Agriculture and Rural Development (DARD) in Son La and Lai Chau provinces. The project activities were implemented through an innovative process in five phases: (i) diagnostic research, (ii) participatory technology development, (iii) outreach strategy research, (iv) development and implementation, and (v) evaluation. The project developed a participatory monitoring and evaluation (PM&E) system including detailed guidelines for meetings with farmers in the planning, implementation, and evaluation of field trials to get community feedback. The project implementers then modified the PM&E guidelines to make them more specific for each research component and to adapt them to the local cultural conditions, the needs of the participating communities, and the institutional settings of the Vietnamese partner organizations. The project developed an impact assessment pathway to track the achievement of impacts, mainly during the project duration.

The CIRAD ADAM Project received funding from the French Development Agency. The project implementer was NOMAFSI, with technical assistance from CIRAD. The project commenced with the signing of a cooperation agreement between CIRAD and NOMAFSI in Nov. 2008 and reached completion in 2014. The project activities concentrated on four districts: Van Chan District in Yen Bai Province, Thanh Ba District in Phu Tho Province, and Moc Chau and Mai Son districts in Son La Province. The initial project proposal designed the research activities and outputs for the tea-based farming system. However, after several field trips by researchers at the agrarian diagnosis stage of the project, the project expanded its focus to include annual crops such as maize and crops intercropped with maize (eg, legumes, buckwheat, and stylo grass). The project had two specific objectives: (i) to design and test innovations based on agroecology for sustainable tea production on sloping lands, and (ii) to promote direct seeding mulch-based cropping (DMC) systems for the sustainable production of annual crops in various biophysical and socioeconomic contexts in the mountainous areas of Vietnam. The project implementation took place in three phases: (i) agrarian diagnosis studies, (ii) on-farm research, and (iii) extension of technologies. Although the project collaborated with DARD and local agricultural extension networks, the purpose of this formal collaboration was to inform the local partners about the research activities rather than to involve them actively in the research process. The NOMAFSI and CIRAD researchers were the main designers and implementers of the research components and on-farm experiments, which they carried out with the use of detailed protocols. The project set up some on-farm demonstration sites and conducted field visits; however, overall, the project adopted a conventional research approach. The project design did not mention impact assessment.

The NOMAFSI Project was implemented by NOMAFSI and funded by the Vietnamese government from 2011 to 2013. The overall objective of the project was to identify appropriate measures for sustainable maize development on sloping lands of the northern mountainous regions of Vietnam. The project implemented research activities in four provinces: Hoa Binh, Yen Bai, Cao Bang, and Son La. The project design focused on the linear links between research activities and research outputs. The NOMAFSI Project proposal followed the template provided by the Vietnamese Ministry of Agriculture and Rural Development (MARD), which identifies direct research outputs as the final products of a research project. Although the NOMAFSI Project proposal mentioned the expected long-term benefits or impacts, the project design did not include an impact pathway or specific impact indicators. The project implementation occurred in three phases: (i) diagnostic research, (ii) on-farm technology development, and (iii) preextension evaluation. Like the CIRAD ADAM Project, the NOMAFSI Project formally collaborated with the provincial and district DARD and an extension network during the research process. The project design included the transfer of direct research products (such as sustainable techniques for maize production on sloping lands and publications) to the relevant DARD and extension networks; however, there were no clear mechanisms for building the capacity of local stakeholders to scale-out the application of the research outputs. The project did not include any scheme for assessing the impact of the research intervention.

10.3.4.2 Applying the Holistic Assessment Framework to AR4D Projects in the Northwest Highlands

Applying the holistic impact assessment framework to these three research projects in the Northwest Highlands helped to gain a better understanding about the contribution of AR4D projects (which are underpinned by participatory processes) in comparison with conventional agricultural research projects (which are underpinned by top-down approaches). The study found that, although the projects involved a small number of farmers in trials and extended the research outputs to a limited scale, some initial human, social, economic, and natural outcomes and impacts were observed. The three research projects took place in similar socioeconomic settings; however, their different use of research approaches and communication strategies resulted in different levels of outcomes and impacts. The ACIAR Northwest Project employed a participatory approach and used participatory processes facilitated by various participatory communication strategies, from the research design to the implementation and monitoring and evaluation stages. It generated higher actual and potential outcomes and impacts on target communities and research partners than the other two projects. Table 10.3 presents a summary of the major outputs and initial impacts of the three projects.

In terms of the projects' direct research outputs, the three projects all aimed to achieve research outputs but varied in the levels of output achievement. The research outputs achieved by the NOMAFSI Project were quite consistent with its initially specified research outputs. Compared to the planned research outputs, all three projects achieved the direct research outputs as they expected. The ACIAR Northwest Project and the CIRAD ADAM Project produced some additional outputs compared to their initial research designs due to adjustments of their research activities to make their projects more adaptive to local conditions. In the NOMAFSI Project, a variation in research outputs occurred because the original project length was extended in two later projects. Through reviewing the projects' documents and

TABLE 10.3 Major Research Outputs and Impacts of AR4D Projects in the Northwest Highlands of Vietnam

Key Indicators	ACIAR Northwest Project	CIRAD ADAM Project	NOMAFSI Project
I. Direct research outputs			
Technology development	– Profile of existing maize- and temperate fruit-based farming systems – Improved soil and crop management practices in maize-based farming systems – An innovative approach to AR4D and PM&E systems	– Identification of cover crops for use in developing DMC systems – An agrarian diagnosis research report of farmers' constraints and opportunities – Adaptive sustainable farming systems for maize production on sloping lands	– Assessment report on maize production in northern mountainous regions – Appropriate maize varieties and planting density recommendations – Eight demonstration sites of appropriate technologies for maize production on sloping lands
Outreach capacity building	– Capacity building for a large number of researchers from research partner institutions participating in the project – Twenty-nine undergraduate students from TBU received scholarships from UQ to conduct and contribute to numerous studies, and four researchers from Vietnamese partner institutions received John Allwright Fellowship scholarships and enrolled in UQ postgraduate courses – Eighteen local extension staff participated in season-long training-of-trainer courses in FFS methodology, sustainable soil and crop management – Co-organization of one international conference on conservation agriculture and series of technical training, planning and innovation workshops – Publication of scientific papers, training materials, Farmer Field & Business School guidelines, participatory video	– Training on conservation agriculture for researchers and local stakeholders (148 researchers, 570 local farmers, and 135 extension staff) – Support for three master's students – Several field visits to experiments organized for stakeholders including local farmers, extension staff and authorities – Co-organization of one international conference on conservation agriculture and a series of technical workshops and field days – Translation of video and guide on conservation agriculture from Madagascar into the Vietnamese language – Publications, training materials, newsletter, brochures, academic degree attainment	– Two advanced technical processes for sustainable maize production on sloping lands submitted to MARD for approval – Six technical training courses and six on-farm workshops at demonstration sites for 360 farmers in three provinces – Publication of two journal articles – One PhD student involved and supported by the project – One evaluation report on maize production on sloping lands of the northern mountainous regions of Vietnam – One final evaluation workshop with the participation of multiple stakeholders – Final evaluation reports examined by MARD

TABLE 10.3 Major Research Outputs and Impacts of AR4D Projects in the Northwest Highlands of Vietnam—cont'd

Key Indicators	ACIAR Northwest Project	CIRAD ADAM Project	NOMAFSI Project
II. Livelihood impacts			
Livelihoods capital: – Human (H) – Social (S) – Economic (E) – Physical (P) – Natural (N)	– H: Improved understanding on sustainable maize-based farming systems and decision-making capacity for farmers in local villages and some innovations made by farmers – S: Development of strong collaboration between local farmers and researchers, and between local farmers and extension staff – E: Maize yield increased by 1.5–2 tons of dried seeds/ha when applying minimum tillage and intercropping with legumes; intercropping pumpkins and rice bean with maize brought significant income increases for farmers – P: Evidence not available due to the short period of time since project completion – N: Minimum tillage had been applied in more than 80% of maize areas and rice bean was being intercropped with maize in large areas, leading to significant improvement in soil quality and reduction in soil erosion	– H: Improved awareness of local communities on sustainable maize production techniques but limited change in decision-making capacity at local villages – S: Established partnerships between local farmers and researchers but a weak collaboration between local farmers and extension staff – E: Maize yield increased by 1.0–1.5 tons of dried seeds/ha when applying minimum tillage and intercropping with legumes; intercropping rice bean with maize brought additional net income but in a limited scale – P: Evidence not available due to the short period of time since project completion – N: Minimum tillage had been applied in more than 70% of maize areas, leading to initial improvement in soil quality and reduction in soil erosion	– H: Improved awareness and capacity for research farmers on sustainable maize production but at limited scale – S: Establishment of partnership between local farmers and researchers but weak collaboration between local farmers and extension staff – E: Maize yield increased by 1.0–1.5 tons of dried seeds/ha when applying minimum tillage intercropping with legumes but farmers were not interested in applying intercropping techniques – P: Evidence not available due to the short period of time since project completion – N: Minimum tillage technique applied in less than 25% of maize areas, leading to initial improvement in soil quality and reduction in soil erosion
III. Institutional impacts			
Structures and processes: – Policies – Institutions – Processes	– Enhanced capacity of field researchers from partner organizations through integrative research approach – Innovative AR4D approach utilized by the project's research partner organizations – Improved capacity on PM&E of trials and technology extension for local extension staff – Major contribution to the formulation and execution of Decision 14/QĐ-UBND of Son La province on maize production on sloping lands	– Enhanced technical capacity for field researchers but lack of community facilitation and PM&E skills – Limited contribution to the formulation and execution of Decision 14/QĐ-UBND of Son La province on maize production on sloping lands	– Enhanced technical capacity of field researchers and local extension staff but lack of community facilitation and PM&E skills – Limited contribution to the formulation and execution of Decision 14/QĐ-UBND of Son La province on maize production on sloping lands

Continued

TABLE 10.3 Major Research Outputs and Impacts of AR4D Projects in the Northwest Highlands of Vietnam—cont'd

Key Indicators	ACIAR Northwest Project	CIRAD ADAM Project	NOMAFSI Project
IV. Impacts in the vulnerability context			
Vulnerability context: – Shocks – Trends – Seasonality	– A large number of farmers shifted from the application of full tillage to minimum tillage for maize production – Increasing number of farmers shifted from one maize crop per year to two crops (two maize crops or complementary crops intercropped with maize) in research areas – Improved market engagement identified by farmers participating in the Farmer Field & Business School group	– Initial changes from maize monocrop to legume intercropped with maize farming system but on a limited scale – Limited evidence of changes in maize-based production on sloping lands by local farmers except the application of minimum tillage	– Limited change in farming practices by local farmers regarding the shift from one maize crop per year to two crops (two maize crops or complementary crops intercropped with maize) in research areas

conducting interviews with farmers and research partners, this study determined that the ACIAR Northwest Project implemented the most innovative research process among the three projects. The project's transdisciplinary and multiinstitutional research approach, which involved a large number of partners and collaborating research institutions, was highly accepted by researchers from different research institutions, local authorities, and extension agencies.

The ACIAR Northwest Project trained a number of local extension staff and farmers in farmer field school (FFS) methodology and crop management practices. This training enhanced the capacity of local research partners to scale-out the research outputs. The CIRAD ADAM Project also conducted various field visits and training for local extension staff and farmers. Although the NOMAFSI Project conducted various technical training courses and some on-farm demonstration sites for farmers and local extension staff, the project applied a top-down approach in both research implementation and training activities. The main short-term objectives of the three projects were direct research outputs, such as developed technologies, a number of publications, and technical processes. The NOMAFSI Project aimed to achieve these outputs to meet indicators, as planned in the research proposal and approved by MARD. In contrast, the other two projects targeted the achievement of these research outputs as the outcomes of research interventions aimed at making wider impacts.

The three selected agricultural research projects were implemented on a small-farm scale, and this study was carried out shortly after the projects were completed. Therefore, in relation to the projects' institutional impacts, it was difficult for this study to capture significant changes in the local livelihood contexts such as the natural environment, human health, or prices for maize, temperate fruits, and other annual crops. However, local extension agents and farmers acknowledged initial institutional impacts through local policy development. For example, in response to the findings of various research initiatives in Son La Province, the provincial government issued Decision 14/QĐ-UBND in early 2014. The decision decreed the piloting of sustainable production techniques for maize on at least 30 ha per commune, with priority for the poorest communes. Twelve districts executed the decision in 2014 on 2800 ha. The three research projects investigated in the present study all carried out activities in Son La Province and all three projects shared the objective of developing sustainable maize production on sloping lands. While it was not possible to determine which projects had the most influence on the decision, local extension staff and farmers reported the view that the ACIAR Northwest Project made a major contribution to the issuance of the decision. The ACIAR Northwest Project collaborated closely with DARD as the provincial center of agricultural extension and with other groups in its extension network in designing the project activities, carrying out trials, and scaling-out technologies. The ACIAR Northwest Project also conducted season-long training-of-trainer (TOT) courses on FFS methodology for local extension staff, and this study's respondents considered these courses to be an important step in institutionalizing the project's technologies.

In terms of the projects' livelihood impacts at a community level, this study observed that there was significant change in the awareness and understanding of local farmers, especially the trial farmers, on sustainable farming on the sloping lands in the three projects' research sites. Both local farmers and extension staff assessed the sustainable agricultural techniques such as minimum tillage, mulching, and intercropping in maize-based systems as positive in reducing the risk of soil erosion. However, the FGDs with farmers involved in the three research projects indicated that, despite understanding the importance of applying sustainable techniques in maize-based farming systems, only the farmers in the ACIAR Northwest Project believed they would continue to apply some of those techniques without further support from the project. The farmers in the CIRAD ADAM Project and NOMAFSI Project areas were not sure about applying the new techniques if they did not get support from the projects.

In addition, the participatory approach of the ACIAR Northwest Project enhanced the active engagement of the community in the project. This helped to build trust between the field researchers and farmers and between the local extension staff and farmers. Farmer researchers in the research sites such as Phieng Luong and Muong Sang communes became more confident in sharing their research experiences in community meetings, which helped to strengthen the social relationships among the research partners. The study found lower levels of farmer participation in research activities in the CIRAD ADAM Project and the NOMAFSI Project, with farmers indicating that they only did what the researchers asked them to do. For example, the results of the semistructured interviews with farmers showed that more than 85% of farmers in Phieng Luong and Muong Sang communes (ie, ACIAR Northwest Project research areas) participated in at least one training course on sustainable agriculture (on topics such as mulching, minimum tillage, and intercropping). In contrast, less than 50% of farmers in Chieng Hac and Chieng Dong communes (ie, CIRAD ADAM Project and NOMAFSI Project research areas) participated in at least one such training course. This variability resulted in different levels of awareness among the farmers about sustainable agriculture. In addition, the CIRAD ADAM and NOMAFSI Projects conducted trials with the limited involvement of local extension staff in either research planning or implementation.

In terms of the projects' economic impacts, the local people reported that the sustainable techniques such as minimum tillage, mulching, and intercropping had helped not only to reduce labor cost but also to increase the maize yield and income for farmers. Farmers in Phieng Luong and Chieng Hac communes (ACIAR Northwest Project and CIRAD ADAM Project, respectively) predicted that applying the techniques of minimum tillage in combination with mulching in maize cropping could help to reduce labor and fertilizer costs by about 30% over 3 years. Farmers in most research sites also reported that, in the 2013 crop season, the maize yield had increased by 1.5–2 tons/ha, leading to an additional income gain for farmers. In Phieng Luong Commune (ACIAR Northwest Project), farmers had intercropped pumpkin with maize since 2013 and reported that this technique had provided additional economic profit. Some farm households in Chieng Hac Commune (CIRAD ADAM Project) had recently intercropped legumes (such as rice bean, black bean, and peanuts) with maize; some households were successful, but on a very limited scale. In Chieng Dong Commune (NOMAFSI Project), several farmers had intercropped complementary crops such as black bean and peanuts with maize and earned additional increased income in the two last two crops.

In regard to the projects' physical and natural impacts, due to the small scale of the three selected projects, this study could not clearly capture changes in local infrastructure systems, vegetation improvements in maize-based farming systems, and changes in soil erosion. However, the shift from maize monocropping to intercropping was likely to improve the soil quality. Farmers and local extension staff agreed with the suggestion that higher soil fertility, reduced soil erosion, and improved vegetation were the likely benefits in fields where farmers had intercropped legumes with maize.

Among the sustainable production techniques developed by the projects for maize cropping, the minimum tillage technique was highly accepted by farmers in the research and technology outreach areas of the ACIAR Northwest Project and the CIRAD ADAM Project. Despite the small number of trials and extension sites and the low number of farmers involved, farmers applied the minimum tillage technique in more than 70% of areas where the two abovementioned projects conducted trials and extension demonstrations. According to farmers in Phieng Luong and Muong Sang communes (ACIAR Northwest Project), the application of minimum tillage technique involving opening a narrow trench to apply seed and fertilizer (either by hoe or harrows pulled by buffalo), coupled with the use of in situ organic material for mulch, had helped to reduce soil erosion by between 40% and 50%. However, several farmers, especially farmers in Chieng Dong and Chieng Hac communes (NOMAFSI Project and CIRAD ADAM Project, respectively), were hesitant to confirm that they would continue to apply any of the sustainable techniques except for minimum tillage without getting further support from the projects. The FGDs and semistructured interviews with farmers identified that the reasons for this hesitance were the lack of mulching materials, the requirement for more labor to source mulching materials, and the prospect of receiving little and unstable additional cash income from the complementary crops.

10.4 CONCLUSION

The application of a holistic impact assessment framework is crucial for understanding and sustaining the impacts of AR4D toward the goal of sustainable agriculture and social change. Using the sustainable livelihoods framework as a lens for identifying multiple livelihood impacts provides a better understanding of the complexities involved in social change and development, while participatory techniques enhance the participation of target stakeholders in impact assessment processes. This study developed a holistic impact assessment framework from a comprehensive livelihoods perspective, which evaluators can apply to assess the impact of AR4D projects in different social contexts. The results of testing the holistic impact assessment framework in three AR4D projects in the Northwest Highlands indicated that the AR4D project that adopted participatory processes to facilitate the immediate use of research outputs for development generated better social, human, economic, and environmental outcomes and impacts for local communities than the other two projects. When applying the participatory techniques in the proposed framework, projects should flexibly use and modify the framework to adapt it to the local context and to more comprehensively assess the impacts of AR4D in culturally diverse regions.

References

ACIAR, 2009. ACIAR Country Profile 2009–2010: Vietnam. Australian Centre for International Agricultural Research (ACIAR), Canberra.

Adato, M., Meinzen-Dick, R., 2002. Assessing the Impact of Agricultural Research on Poverty Using the Sustainable Livelihoods Framework. In: FCND Discussion Paper 128, EPTD Discussion Paper 89. International Food Policy Research Institute (IFPRI), Washington, DC.

Campbell, J.R., 2001. Participatory rural appraisal as qualitative research: distinguishing methodological issues from participatory claims. Hum. Organ. 60 (4), 380–389.

Carpenter, D., McGillivray, M., 2012. A Methodology for Assessing the Poverty Reduction Impacts of Australia's International Agricultural Research. ACIAR Impact Assessment Series Report No. 78, Australian Centre for International Agricultural Research (ACIAR), Canberra (pp. 46).

Castella, J.-C., 2012. Agrarian transition and farming system dynamics in the uplands of South-East Asia. In: Paper Presented at the 3rd International Conference on Conservation Agriculture in South East Asia: Conservation Agriculture and Sustainable Livelihoods, Innovations for, with and by Farmers to Adapt to Local and Global Changes, 10–15 December 2012, Hanoi.

Catley, A., Burns, J., Abebe, D., Suji, O., 2008. Participatory Impact Assessment: A Guide for Practitioners. Feinstein International Center, Tufts University, Boston, MA.

Cramb, R.A., Purcell, T.D., 2001. How to Monitor and Evaluate Impacts of Participatory Research Project: A Case Study of Forages for Smallholders Project. CIAT Working Document No. 185. Centro Internacional de Agricultura Tropical (CIAT), Cali, Colombia (pp. 56).

Cramb, R.A., Prucell, T., Ho, T.C.S., 2003. Participatory assessment of rural livelihoods in the Central Highland of Vietnam. Agric. Syst. 81, 255–272.

Cromwell, E., Kambewa, P., Mwanza, R., Chirwa, R., 2013. Participatory impact assessment: the "Starter pack Scheme" and sustainable agriculture in Malawi in drought policy contexts: lessons from southern Ethiopia. In: Holland, J. (Ed.), Who Counts?: The Power of Participatory Statistics. Practical Action Publishing, Rugby, pp. 163–180.

DFID, 1999. Sustainable Livelihoods Guidance Sheet. Department for International Development (DFID), London.

GSO, 2014. Statistical Year Book of Vietnam 2013. Statistical Publishing House, Hanoi.

Ha, D.T., Le, Q.D., Nguyen, Q.T., Dam, Q.M., 2010. Sustainable techniques for sloping land agriculture in North-West of Vietnam. In: Bo, N.V., Tuat, N.V., Viet, N.V., Liem, P.X., Hoang, N.H. (Eds.), Results of Science and Technology Research 2006 to 2010—Proceedings of Science and Technology Conference, 5–6 November 2010, Hanoi. Agricultural Publishing House, Hanoi.

Harrison, E., 2002. 'The Problem with the Locals': partnership and participation in Ethiopia. Dev. Chang. 33 (4), 587–610. http://dx.doi.org/10.1111/1467-7660.00271.

Krall, S., Baur, H., Poulter, G., Puccioni, M., Lutzeyer, H.-J., 2003. Impact assessment and evaluation in agricultural research for development. Agric. Syst. 78 (2), 329–336.

Lilja, N., Ashby, J.A., Sperling, L., 2001. Assessing the Impact of Participatory Research and Gender Analysis. CGIAR Systemwide Program on Participatory Research and Gender Analysis, Cali, Colombia.

Marasas, C.N., Anandajayasekeram, P., van Rooyen, C.J., Wessels, J., 2001. A comprehensive conceptual framework for assessing the impact of agricultural research and development. Agrekon 40 (2), 201–214. http://dx.doi.org/10.1080/03031853.2001.9524945.

MoLISA, 2014. Decision 529/QD-LDTBXH Dated 6 May 2014 on Releasing the Results of General Census on Poor Households and Marginally Poor Households in 2013. Ministry of Labour and Social Affairs (MoLISA), Hanoi.

Neef, A., Neubert, D., 2011. Stakeholder participation in agricultural research projects: a conceptual framework for reflection and decision-making. Agric. Hum. Values 28 (2), 179–194. http://dx.doi.org/10.1007/s10460-010-9272-z.

Nguyen, H.N., Van de Fliert, E., Nicetic, O., 2015. Towards a holistic framework for impact assessment of agricultural research for development—understanding complexity in remote, culturally diverse regions of Vietnam. Australas. Agribusiness Rev. 23, 12–25.

NOMAFSI, 2012. Study on Agricultural Forestry Technology Transferring for the Northwest Highlands—Nghiên cứu chuyển giao lĩnh vực Nông lâm nghiệp cho vùng Tây Bắc. Northern Mountainous Agriculture and Forestry Science Institute (NOMAFSI)—Viện Khoa học kỹ thuật nông lâm nghiệp miền núi phía Bắc, Viện Khoa học nông nghiệp Việt Nam, Phu Tho (pp. 80).

Pretty, J.N., 1995a. Participatory learning for sustainable agriculture. World Dev. 23 (8), 1247–1263. http://dx.doi.org/10.1016/0305-750x(95)00046-f.

Pretty, J.N., 1995b. A Trainer's guide for participatory learning and action. Sustainable Agriculture Programme, International Institute for Environment and Development, London.

Robinson, L., 2002. Participatory rural appraisal: a brief introduction. Group Facilitat. 4, 29A.

Ruane, J., 2014. An FAO E-mail Conference on Approaches and Methodologies in Ex Post Impact Assessment of Agricultural Research: The Moderator's Summary. FAO, Viale delle Terme di Caracalla, 00153 Rome, Italy. http://www.fao.org/nr/research-extension-systems/res-home/news/detail/en/c/217706/.

Stur, W., Le, T.H.S., Lienhard, P., 2013. Final review report for the ACIAR AGB 2008/002 Project "Improved market engagement for sustainable upland production systems in the north-western highlands of Vietnam". Australian Centre for International Agricultural Research (ACIAR), Canberra, Australia.

Thai, T.M., Neef, A., Hoffmann, V., 2011. Agricultural knowledge transfer and innovation processes in Vietnam's northwestern uplands: state-governed or demand-driven? Southeast Asian Stud. 48 (4), 425–454.

Van de Fliert, E., 2008. Project Proposal AGB/2008/2002: Improved Market Engagement for Sustainable Upland Production Systems in the North West Highlands of Vietnam. Australian Centre for International Agricultural Research (ACIAR), Canberra.

Van de Fliert, E., 2010. Participatory communication in rural development: what does it take for the established order? Ext. Farm. Syst. J. 6 (1), 96–100.

CHAPTER

11

Changes in the Nature of the Cat Ba Forest Social-Ecological Systems

A.T.T. Nguyen[*,‡], C. Jacobson[†], H. Ross[*]*

[*]The University of Queensland, Brisbane, QLD, Australia [‡]Vietnam National University of Forestry, Ha Noi, Vietnam [†]University of the Sunshine Coast, Sippy Downs, QLD, Australia

11.1 INTRODUCTION

It is undeniable that human activity is a major force of changing ecosystem dynamics at various scales. To study these processes, a comprehensive conceptualization that recognizes social and ecological systems as "coupled" has emerged (Berkes et al., 1998; Gunderson and Holling, 2002; Anderies et al., 2004; Hahn et al., 2006; Ostrom, 2007). Coupled social-ecological systems (SESs) are a form of complex adaptive system (Levin, 1998; Rammel et al., 2007); self-organization and coevolutionary dynamics are the main characteristics of such systems (Rammel et al., 2007). Complex adaptive systems offer a conceptual framework to understand complex interactions in SESs including the interactions of ecosystem dynamics and institutional change over spatial and temporal scales. However, a greater understanding of coevolving SESs is first needed (Berkes et al., 2003).

The purpose of this chapter is to trace the dynamics of forest SESs in Cat Ba Island, focusing on linkages between access to the forest ecosystem, local ecological knowledge (LEK), and local livelihoods. In doing so, we highlight the importance of multiple drivers of change and disturbances in the system. The analysis of system changes over the past decades provides a greater understanding of adaptation processes, and informs the interpretation of responses to future changes on Cat Ba, including the more recent emphasis on conservation.

11.2 CASE STUDY CONTEXT AND METHODS

Cat Ba Island, with an area of roughly 180 square kilometers, is the largest of the 366 islands that comprise the Cat Ba Archipelago, which makes up the southeastern edge of the Ha Long Bay UNESCO World Heritage site in northern Vietnam. In 2004, almost all of the Cat Ba Archipelago was

Redefining Diversity and Dynamics of Natural Resources Management in Asia Volume 2
http://dx.doi.org/10.1016/B978-0-12-805453-6.00011-5

designated as a UNESCO Biosphere Reserve, called the Cat Ba Biosphere Reserve (CBBR), with a total area of 26,140ha, the majority of which is the national park (HPPC, 2005).

CBBR has rich biodiversity, with many unique ecosystems and fauna and flora such as tropical rain forest on limestone hills, mangroves, coral reefs, sea grass beds, caves and grottos, cliffs and valleys. Moreover, the island is home to numerous rare and endangered species of plants and animals, such as the golden-headed langur. This species is categorized in the world's top 25 most endangered primates and is now one of the world's rarest primate species (Cat Ba Biosphere Reserve Management Unit, 2007). Because of its scenic beauty and importance to biodiversity conservation, CBBR has become a popular tourist destination and attracts numerous national and international interests. Today, the island is home to over 16,000 people with the majority (10,000) residing in the district center—Cat Ba Town. The remaining 6000 people are distributed across the island in six communes—Viet Hai, Gia Luan, Phu Long, Tran Chau, Xuan Dam, and Hien Hao. Before 1979, Cat Ba Island had a small population but the population has increased significantly because of a large number of migrants from the mainland. The ethnic groups are Kinh Vietnamese (96%) and Chinese-born Vietnamese (4%), living in the same areas (Hai Phong People's Committee, 2006).

The research described in this chapter was based on three communes in close proximity to Cat Ba National Park (Gia Luan, Viet Hai, and Tran Chau) who were previously dependent on livelihoods strongly associated with the forests, combined with some agriculture. Each commune had distinct characteristics of participation in conservation initiatives, biophysical condition, and varied preference for agricultural cultivation. Each of these communes has also been engaged in a series of Integrated Conservation and Development Projects (ICDPs) to support local people's livelihoods and biodiversity conservation. Fig. 11.1 shows the location of the research communes.

Research conducted between June 2009 and January 2012 includes

- A 2-week period of participant observation in each commune to observe how the communities operated in their daily lives, and how the livelihood system of the local people functioned.
- A series of 51 key-informant interviews, 18 with people of Gia Luan Commune, 17 interviews in Tran Chau Commune, and 16 in Viet Hai Commune, addressing issues including changes in the focal SESs over the past decades: how they have changed; what drivers of change have influenced the systems; and how management practices of the forest SESs changed. Due to the nature of the study, interviewees needed to satisfy several criteria, such as knowledge about the forests, age, gender, or belonging to a certain organization.
- Four focus groups: one mixed gender group held in each commune, except in Tran Chau where two focus groups were held, one in Hai Son Hamlet, and one with the rest of commune, because of the specific conditions of Hai Son Hamlet.

The focus group discussion and key-informant activities were used to explore the context, refine the study, and establish the key study terms. This technique was very useful for the study, particularly in the case of any conflicting answers obtained in individual interviews. The purpose of the meetings was explained in each group activity, as were the other tasks and goals of the research team. The historical time line and seasonal calendar were obtained as the same time from focus group discussion. All focus group discussions were held at an informal

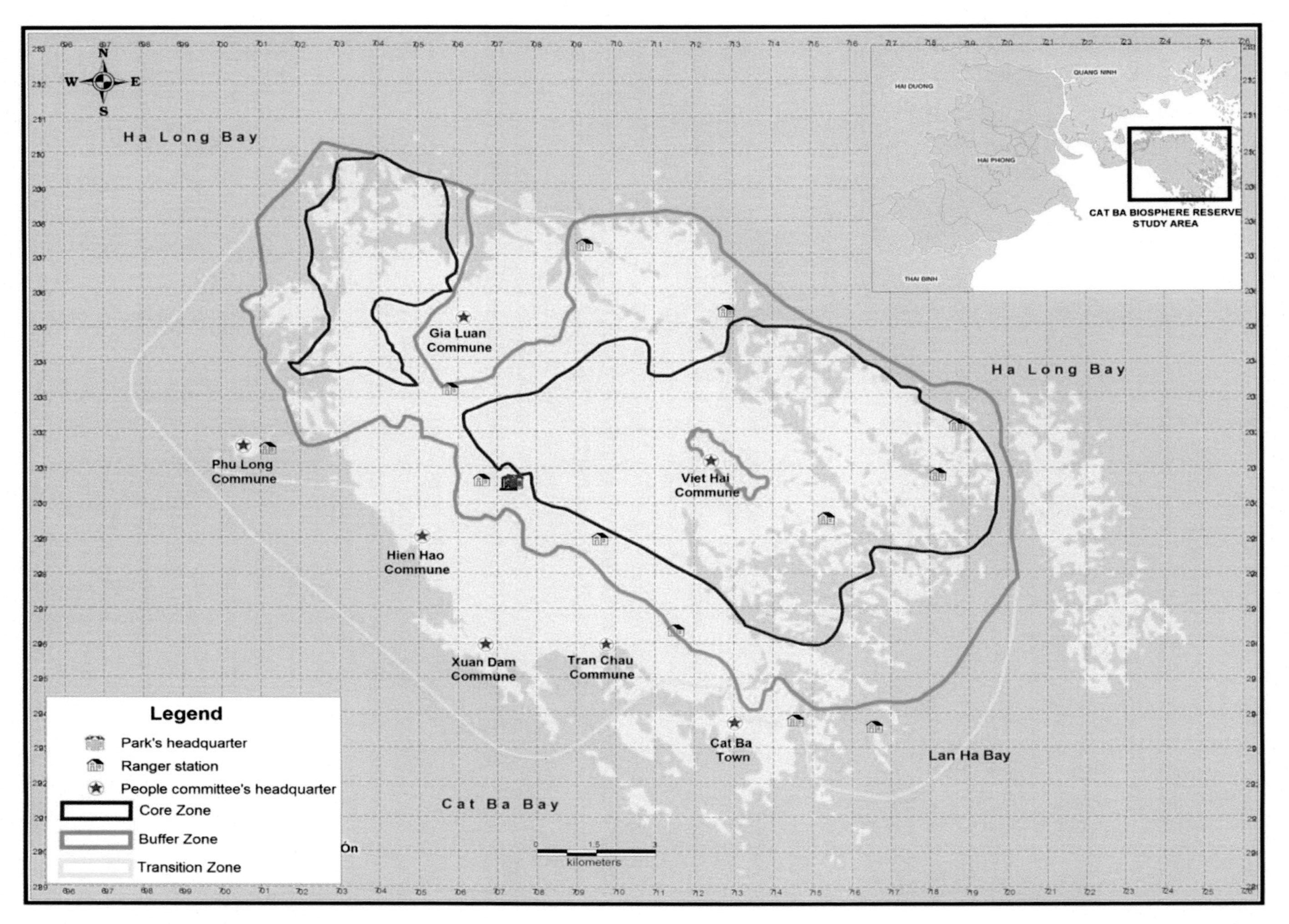

FIG. 11.1 Map of Cat Ba Biosphere Reserve.

atmosphere so that local people could feel free to express their personal understandings, perceptions, and knowledge. The length of each focus group discussion varied from 3 to 6 hours.

Data analysis emphasized manual content selection. Excerpts or direct quotations were chosen as examples of widely expressed themes in the research. Analysis of documents to triangulate against data from key informants, including secondary data, such as reports, legal documents, statistics, and research related to the research area and topic was also undertaken.

11.3 CHANGES IN SOCIAL-ECOLOGICAL SYSTEMS

The history of human-nature relationships in Cat Ba Island is relatively short and poorly documented compared to other areas in Vietnam. The social-ecological change taking place in Cat Ba Island communes is discussed below in terms of four major periods: from 1948 to 1978, from 1978 to 1986, from 1986 to 2000, and from 2000 to now.

11.3.1 From 1948 to 1978: Traditional Agriculture Practices, Forest Extraction

This period was distinguished by the mass migration of refugees from the mainland city of Hai Phong, during the extended war with French colonial forces (Dawkins, 2007). The first *Kinh* people settled in the Viet Hai Commune in 1948; Kinh is one of 54 ethnic groups in Vietnam and accounts for approximately 86% of the country's population. The elders interviewed only referred to two communes existing in this period: Viet Hai and Gia Luan. The formation of Hai Son Hamlet in 1978 is taken to mark the end of this period. The period was characterized by traditional agricultural practices and forest extraction. The three communes studied combined lowland rice production with the practice of slash-and-burn agriculture (swidden), in addition to using forest resources in the mountainous areas. Land use was relatively extensive, with a single rice crop per year in the lowlands and swidden agriculture with long fallow periods in the uplands.

According to interviews with elders, the population of Viet Hai Commune fluctuated during this period while the case of Gia Luan Commune was relatively simple. In the early 1950s Viet Hai Commune was officially formed, consisting of Viet Hai Hamlet and Tra Bau Hamlet. By the end of 1959, the population of the commune was 28 households, all of them Kinh people. In 1965, the population of the Viet Hai Commune rose significantly to 57 households due to a government resettlement program that aimed to reduce the stress of overcrowding on agricultural land (Fig. 11.2). Gia Luan Commune had a total population of 50 households in the 1960s, and this commune was not divided into hamlets. In 1948, when Gia Luan was first established, there were 12 Chinese households and only 2 Vietnamese households in the area.

Traditional agricultural practices and forest exploitation dominated the livelihood activities of locals. As in other forested areas in Vietnam, slash and burn was one of the methods of agricultural cultivation, but this method was limited in these areas due to the landscape being dominated by limestone karst, with steep limestone hills. Nevertheless, some forested land was converted to agricultural land. Slash and burn cultivation provided one crop per year with very low yields, which were highly dependent on natural conditions (water, weather, and no disease). These low yields led to food shortages. In this historical period the common

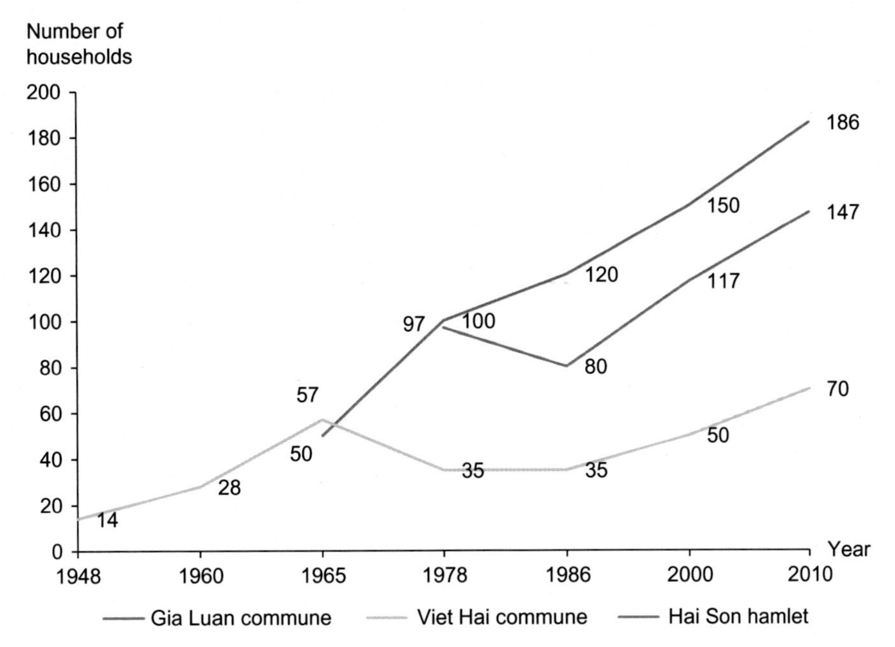

FIG. 11.2 Population changes through time in three selected communities.

food crops in Viet Hai were rice and maize, while maize was the main crop in Gia Luan. Home gardens were maintained as well as household livestock such as poultry and pigs.

According to focus group discussions and interviews with older villagers, in this period forest resources were abundant, and exploited by the then-small population. The forests were described as having plenty of large, old growth trees and a diversity of wildlife. For example, the Cat Ba langur population was very abundant, estimated at 2400–2700 individuals in the 1960s (Nadler and Long, 2000). Forests were open access for local people. They were utilized for collecting wild foods, wildlife for meat, timber for building materials and other needs, and other nontimber forest products (NTFPs). The exchange of forest products for food or money in this time was very limited. Because people's livelihoods were forest-based, close relationships were formed with the forests and LEK was created and built up through daily practices.

Living conditions of local people in this time were very difficult. The economic activities in this period were village-oriented, with very limited trade extending outside the boundaries of the commune. There was no infrastructure in the areas, such as a road system, primary schools, irrigation system, or electricity. The remote areas of Viet Hai Commune were isolated from the rest of the island. It was not uncommon for women to travel out of the boundaries of their commune less than once a year.

11.3.2 From 1978 to 1986: Significant Reduction of Forest Resources, Local Ecological Knowledge, and Social Institutions

In 1978, Hai Son Hamlet was officially formed, on the main road running from Cat Ba Town to the national park's headquarters; at first it was known as Khe Sau Hamlet. Around this time Chinese people returned to China and more Vietnamese people from the mainland (Thuy Nguyen—Hai Phong) migrated to Cat Ba, this time under government programs. The population of the hamlet at the time was 97 families with 420 people. This included many extended and large families with more than three children. Despite being a young community, people living within the park border were still dependent on the forest resources.

For Hai Son Hamlet, this time was considered the wealthiest with regard to natural resources. The forests provided villagers with plenty of large trees, wildlife, and other NTFPs. The new arrivals to the village cut down many large trees for building houses and making furniture. Wildlife at this time included "black monkeys" (the local name for Cat Ba langur), "yellow monkeys" (Rhesus Macaque—*Macaca mulatta*), and some other small animals that went down to home gardens to find food. As in the older communities, the local people depended heavily on the forest resources. Meanwhile, social economic conditions were very difficult. There was nearly no infrastructure except for a primary school, which was not a solid building. There was no electricity, road systems, health care, or communication methods.

This was a postwar period, and the study communes experienced opposing population trends. After the end of the American war, in 1975, refugees who had fled from the war moved back to their hometowns. In the Viet Hai Commune after 1978, the population reduced from 57 to 35 households due to the commune's location being isolated from other parts of the island. In contrast, the population of Gia Luan Commune increased significantly from 50 to 100 households over the same period due to increasing population, resulting in division of households. However, the commune did not divide into hamlets (Fig. 11.2).

The economic activities during this period were characterized by collective agriculture and agricultural subsidies under new government policies. This movement spread across the isolated communes on the island. The collectivization created a series of work-exchange teams (*Tổ đổi công*) that were converted into agricultural cooperatives. The locals reared livestock in home gardens for family consumption. Another prevalent livelihood of Viet Hai Commune and Hai Son Hamlet was to cultivate Hương Nhu (*Ocimum gratissimum*) for the extraction of oil. Local people could exchange oil for rice; one kilogram of oil was worth 60 kg of rice. Interviewees, who were village elders, recalled that during the period between 1978 and 1986, there was more than enough rice to eat, but nothing else because of the lack of markets. In addition, the pattern of natural resource extraction continued to be more intensive in terms of the frequency and types of forest products extracted. People exploited the forest resources to meet their livelihood needs. For many families at the time, logging was the main source of income.

Those practices substantially changed when the Vietnamese government nationalized all forests in the country after independence. To control the forest, the government established a system of state forest enterprises (Mai and To, 2015). In Cat Ba Island, Cat Ba State Forest Enterprise was formed and operated on the island until 1986. The main management activities of this organization were forest extraction (mainly logging) and forest plantation. The enterprise felled a large amount of timber each year to fulfil government production quotas. Forest cover on foot slopes and in valleys declined during this period as a result of excessive logging operations in the area combined with the subsequent conversion of cleared forested land to agricultural land (Nadler and Long, 2000). The local residents continued to maintain and build up ecological knowledge through daily forest use. The increasing population, coupled with inappropriate agricultural policies (causing food shortages), led to intensive forest extraction in all of the communes of this study. By the end of the period, the forest resources started to show signs of depletion.

11.3.3 From 1986 to 2000: Beginning of Protected Area Management

In the period from 1986 to 2000, Vietnam adopted the global discourse on biodiversity conservation. The forest sector began to shift from utility to preservation and development. The Cat Ba Protected Area was established in 1983 by the Ministry of Forestry. Three years later, Cat Ba National Park was formed under the Council of Minister's Decision 79—CT (Furey et al., 2002). The formation of these protected areas was symbolic of international trends in protected areas, including commitments under the Convention on Biological Diversity. However, locals at the time were not made aware of the existence of the protected areas. This indicates that the protected areas were "paper protected areas" rather than introducing any real protection to the areas. On the other hand, the establishment of Cat Ba National Park had a significant impact on the local people and communes whose livelihoods had depended heavily on the forest resources. Local people were forced to shift away from free access to the forests due to the new prohibition of collecting forest products. Furthermore, Viet Hai Commune members complained that their agricultural land had been reduced by 28 hectares due to the conversion of land for the cultivation of *Huong Nhu* (for oil extraction) under the management of the national park for conservation purposes. In the period, significant events occurred in each of the communes.

11.3.3.1 Viet Hai Commune

In Viet Hai Commune, 1991 represented a very important milestone. Local people moved from traditional agriculture with low productivity and forest-based subsistence to more intensive agriculture activities such as rice cultivation, home gardens, and raising livestock. New agricultural technologies were introduced to the commune such as pesticides, fertilizers, new breeds of rice, and other crops to improve agricultural production. The population of the commune remained at 35 households during this time. However, despite the increases in productivity, agricultural production was still insufficient to meet all of their food and livelihood needs all year long. The second turning point was that the agricultural cooperatives were disbanded, and there was a shift away from state subsidies and central planning. Relations between the state and the peasants in the commune became more in line with free market principles. Under the Communist Party-issued Resolution 10, or *Khoán 10*, agricultural collectives were obliged to contract land to households for 15 years for annual crops and 40 years for perennial crops. The implementation of Resolution 10 began in Vietnam in 1988; however, it had no significant impacts on the commune until 1991 (Kirk and Nguyen, 2009).

The third important point is that a boat was introduced to improve the connection between the commune and Cat Ba Town. Residents of the commune stated that this event was a critical point in their lives, both in terms of economic and cultural connections. The locals could access Cat Ba Town daily, which meant more access to education for children and cultural exchange with the rest of the island and the world. The most important impact was the increased opportunity for economic development. With their agricultural products able to circulate in a larger market, locals were able to push for greater agricultural development. In contrast, it was an unsuccessful time for biodiversity conservation. Elders recalled that this was a peak time for the extraction and consumption of forest products from the area. Many boats waited to pick up lumber and other forest products from the local commune to sell in Cat Ba Town.

In 1999, the basic infrastructure of Viet Hai Commune developed significantly. Construction work began on the main commune road that connects the commune center to the ferry station. At the time this was only a dirt track. It was also the first time that the commune had an electricity supply, which was produced from a communally owned generator but only operated nightly from dusk to 11 p.m. A landline phone was also set up. This year was recorded as the first time that tourists visited Viet Hai, albeit a very small number as there were no services for tourism in the commune. All of these new socioeconomic improvements enabled local people to communicate with the wider world, and assisted local people to improve their standard of living. In addition, local people experienced a shift in thinking about the forest, from forest extraction to gradual acceptance of forest conservation, due to the introduction of other alternative livelihoods such as tourism. The population of the commune increased significantly, to 50 households. The 15 new families settling in the commune were a spontaneous migration via families and relative networks.

11.3.3.2 Hai Son Hamlet

The establishment of Cat Ba National Park in 1986 affected Hai Son Hamlet as well as other communes located within the park boundaries. The local people were given directions not to collect any of the resources that they were accustomed to taking from the forests, without being provided with adequate compensation. They stopped cultivating *Hương Nhu* to exchange for rice. The economic condition of local people after the establishment of the national park was described as a most difficult time, causing hunger and poverty. People shifted from having plenty of rice to eat and free access to the forest for their needs, to not enough food and no access to the forests. As a result, the population of the hamlet decreased by 20 households to approximately 80 families totaling less than 300 people. All of the families, especially the larger families, desired to move away during this time of hardship. In the same year, the main road of the island was under construction, which led to some major changes in the socioeconomic conditions of local people over the following periods.

Although activities that impacted the forest were forbidden, due to low law enforcement levels and few park rangers the illegal forest extraction continued around this commune. Hai Son Hamlet was made up of a young community of migrants who were quite well-educated compared to other communes on the island at that time. Their community location (next to the core zone of the national park, surrounding the park headquarters) combined with frequent contact with the forest allowed them to acquire LEK very quickly. This knowledge was maintained in the commune for the purpose of supplementing livelihood activities.

Toward the end of the 1990s, basic infrastructure of the local community developed considerably, with the construction of a health station and a primary school. The local people reported that their living conditions had improved and that they had gradually became less reliant on forest resources to sustain their livelihoods. The households in the commune were now able to buy basic electronic goods such as radios and black and white televisions (run by battery as there was still no electricity network). The population of the community increased to 117 families, consisting of 420 people. The increase in population was mainly due to newly formed families separating from extended families.

11.3.3.3 Gia Luan Commune

Gia Luan shared the same trend with two other communes that were affected by *đổi mới* and the existence of the national park. The living conditions of local people at this time were very difficult with not enough food to eat, most notably a shortage of rice. The most remarkable economic event was the abolition of the system of the subsidies. All agricultural activities operated under the management of an agricultural cooperative alongside traditional agricultural practices. Planting and harvesting bamboo was introduced as an alternative livelihood. The only road access to the commune was an unsealed road. The population grew to 120 families through natural increases only, with no migration.

After a long period of extraction, the forest resources became depleted. Forest resource extraction by local people was controlled by district rangers as part of the management of the Cat Ba National Park. However, the level of law enforcement was very low. The local people continued poaching and collecting NTFPs at a high rate. Gia Luan Commune is an old community, with a longer history and relationship with the forest. The Gia Luan people were said to have a very good knowledge of wildlife, developed through their practices of hunting and logging. It was not an easy task to tell people not to go to the forests while they had an abundance of free time, labor, and ecological knowledge. However, as a result of the new regulations the people of the commune reduced their interactions with the forest, and their LEK was also reduced.

11.3.4 From 2001 to Now: Social Development, Livelihood Diversification, and Environmental Issues

The early years of the 21st century were characterized by rapid economic development and forest protection around all communes. The acknowledgment of the high biodiversity values of the island, especially the endangered endemic primate species the Cat Ba langur, coupled with the poverty of the communes, led to the establishment of small-scale, foreign-funded ICDPs from 2000 (Brooks, 2010). The local communes were aware of the existence of some ICDPs and expressed positive attitudes toward environmental education aspects that changed their thinking about ecosystems and their livelihoods. The introduction of a new institutional framework enhanced ecosystem management. In 2004, almost all of the Cat Ba Archipelago was designated as a UNESCO Biosphere Reserve.

As a result of strengthened law enforcement and forest depletion, local people recounted that they stopped logging at this time, but continued their poaching practices. However, compared to the previous period, there were now fewer poachers and they only caught small wildlife. According to the estimation of local people, approximately 80% of the native forests have recovered. Valuable timber species have regenerated. The protection of the forest has allowed them to grow steadily. People have still collected some NTFPs such as honey, and wood for fuel. By the end of the 2000s, there had been very significant changes in the communes studied. Local people adopted livelihood diversification to raise their income levels and reduce their dependence on forest resources. As a result, there has been further reduction of LEK caused by a lack of practice. However, there have been some opportunities to apply LEK through ICDPs, such as beekeeping projects, and langur conservation projects.

The local people said that they still depended on the forest resources but in more sustainable ways. Local people could benefit from ecosystem services of the forests and stated that forests could bring tourism to Cat Ba Island. In addition, some of them could benefit directly through the forest resources by beekeeping. Bees were kept in home gardens and fed by forest flowers. Cat Ba forest-based honey has been very popular with tourists and sells for a very good price. However, the number of households with commercial beekeeping as their main source of income is limited: less than 10 households in all three research communes. The general trend is that many households have a few bee colonies that allow them to earn additional income. As one person stated: "Every family in our commune has at least one to two bee colonies, my family has four. It helps us to earn some extra money. For example, that money could pay for clothes, books, school fees for kids at the beginning of school year" (Galuan interviewee). Another example of forest-based alternative livelihood is to keep a medicinal home garden. People stated that some households in Tran Chau Commune, who were knowledgeable in medicinal plants, took part in the medicinal home garden projects. However, it was more a good idea rather than a real alternative livelihood. As there was no stable market for medicinal products the locals became frustrated and went back to collecting medicinal plants from the forests.

The first decade of the 21st century was very important for all communes in terms of infrastructure development. Both Viet Hai and Gia Luan communes were considered extremely poor communities and eligible for Program 135 assistance from the government. These programs invested in developing basic infrastructure for the island communes, such as fresh water supply (only in Gia Luan); irrigation systems; transport systems; schools; a cultural house; a post office; and a health station. In Viet Hai, during 2003 the road to the ferry station was upgraded with a concreted surface, and other projects included a permanent irrigation and drainage system and a reservoir. In 2010, the national electricity network reached the commune. Similarly, basic infrastructure has been developed in Hai Son Hamlet, which allowed local people to connect to the broader society. The national electricity network and power supply network has supplied power constantly to the community since 2001. The main road was upgraded and extended to connect all hamlets. In Tran Chau a new post office, a culture house, and a health station were built. The primary school was rebuilt as a two-level building. The irrigation and drainage systems were concreted. As a result of this infrastructure development, the local people said happily that their standard of living had improved considerably. They could now build solid houses with brick walls, concrete flat roof and multistoried houses. They could also buy household electrical goods and valuable vehicles such as tractors and scooters.

Following the growth of infrastructure in the island communes, agriculture has also developed significantly. Locals have applied some modern technology to their agricultural cultivation practices such as using machinery for land preparation, introducing new high-yielding rice varieties, and intensive use of fertilizers and pesticides in cultivation leading to an enormous increase in agricultural yields, turning from not enough food crops for households' consumption to selling to local and inland markets. Viet Hai Commune cultivates two rice seasons a year, mixed with other crops and vegetables. In the past, farmers cultivated only one paddy rice season a year with very low productivity. Like a double-edged sword, intensive utilization of fertilizer and pesticides in cultivation could cause negative effects to the people's lives such as health issues, and the disappearance of some species. For example, in the case of Viet Hai Commune, the failure of the beekeeping projects was caused by intensive use of pesticides and fertilizers (Dawkins, 2007). In the case of Hai Son Hamlet, agricultural development brought economic development for most of the locals. However, the lack of water has remained a constraint to agricultural development. The locals complained that they could gain more benefits if they had enough water for irrigation.

Tourism has been an increasingly important driver for the Cat Ba economy and local people's livelihoods and culture. However, each commune has been impacted by tourism in different ways. For example, in Viet Hai ecotourism or community-based tourism has strongly influenced the community. The local people stated that the number of tourists to the commune has grown significantly but most were foreigners. They have benefited directly from tourism services and others have benefited indirectly due to increased demand for agricultural products and through services such as motorbike taxi, or *Xe Om*. The people from Hai Son Hamlet are able to gain some income from tourism activities in both direct and indirect ways. Military Cave, or *Hang Quan Y*, one of the popular tourism places, is located in the hamlet area. Some households benefit from tourism activities such as selling entrance fees, protecting the cave, and some selling of food and drinks. In addition, some locals are involved in tourism activities in Cat Ba Town, such as being employed as tour guides or working in hotels and restaurants. Due to the proximity of the hamlet to the larger Cat Ba Town, their agricultural products, especially forest-based honey, are in high demand. Fewer benefits from tourism were evident in the Gia Luan Commune compared to other communes; except for people who were working in Cat Ba Town, no direct benefits from tourism were declared by the locals. They even considered tourism as having a negative effect on the local living standard, such as increasing the cost of living compared to their income. Other complaints included an increase in traffic noise, dust, and litter caused by tourists.

The cultural benefits of tourism are understood by local people. In the case of Viet Hai Commune, due to economic development, the living standard of the locals has increased significantly. They have been able to buy modern furniture and electric household goods, as well as repair old houses. However, some locals said sadly that traditional houses have been gradually replaced by new brick houses. These new houses are not what tourists expect to see in the Viet Hai Commune. To maintain the number of tourists visiting the commune, the local people said that the old style house using the old architectural style with natural materials should be preserved for tourism purposes. This is considered a cultural loss. The locals suggested that the local government should consider conservation not only for biodiversity but also for cultural values of their commune. Besides the preservation of cultural values, the local people have admitted that they are learning new things from tourism. For example, in two focus groups and interviews in Tran Chau Commune, people discussed the issue of litter. They learned that they could gather litter to put in a designated area rather than freely discharging it into the environment.

There was a similar upward trend in population at all of the communes studied. The total number of households has increased steadily since the early 2000s. The local people explained that Cat Ba Island is now a very good place for settlement both in terms of economic opportunities and the environment. In 2010, the population of Viet Hai Commune increased to 70 families, arising from newly formed families when children married. Hai Son Hamlet had 147 households consisting of 545 people. Similarly, Gia Luan Commune has remained as two hamlets with 186 households consisting of 568 people. Tran Chau comprises six hamlets, equivalent to 1508 heads. However, the growing human presence on the island caused by both local population growth and a boom in tourism has resulted in many negative impacts for the environment. The long history of exploitation of natural resources on the island coupled with the growing human presence has led to some environmental issues such as water shortages and environmental pollution. In the cases of Viet Hai and Gia Luan communes, litter has been identified as the main concern by local people. Tourists (mainly domestic) leave litter everywhere: on the road, in the forests and in the sea. They explained that local people's awareness of environmental issues had increased and that most of them have started collecting rubbish and dumping it in designated areas. However, some people maintain bad habits of dumping rubbish in the garden or in common areas. In terms of environmental pollution, Tran Chau suffered more than other communes. They complained of bad odors and flies caused by dumping rubbish in inappropriate landfill near the commune. Every day, hotels, restaurants, and other tourism facilities produce an enormous amount of solid waste. In Cat Ba Town, this solid waste is collected and transported to the nearby landfill without any treatment.

Water shortages were a big issue listed by local communes, especially Tran Chau commune. Groundwater is the main source of fresh water for domestic consumption in Cat Ba Town. One pumping station is located in Cat Ba Town, and two others are in Tran Chau Commune. Every day, water is pumped into the water pipeline system to a storage dam for preliminary treatment, and then to hotels and households. In the peak tourism season, the fresh water supply meets only about 30% of the total water demand (Mai, 2012). This means that to supply fresh water for Cat Ba Town, Tran Chau is forced to restrict its own water use. Two focus group discussions in Tran Chau indicated that they had suffered increasingly from water shortages both for domestic use and for agriculture. The examples given were cases of reduced water levels in all wells in the commune. The locals face the issue of insufficient fresh water to meet their daily needs. The stream near the commune had no water for many months. Due to this water issue, there has been a negative impact on the development of agriculture in Tran Chau Commune. The locals explained that they practiced dry land agriculture, but if the water supply was enough to support irrigation, they could cultivate two rice seasons a year and increase their agricultural yield.

Correlated with these environmental issues is the current social and institutional instability. Social conflicts associated with resource use and management decisions have escalated over the last decade. For example, the local people of Tran Chau Commune have had to use buckets to carry water for daily use, while the local government has decided to drill more wells in their commune to sell more water to Cat Ba Town. Some local people have destroyed the wells as a response to this conflict over water use. Other conflicts could be observed in the case of law enforcement; between local people and rangers who have tried to keep them away from the forests. The social conflicts could be seen within and between communes in terms of conservation benefits. The criteria used to select representatives for some conservation associations was not equitable or transparent. More people and fewer opportunities have caused tension in some communes. In addition, the money for forest protection contracts was developed with the intention to compensate households for income loss from forests, but this was another source of conflict. The households that did not hold contracts felt disadvantaged compared to those with contracts.

11.3.5 Synthesis

This historical social-ecological analysis has shown that there has been significant change in the livelihood activities of local people, and hence their relationship with the simultaneously changing forest ecosystems. Until the late 1980s, most of the people in the study communes still hunted, trapped wildlife, practiced slash and burn agriculture, and gathered NTFPs such as medicinal plants, honey, and wild fruits for subsistence. Their traditional livelihood practices have been eroded due to a set of social, economic, and institutional factors. The social-ecological drivers of change such as the establishment of protected areas brought about the economic displacement of local people through loss of their access to and use of forest resources. Such displacement, combined with the collapse of traditional livelihood options has resulted in adaptations in livelihood systems as well as LEK. The communes moved from a self-sufficient economy with much dependence on the forest products to a market-based economy of cash cropping, the sale of NTFPs in markets, and tourism. The changing socioeconomic conditions accompanying the formation of CBNP forced local people into difficult situations through the loss of access to forest resources, simultaneous reduction of agricultural lands, and failure of agricultural markets. To address subsistence and economic needs in their communities, local people have made use of their LEK to illegally harvest forest resources. Such illegal activities have been associated with a decline of wildlife populations on Cat Ba Island (Timko, 2001; Brooks, 2010; Schneider et al., 2010).

This historical analysis of the SESs on Cat Ba Island shows that the level to which people rely on biodiversity for subsistence has declined over time with moves from hunting and gathering dependency toward diversified forms of livelihood, placing less pressure on forest resources. In the meantime, the forest resources were severely depleted, and while forest cover is recovering, wildlife populations are not (Thuc et al., 2014). There are some major drivers of change that have affected all systems of the island. The local forest systems have presented the typical characteristics of SESs such as a complex and adaptive system, with nonlinear relationships and cross-scale interactions. For instance, changes in agriculture and development policy mediate the people-forest relationship and exploitation of wildlife. The many key drivers of change in the three communes arise at different scales from international to local. International-scale interactions are seen in the formation of CBNP and CBBR, as national responses to international influences in favor of biodiversity conservation. The significant biodiversity values of the island have attracted much attention from international organizations, both government and nongovernment, toward conservation initiatives. Economic factors are another of the international- and national-scale interactions. The number of international tourists and the increasing demand for wildlife trade to China are examples. The local-scale interactions can be seen in the case of the road system development that affects other aspects of the systems. The ways that the systems change are very complicated. Therefore, interventions both for biodiversity conservation and community development should be considered carefully in anticipation of their systemic effects and likelihood of being sustained in local conditions.

11.4 DISCUSSION AND CONCLUSIONS

Over time, there have been significant changes in SESs, which have moved from a natural forest extraction to a conservation basis. In the past the SESs of Cat Ba Island were characterized by human exploitation of natural resources. Local people interacted with forest resources as a means of livelihood, leading to the development of local knowledge; they knew what kind of forest products to harvest, and which included the identification, distribution, ecological interaction, and techniques to harvest these products. The local forest resources were then depleted under overexploitation by local people, especially as populations increased. As a response to the global biodiversity conservation discourse, the national protected area was established through the CBNP. The CBBR is another international response to biodiversity conservation values on the island. However, it placed greater emphasis on the role of local communities to move the forest SESs from exploitation to biodiversity conservation.

However, the shifting paradigm has been influenced by many drivers across scales. The local people possess rich knowledge of wildlife and tree species, as well as an understanding of ecosystem interactions. To comply with the regulations and law, local people have had to give up their habits and traditional livelihoods such as hunting, trapping, and gathering NTFPs. The main issue for them has been how to change their livelihood systems and still address their subsistence and economic needs. Through the use of a SESs framework, mutual feedbacks between environment and people have been understood by analyzing drivers of change such as conservation policies, *đổi mới* policy, the tourism boom, and the introduction of ICDPs, this research has gained a better understanding of how the local SESs, including people's livelihoods and knowledge, have adapted over time. The local people have been forced to leave their incomes in forest extraction. They have adopted some adaptive livelihood strategies such as agricultural intensification, and livelihood diversification. The results indicate that agricultural intensification has played an important role in local livelihood systems. However, there are some factors that have hindered this process such as uncertainty in the markets for agricultural products, and environmental stressors such as water availability. Local people have tended to adopt varied livelihood activities as a strategy to reduce risk. The tourism industry has been an important economic sector of Cat Hai District and Hai Phong City, but has shown fewer benefits for the locals within the national park. The benefits that local people did gain were in the form of more off-farm job opportunities.

The dynamic linkages between livelihoods, forests, and governance have been, and will continue to be, major elements of the island's state. The integrity of native ecosystems and the services that they provide will affect local livelihood activities such as agriculture and tourism. The future conservation of the unique biodiversity and ecosystems of Cat Ba will largely depend on local residents, who must ultimately use their knowledge and understanding in an ecologically responsible way to maintain positive economic and social practices.

The main issue facing Cat Ba in the future is how the local communities cope with changes in forest SESs. In general, they have positive attitudes toward changing their long-held livelihood strategies and land management practices. The loss of local knowledge is acceptable for local people if they have adequate other sources of income that could pull them away from the direct extraction of forest resources. However, changing people's habits as well as behaviors that have been maintained for decades is not an easy task. Illegal forest extraction has been documented in all the communes studied and in the annual records of Cat Ba National Park. There are many socioeconomic conditions that affect their behaviors such as the high value of forest products and low level of law enforcement. To guide the communities' responses in a way that increases the sustainability of social ecological systems, it is important to understand their capacity to generate and apply knowledge to current and future changes in the conservation dilemma (Gómez-Baggethun and Reyes-García, 2013).

The relevance of SESs analysis is as a multiperspective approach that combines theories of knowledge, livelihoods, human perception and attitude, as adopted in this study, cannot be overemphasized. The approach has proven to be a worthy tool for revealing and understanding holistically the relationships between conservation and development, the processes that drive these relationships, and their ultimate socioeconomic and environmental implications. The introduction of some conservation development projects has led to more favorable livelihood outcomes and heightened environmental awareness coupled with conservation-oriented practices. Overall, the approach has shown that if people are realistically enabled to have control over park resources, not only will there be improvements in livelihoods but also a sense of resource ownership and motivation toward more conservation-oriented behaviors. Thus, behavior is a fundamental factor in exploring the link between conservation and humans. As Schultz (2011) stated, "Conservation is a goal that can only be achieved by changing behavior."

References

Anderies, J.M., Janssen, M.A., Ostrom, E., 2004. A framework to analyze the robustness of social-ecological systems from an institutional perspective. Ecol. Soc. 9 (1), 18.

Berkes, F., Folke, C., Colding, J., 1998. Linking Social and Ecological Systems: Management Practices and Social Mechanisms for Building Resilience. Cambridge University Press, Cambridge, UK.

Berkes, F., Colding, J., Folke, C., 2003. Navigating Social-Ecological Systems: Building Resilience for Complexity and Change. Cambridge University Press, Cambridge, UK.

Brooks, M.A., 2010. Constraints and Enabling Factors for Effective Conservation in Vietnam: Cat Ba Island Case Study. Unpublished Doctoral Thesis, The University of Queensland, Brisbane, Australia.

Cat Ba Biosphere Reserve Management Unit, 2007. Information on the Biosphere Reserve 'Cat Ba Archipelago' for the Hon Tom Burns-Queensland Government Special Representative to Vietnam. Presentation to University of Queensland, Australia.

Dawkins, Z., 2007. The Social Impact of People-Oriented Conservation on Catba Island, Vietnam. Australia National University, Canberra, Australia.

Furey, N., Le, X.C., Fanning, E., 2002. Cat Ba National Park Biodiversity Survey 1999. Ministry of Agriculture and Rural Development Forest Protection Department, Hanoi, Vietnam.

Gómez-Baggethun, E., Reyes-García, V., 2013. Reinterpreting change in traditional ecological knowledge. Hum. Ecol. 41 (4), 643–647.

Gunderson, L.H., Holling, C.S., 2002. Panarchy: Understanding transformations in human and natural systems. Island Press, Washington, DC.

Hahn, T., Olsson, P., Folke, C., Johansson, K., 2006. Trust-building, knowledge generation and organizational innovations: the role of a bridging organization for adaptive comanagement of a wetland landscape around Kristianstad, Sweden. Hum. Ecol. 34 (4), 573–592.

Hai Phong People's Committee, 2005. Regulations on Operation of Cat Ba Archipelago Biosphere Reserves Management Unit. Hai Phong People's Committee, Hai Phong, Viet Nam.

Hai Phong People's Committee, 2006. The Master Plan for Cat Ba National Park (2006–2010) and Vision 2020. Hai Phong People's Committee, Hai Phong, Viet Nam.

Kirk, M., Nguyen, D.A.T., 2009. Land-tenure Policy Reforms: Decollectivization and the Doi Moi System in Vietnam. The International Food Policy Research Institute (IFPRI). Retrieved from website, http://www.ifpri.org/sites/default/files/publications/ifpridp00927.pdf.

Larsen, P.B., 2008. Linking livelihoods and protected area conservation in Viet Nam: Phong Nha Ke Bang World Heritage, local futures? In: Galvin, M., Haller, T. (Eds.), People, Protected Areas and Global Change: Participatory Conservation in Latin America, Africa, Asia and Europe. NCCR North-South, Swiss National Centre of Competence in Research North-South, University of Bern, Germany, pp. 431–470.

Levin, S.A., 1998. Ecosystems and the biosphere as complex adaptive systems. Ecosystems 1 (5), 431–436.

Mai, T.V., 2012. Sustainable Tourism - Systems Thinking and System Dynamics Approaches: A Case study in Cat Ba Biosphere Reserve of Vietnam. Doctor of Philosophy thesis, The University of Queensland.

Mai, T.V., To, P.X., 2015. A systems thinking approach for achieving a better understanding of swidden cultivation in Vietnam. Hum. Ecol. 43, 169–178.

Nadler, T., Long, H.T., 2000. The Cat Ba langur: Past, Present and Future. The definitive report on *Trachypithecus poliocephalus*, the world's rarest primate, Endangered Primate Rescue Centre, Hanoi, Viet Nam.

Ostrom, E., 2007. A diagnostic approach for going beyond panaceas. Proc. Natl. Acad. Sci. 104 (39), 15181–15187.

Rammel, C., Stagl, S., Wilfing, H., 2007. Managing complex adaptive systems – a co-evolutionary perspective on natural resource management. Ecol. Econ. 63 (1), 9–21.

Schneider, I., Tielen, I.H., Rode, J., Levelink, P., Schrudde, D., 2010. Behavioral observations and notes on the vertical ranging pattern of the critically endangered Cat Ba Langur (*Trachypithecus poliocephalus poliocephalus*) in Vietnam. Primate Conserv. 25, 111–117.

Schultz, P., 2011. Conservation means behavior. Conserv. Biol. 25 (6), 1080–1083.

Thuc, P.D., Baxter, G., Smith, C., Hieu, D.N., 2014. Population status of the Southwest China Serow *Capricornis milneedwardsii*: a case study in Cat Ba Archipelago, Vietnam. Pac. Conserv. Biol. 20 (4), 385–391.

Timko, J., 2001. Conservation at What Cost? The Case Study of Social Implication of Protected Areas and the Role of Local People in Vietnam. Master of Science Thesis, The University of Bristish Columbia.

SECTION VI

DECENTRALIZATION

CHAPTER

12

Decentralization in Forest Management in Vietnam's Uplands: Case Studies of the Kho Mu and Thai Ethnic Community

T.D. Vien, M.V. Thanh*[†,‡]

[*]Vietnam National University of Agriculture, Hanoi, Vietnam [†]International Centre for Applied Climate Sciences, University of Southern Queensland, Toowoomba, QLD, Australia [‡]Center for Agricultural Research and Ecological Studies, Hanoi, Vietnam

12.1 INTRODUCTION

Decentralization is commonly viewed as the process of transferring power from the central to lower levels of government (Agrawal, 1999). This transfer has been widely adopted in different fields by many countries around the world, as it reduces government bureaucracy, democratizes decision making, and improves efficiency of public service delivery (Omar et al., 2001; Lungisile, 2002). Nevertheless, the decentralization process is not simple on the ground implementation and its success depends on a diverse array of factors at different levels of government and their interactions among these levels (Pacheco, 2004). Some of them are actors' decision-making capability, the conditions powers are allocated, and the accountability maintained (Larson and Soto, 2008).

In the forestry sector, there are several arguments that support the transfer, due to the fact that decentralization potentially distributes benefits from forest resources more equitably between forest users. It also facilitates regulating forest utilization more effectively (Ribot, 2001a,b). Furthermore, it supports the collective action of institutions in governing forest resources (Andersson, 2002). The combination of these factors could result in positive implications for conserving forest resources while improving livelihoods of forest dependents (Kaimowitz and Ribot, 2002).

In many developing countries, particularly in mainland Southeast Asia, rapid degradation of forest resources under state management together with high costs of protection

Redefining Diversity and Dynamics of Natural Resources Management in Asia Volume 2
http://dx.doi.org/10.1016/B978-0-12-805453-6.00012-7

has led to increase decentralization in forest management (Fisher, 2000). The government of Vietnam has begun transferring rights and management responsibilities over forest resources to the local governments since the early 1990s by means of forestland allocation (FLA). By implementing FLA, forest tenure rights have been transferring to local organizations, individual households, and communities. The government believes that access to forestland and the rights to make productive use of the land will motivate local people to use and manage the land in an economical and environmentally friendly manner, and in return it will increase the country's forest cover and improve livelihoods of the local people.

There have been diverse local responses to decentralizing FLA since its implementation. Several studies have been conducted to understand how geographical factors contribute to the diverse responses and environmental results, but no studies have been conducted to understand how ethnicity and ecological conditions contribute to the issue. The purpose of this chapter is to fill these gaps through case studies in different ethnic groups residing in the same ecological condition (*ecological factor*), and the same ethnic group residing in different ecological condition (*ethnicity factor*).

12.2 CASE STUDIES CONTEXT

This study was conducted in three forest-independent communities of Kho Mu and Thai ethnic community living in Nghe An and Son La provinces (Fig. 12.1). These are the leading provinces in implementing the decentralizing FLA in Vietnam. The selected hamlets include Na Be, Xieng Huong, and Huoi Toi hamlets.

FIG. 12.1 Location of the study sites.

12.2.1 Na Be

Na Be Hamlet is a part of Xa Luong Commune, Tuong Duong District, Nghe An Province, and belongs to the North Central Region of Vietnam. The hamlet is about 8 kilometers (km) from its commune center. It lies 6 km from national highway No. 7 and close to the Lao border. The hamlet connects to the outside world via a road that is in bad condition and it is difficult to access the hamlet in rainy seasons. The total population of Na Be is 726 people living in 132 households; most of them are Khomu ethnic minority. The locals' livelihood heavily relies on swidden cultivation, animal husbandry, and exploiting forest resources. Around 60.6% of households suffered from food shortages between 2 to 6 months per year. There is no marketplace in the hamlet and surroundings.

12.2.2 Xieng Huong

Xieng Huong Hamlet is also a part of Xa Luong Commune. It is about 3 km from the commune center and lies along the Nam Non River. The river separates Xieng Huong from national highway No. 7 and there is road that is in good road condition running through the hamlet. However, inner roads are not in good condition and it is difficult to move around during the rainy season. The total population in Xieng Huong is 394 people living in 81 households. Most of them are Thai ethnic minority. Similar to Na Be Hamlet, the main local livelihoods in Xieng Huong are from swidden cultivation, animal husbandry, and exploiting forest resources. Some households earn extra income from cage-raising fish. The local people suffered from food shortages between 2 and 3 months per year. There is no market in the hamlet; nevertheless, the local people are able to easily access markets in the commune center.

12.2.3 Huoi Toi

Huoi Toi Hamlet is in Chieng Hac Commune, Yen Chau District, Son La Province, and belongs to the Northwest Region of Vietnam. It is about 1 km from its commune center. There are 41 households with 192 people. All of the Huoi Toi Hamlet residents are Thai ethnic people who have long been settled in the hamlet. Almost all households in the hamlet engage in agricultural and forest-related activities. Owing to the small area of paddy field, the local people mainly rely on slopping land-based farming or swidden cultivation and animal husbandry. Unlike Na Be and Xieng Huong, farming systems in Huoi Toi are more market-oriented. Thai people in Huoi Toi had much better and diverse cash incomes from various sources: maize, livestock, fruits (litchi, longan, and mango), and fishes, in which, maize was their main income source.

12.3 METHODS

Several methods were employed for data collection. First, in-depth interviews were conducted with officials from the provincial to the commune level and knowledgeable villagers. These interviews provided us with insight into how the government policies were implemented in selected districts and communities and its impacts on forests and the local livelihoods. Second, group discussions were organized in each selected hamlet with male and female villagers from different age groups. The group discussions helped us gather information on the socioeconomics of households, land-use changes, and livelihood patterns in the study sites. This step also involved observation through transect walks in different directions in the study areas. Finally, a structured questionnaire was used to survey 117 households in three hamlets. The survey helped derive important data on demographic, socioeconomic, land use, and livelihood patterns of the local people before and after implementation of FLA. The secondary data, such as maps, reports, and statistical figures, were also gathered during field data collection.

12.4 MAIN FINDINGS

12.4.1 Decentralization in Forest Management and Its Effectiveness

FLA is seen as a measure of decentralization in forest management in Vietnamese uplands. The measure was started with Decree 02/CP in 1994 to allocate forestland to local organizations, households, and individuals. It was followed by Decree 01/CP in 1995, focusing on contracting of land for agriculture, forestry, and aquaculture purposes. In 1999, the government made another push by issuing Decree 163/CP, emphasizing leasing of land for forestry purposes. These policies were supported by several supplemental forest development and protection programs, namely 327 (known as regreening bare hills and uncultivated lands) and 661 (so-called five-million-hectare afforestation program), which paid forestland recipients for their work in forest development and protection.

The decrees have been implemented in both Nghe An and Son La provinces; however, time of implementation of these decrees in the two provinces is different. Regarding Decree 02/CP, Nghe An implemented it right after it took effect in 1994 in the whole province. While in Son La, it was implemented in 1995 in only seven communes of Yen Chau District where one of the study sites is located. As for Decree 163/ND-CP, Son La started implementing in July 2001, while Nghe An started in June 2002. In fact, however, even before Decree 163/ND-CP officially took effect in Son La, in 1998, the provincial authorities themselves had allocated parts of forestland to their forest-adjacent people to avoid the haphazard activity of clearing and burning forests for shifting cultivation. This initiative was accepted by the government and was considered as a breakthrough step of decentralization of Son La authorities over forest resources, and it had been taken into account by the policy makers for forming Decree 163/ND-CP.

The implementing process of both decrees was the same: central government—province—district—commune—local households. However, when implementing decree 02/CP, Son La Province did not allocate forestland to individual households, but to communities or hamlets, while Nghe An authorities allocated forestland to individual households. And provincial authorities made their own circulars/decisions to appoint their lower administrative levels to implement the decrees.

To allocate forestland to recipients, both provinces established FLA management boards at different levels. Members of these boards are officials from several organizations at the same level: land administration, agriculture and rural development, forest protection unit, representative of people's committee, police, finance, and price department. Unlike Son La, the province that set up the boards in all its levels, Nghe An Province did not set up a board at the provincial level. Interestingly, although members of a board are quite various, the ones who really worked in the field were very few: (1) following Decree 02/CP, District Offices of Land Administration (DOLA) was the office that directly allocated forestland to local people; and (2) following Decree 163/ND-CP, not only DOLA was appointed to allocate forestland, but District Forest Protection Unit (DFPU) was also assigned.

When implementing Decrees 02/CP and 163/ND-CP, the government allowed two provinces, Nghe An and Son La, to flexibly adjust things to be feasible for the actual situation in each area, and to ensure local people have enough land for agricultural purpose. Therefore, although Decree 02/CP and 163/ND-CP did not mention allocation of forestland for recipients to do farming, the two provinces still allocated a fixed area for local people to cultivate crops. This land is called "rotational swidden or rotational shifting cultivation area."

As for the funds to implement the FLA, this was managed by provincial authorities, while district and commune authorities are the ones who implement the allocation in practice. They find it is difficult to implement the allocation and to issue land-use certificates. In fact, the not-in-time payment led to the tardiness of FLA in both provinces. In Son La, according to the provincial plan of FLA under Decree 163/ND-CP, by 2003 all the forestland in Son La Province should have been completely allocated to households and individuals, and those recipients were to be titled as owners with a "Red book" or forestland certificate. Nevertheless, as of March 2004 the fund for implementing the decree still lacked the huge amount of 4 billion VND from a total of 15 billion VND (Son La Provincial Forest Protection Unit, 2004; Son La Provincial Department of Natural Resources and Environment, 2003). The same thing occurred in the Nghe An case, as sometimes the implementers needed financial support to work but they had to wait for a long time.

Diffusion of the decree contents to local people was not thoroughly done, so the recipients did not know clearly what and how much they would get for their work on forestland, so they did not seriously work on the land. In many other cases, recipients thought that the allocated forestland became their own land immediately, thus they started to extract as much as possible in the area they received. The overload of forest allocation cadres (in forest protection unit and land administration offices) was also a reason of tardiness of forest allocation following Decree 163/ND-CP. In Son La, for example, one implementer had to allocate 2000 hectares, while normally one could manage only 1000 ha.

12.4.2 Forestland Allocation Policies and the Case Study Sites

The implementation of FLA has different impacts in different places. In Huoi Toi Thai hamlet of Son La province, positive impact appeared as swidden area decreased significantly and forested areas increased accordingly. In another Thai hamlet, Xieng Huong in Nghe An province, however, land use pattern in different time was almost the same. Interestingly, forest of Na Be hamlet of Khomu people got worse after the implementation of the Decrees 02/CP and 163/ND-CP: more swidden area appeared and the natural forest area shrank. The forest cover changes of the three hamlets—Xieng Huong, Na Be, and Huoi Toi—before and after the implementation of the decentralization policies, are shown in the following participatory maps (Figs. 12.2–12.4).

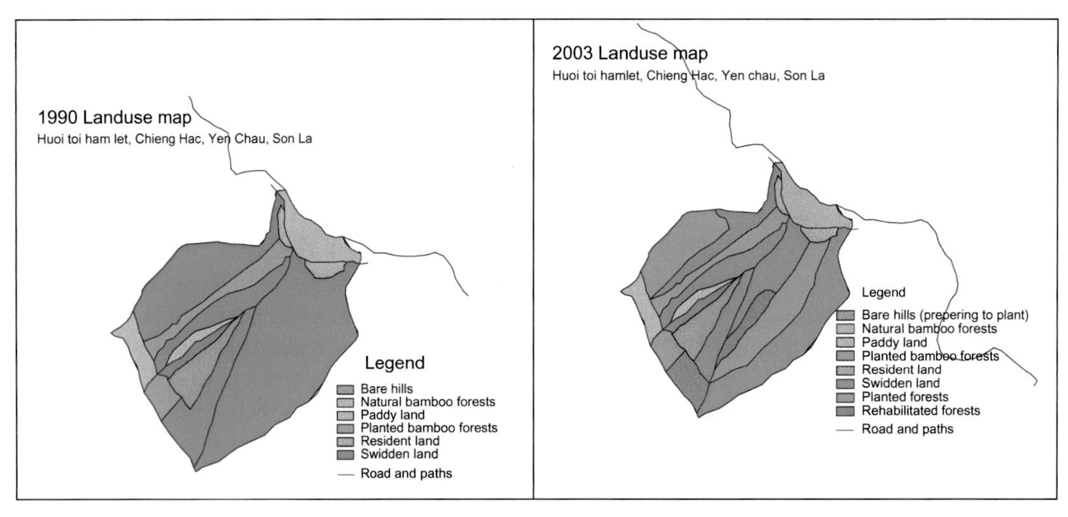

FIG. 12.2 Maps of Huoi Toi Hamlet in 1990 and 2003. *Source: From Group discussion in 2003.*

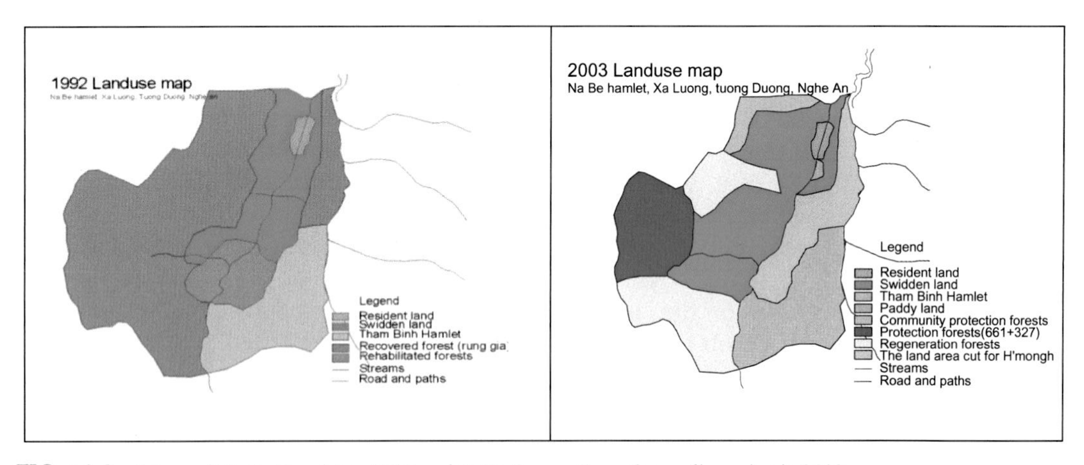

FIG. 12.3 Maps of Na Be Hamlet in 1990 and 2003. *Source: From Group discussion in 2003.*

The heavy destruction of the forest is still going on as people in Na Be continuously cut valuable timber in the forest with a huge volume offered for sale (about 20 cubic meters per month), and every household extracts firewood not only for its own consumption but also for sale. Other nontimber forest products are also gathered.

Thai people understand the Kinh language very well and this helps them better understand state policies compared to the Khomu people. The number of Thai people who answer "understand policy" is higher than that among the Khomu ethnic group. And similarly, the number of Thai people applying to have forestland is higher than that of the Khomu people.

Under the effects of FLA, farmers in the three hamlets have adopted different strategies for their agricultural production as well as forest management in accordance with their particular

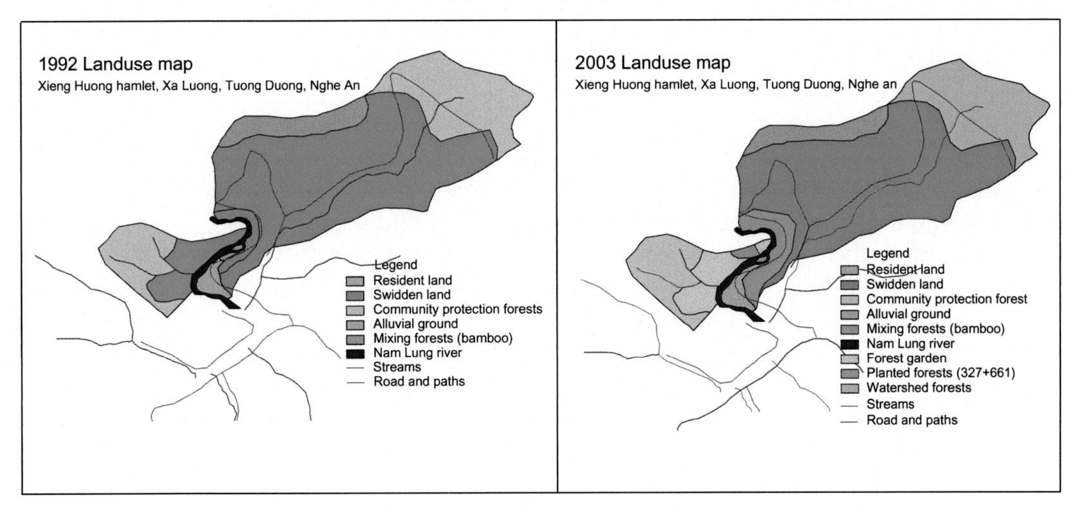

FIG. 12.4 Maps of Xieng Huong Hamlet in 1990 and 2003. *Source: From Group discussion (2003).*

socioeconomic and natural conditions. While the Thai and Khomu groups in Xieng Huong and Na Be hamlets still cling to food production for their own consumption, the Thai in Huoi Toi have mostly produced agricultural products for market. Having better access to markets and better knowledge may account for the case of the Thai in Huoi Toi.

12.5 DISCUSSION

12.5.1 Decentralization Elements Embedded in Forestland Allocation Policies

Some decentralization elements are embedded in FLA policies (01, 02, and 163 Decrees). These policies have made a significant change in the forestry sector, specifically the way in which forestland is used and managed. Under the policies, the government allocated a part of forest and forestland to local organizations, households, and local communities for a long, fixed period through granting them a land-use certificate, in place of the previous government monopoly over forests. To implement these policies effectively, the government has created a legal framework for realizing decentralized management of land and forest in the upland. This provides more scope for the local authorities and people to participate in the decision-making process with regard to forestland use and management than before. Thanks to the government policies, the forest now has its real owners and local people have been given responsibility over their allocated forest and land. This encourages them to make sound decisions on long-term and more sustainable forest management.

Some decentralization elements are found under the implementation of FLA policies; that is, participation and transparency. In both Nghe An and Son La provinces, there is close collaboration between different administrative levels (provincial, district, and commune). In implementing Decrees 01 and 02 at the grassroots level where FLA teams are formed, in which commune and village leaders are also involved. Before allocation of forest and land to households, village meetings are held for the villagers to decide how to allocate land and forest in their village on the basis of equity and convenience among land users in each village/hamlet. State forest enterprises are also assigned to do forest allocation based on Decree 01 to households for protection and a management cost is paid to them. The district FLA steering committee together with the commune FLA team is responsible for any conflicts that happen or complaints that arise from villagers during the land allocation process. Theoretically, if the land allocation process were done properly as planned, participation from local people and authorities in this process would have been more active and meaningful.

Besides FLA, the government has adopted supporting programs, such as the 327 program, and the 5-million-hectare reforestation program. These reforestation programs aim at supporting and encouraging land users to invest in agroforestry production and forest regeneration. Although the 661 program only started in 1999 in some places, to some degree decentralization in its implementation is obvious through the government and provincial related documents, where the important role of the district project management board is stressed. The provincial steering committee and project management board are responsible for project design and planning, then available funds granted by the government are allocated to districts to implement the provincial plans. In turn, each district allocates its plans to its communes based on the provincial planning. On the government side, annually it provides grants to each province based on their plans and the government's available budget. With these reforestation programs first priority is given to the land users to participate in the project.

Some decentralization elements such as accountability can also be found in other development programs, such as the poverty alleviation and hunger eradication program, and the fixed cultivation and sedentarization program. Before 1992 both of these programs were implemented based on a top-down and subsidy approach, but since 1992 a project approach in the implementation of these programs has replaced the ineffective top-down approach, and in many places it has yielded better results due to improving the accountability of project management boards and project staff. According to this approach, small projects under these programs are designed and planned at the village and commune levels, then they are approved by the district and province levels for funding and preferential loans. The most successful projects are the home garden, livestock, and agroforestry gardening development project.

Decentralization in forest management also promoted through official recognition of customary laws that encourage forest resource protection and village regulations according to Ministry of Agriculture and Rural Development (MARD) Guideline No. 54 issued in 1999. Under this guideline local upland communities are encouraged to revise their customary laws or set up new village regulations in natural resources management (NRM) (called *Huong Uoc*) in accordance with the socioeconomic situation and environmental setting of each village and aligned to the state legalistic framework. The village forest management regulations and customary laws provide village and commune leaders some power to adjudicate violations of the village regulations at a certain level defined by the state laws. This policy aims at enhancing accountability and wider participation of local people and authorities in protecting the natural resources in their home villages. In both Nghe An and Son La, most of the upland villages have formulated their *Huong Uoc* and they are approved by the district governments. However, in fact the success of the implementation of a village's *Huong Uoc* depends upon the awareness of the community members and accountability of local leaders, and each community is different. The only point that matters is how the *Huong Uoc* are implemented in daily community life.

12.5.2 Problems in Implementing Decentralization Forest Management Policies

12.5.2.1 Power Devolution

It is obvious that the state perceives the FLA as decentralization in forest management. However, many important powers are still made by the central government, such as the powers of control over exploitation of forest resources, and power of use of allocated forest, which is mainly transferred to the forest recipients, but this power is still subject to state laws. To support local governments to realize forest management policies, a certain amount of funds, which is often not enough to cover all implementation costs, is annually allocated to them. Moreover, having no other source of funds, Nghe An and Son La, two of the poorest provinces in the country, have to rely on government grants. This created the financial dependency of local governments on the central government. The whole administration order creates an upward financial dependency of the lower levels on the upper levels. To a certain extent it represents the top-down subsidizing approach to forestland, and helps to explain why it is difficult to mobilize active participation from local governments. This partly explains the slow FLA process and poor implementation success of other reforestation and development programs in the province.

In addition, when power is transferred to the hamlet and commune leaders, they do not have the financial resources, and they are hardly able to exercise their decision-making power. In reality, as there is no mechanism for commune and hamlet management boards to generate financial resources to maintain their activities, they rely on small budgets contributed by the villagers; therefore, they are not able to make meaningful independent decisions on forest management.

Another aspect of decentralization is the way powers are devolved. Government staffs involved in the FLA work are also concurrently assigned some other tasks besides their main job. Sometimes, some of them cannot find enough time to fulfill all their assigned tasks. Moreover, the situation often becomes more difficult and complicated when project staff, especially leaders, change their workplace or move to another position, and they have to transfer their duties to others. This transferring needs time to take place and it also takes time for the new staff to get up to speed. This often causes many problems in the implementation process, not only for FLA but also for other projects and programs as well.

One of the vice-chairmen of Tuong Duong District complained to us about his work overload. At the same time he was in charge of 15 sections and most of his time was spent on attending different kinds of meetings and signing different kinds of documents. He hardly finds enough time to go to the field to meet and have discussions with local people, or to see how the work is going in the communes and hamlets. This is an obvious example of how poor supervising and monitoring can be an outcome of power devolution if it is not carefully planned and implemented. Moreover, it may lead to power concentration in a few local leaders and it can also easily lead to increased bureaucracy.

Benefit sharing is another issue that has attracted special attention of land users and project staff. From the land user side, long-term benefits for them in FLA policies are not clearly defined. Small government payments for forest protection and limited initial loans for forest planting do not create incentives and yet foster responsibility among land users to actively implement reforestation programs. From the project staff side, low salaries for their workload also do not motivate them to fulfill their assigned tasks with the necessary accountability.

12.5.2.2 Accountability

As mentioned earlier, success in the implementation of FLA also depends on the accountability of the people/organizations to whom power is devolved. Poor planning and supervision and inadequate monitoring from concerned people/organizations are common in our study sites, and it is considered one of the main causes of the poor implementation success of these policies. This is also the main reason behind the failure in implementing 327 projects in Tuong Duong. Some examples of what happens when accountability is lacking is that there can be a waste of seedlings, low survival rate of seedlings, poor-quality fruit gardens and other perennial plantations, all of which are the inevitable outcome of the supply of low-quality seedlings regardless of local people's demand, and the lack of necessary technical guidelines before and after planting.

Another example of problems caused by the lack of accountability was found in FLA. Many places where the area of allocated forest and land was estimated did not have clearly defined borders in the field. Because of this, the land users do not know exactly where the borders are of their allocated forests and land. This in turn has led to disputes among land users in the same hamlets. In many cases, the consequence of this is that there are critical disputes among land users that until now have not been resolved. Thus, in fact there exists only upward accountability, and there is no clear mechanism for downward accountability. The lack of downward accountability has led to a number of unexpected outcomes occurring during implementation of FLA policies. Those mentioned above are only some evidence; there are many more similar cases. Supervision and monitoring by an independent institution may be needed to increase downward accountability of power devolved to people/organizations as they exercise that power.

12.5.2.3 Participation

Participation may take place through the direct participation of local people or through their representatives, local authorities, who are elected by the local people to make decisions regarding how natural resources should be used. Our research revealed that participation took place at different levels depending on the approaches applied by each project/program. Clear-cut differences in participation were found between the top-down approach applied by 327 projects and the bottom-up approach employed by the Social Forestry and Nature Conservation Project (SFNC) in Nghe An.

With regard to participation in the 327 projects, the villagers and local leaders agreed that they had no voice in the project planning activities and all decisions were made regardless of the local settings and people's needs and preferences. Moreover, the free-of-charge supply of seedlings, together with poor access to information, led part of the project beneficiaries to underevaluate and misunderstand government supports and the objectives of the 327 program. This in turn often yielded undesirable outcomes such as waste of seedlings (by not using them) and the serious damaging of newly established forests and fruit gardens by free-grazing livestock.

Even when participation takes place, there is poor participation from disadvantaged households (eg, poor households, female-headed households) in hamlets due to their weak voice in decision making at the hamlet level. This poor participation is common in the study sites for the disadvantaged groups in all hamlet activities and it creates favorable conditions for the elite group in hamlets to participate in the decision-making process at the commune and hamlet levels. Consequently, the elite households are the ones who benefit most from land allocation and other development programs in the upland.

FLA based on household demand is another example. At first, the forestland seemed equitably allocated among households because it satisfied local people's needs and their investment capability. However, this FLA has revealed inequities in forest management. In our study sites, almost all households of the high-income group have larger forestland than other households in the hamlet, whereas poor households have small areas under forest or none at all. Those who do not have forestland are always in a very difficult situation because they rely purely on farming activities, which only provide food to cover daily needs, and they do not have any income from forest products. They do not even have places to collect firewood for family needs because the forests surrounding their hamlet already have owners. In this case they have to ask for the forestland owners to allow them to collect some firewood from their forests.

12.6 CONCLUSION

Although government policies are important steps toward a framework for a sustainable environment and have created the new opportunities for the people to be involved in forest management, there are some shortcomings in the design and implementation of forest management policies. Both the FLA and the programs (327/661 program) are designed to be nationwide and thus do not take into account different population groups, farming traditions, and physical conditions. Although intended to benefit all population groups, the policies are in fact more suited to market-oriented communities than to, for example, traditional hamlets of H'mong or Kho Mu people in the remote areas.

The top-down approach, which does not create room for active participation of local authorities and communities in forest management, together with poor management, passive planning, and the financial dependency of local governments on the central government, are the main forces that restrict the success of the implementation of government policies and programs in both provinces.

Local government should not only have implementing power and tasks devolved to it, but also meaningful decision-making power over forest resource usage. Besides this, they should be guaranteed the necessary resources, access to information, and a favorable legal framework for realizing their assigned tasks and using their devolved power.

To attract active participation in forest management, benefit sharing among actors must be clearly defined, including both to whom powers are to be devolved to and the local communities' rights and responsibilities. This will create a motivating force for participation.

Local people need to have access to information about government policies and programs so that they can make sound decisions regarding their participation in carrying out these programs.

An appropriate mechanism for supervising and monitoring activities of the actors who have been given power, while they are exercising their powers and assigned tasks, need to be applied so that they can be properly held accountable with regard to the use of their power and the results of their assigned tasks.

References

Agrawal, A., 1999. Greener Pastures: Politics, Markets and Community Among Migrant Pastoral People. Duke University Press, Durham and London, CA.

Andersson, K., 2002. Explaining the Mixed Success of Municipal Governance of Forest Resources in Bolivia: Overcoming Local Information Barriers. Center for the Study of Institutions, Populations and Environmental Change, Indiana University, Bloomington, USA.

Fisher, R.J., 2000. Decentralization and Devolution of Forest Management in Asia and the Pacific. FAO, Bangkok, Thailand, pp. 3–10.

Kaimowitz, D., Ribot, J., 2002. In: Services and Infrastructure versus Natural Resources Management: Building a Base for Democratic Decentralization. Submitted for Conference on Decentralization and the Environment. World Resources Institute, Bellagio, Italy.

Larson, Anne M., Soto, F., 2008. Decentralization of natural resource governance regimes. Annu. Rev. Environ. Resour. 33, 213–239.

Lungisile, N., 2002. Decentralisation and Natural Resource Management in Rural South Africa: Problems and Prospects. Institute of international studies, university of California, Berkeley, US.

Omar, A., Kahkonen, S., Meagher, P., 2001. Conditions for Effective Decentralized Governance: A Synthesis of Research Findings. Center for Institutional Reform and the Informal Sector, University of Maryland, Maryland, College Park, USA.

Pacheco, P., 2004. What lies behind decentralisation?: forest, powers and actors in lowland Bolivia. Eur. J. Dev. Res. 16 (1), 90–109.

Ribot, J., 2001a. Local actors, powers and accountability in african decentralizations: a review of issues. In: Paper Prepared for IDRC, Assessment of Social Policy Reforms Initiative. World Resources Institute, Washington, DC.

Ribot, Jesse C., 2001b. Decentralized Natural Resource Management. Washington, DC: World Resources Institute.

Son La Provincial Department of Natural Resources and Environment, 2003. Reports and data on land use and forest situation in Son La [in Vietnamese], Son La, Vietnam.

Son La Provincial Forest Protection Unit, 2004. Reports and data on forest situation and forest protection in Son La Province [in Vietnamese], Son La, Vietnam.

CHAPTER

13

Institutions for Governance of Transboundary Water Commons: The Case of the Indus Basin

M.A. *Kamran**, A. *Aijaz*[†], G. *Shivakoti*[‡,§]

[*]Nuclear Institute for Agriculture and Biology, Faisalabad, Pakistan [†]Indiana University Bloomington, Bloomington, IN, United States [‡]The University of Tokyo, Tokyo, Japan [§]Asian Institute of Technology, Bangkok, Thailand

13.1 INTRODUCTION

The question of water wars has been quite frequently raised in academic debates recently. Water optimists claim that instead of being a cause of conflict, water has been more often a reason for cooperation (Dolatyar, 2002; Dinar, 2002), while water pessimists think that impending scarcity, in the wake of a burgeoning global population, will ultimately result in an intensified struggle for water resources resulting in wars. The "water war hypothesis" is dismissed by water optimists as lacking in substance empirically and theoretically as it generally relies on a very limited number of case studies (Katz, 2011). Water realists stress that "when it comes to water, past will not be a reliable guide to the future" (Wolf et al., 2006). Although the last full-scale war over water was fought over 4500 years ago between the two Mesopotamian cities of Lagash and Ummah, they confidently assert that such wars are certain and only a matter of time (Bulloch and Darwish, 1993; Ward, 2002). So, the academic world seems to be clearly divided on the question of water wars. However, commenting on existing water politics, Selby (2003) opines that it takes water and society as ontologically distinct and separate and ignores the mutually creative interaction between the society and water. Water politics literature is characterized by a strong state centric approach that tends to undermine the dynamics of power politics at the subnational level (Furlong, 2006). Institutional setup is based on principles of flexibility to allow changes based on emerging needs, basin priorities, new information, and technologies (Wolf, 2007). Keeping in view the increasing human impact on ecosystems, it has become necessary to understand governance of large-scale systems to avoid large-scale tragedies (Dietz et al., 2003).

Redefining Diversity and Dynamics of Natural Resources Management in Asia
Volume 2
http://dx.doi.org/10.1016/B978-0-12-805453-6.00013-9

13.2 TRANSBOUNDARY WATER ISSUES IN THE INDIAN SUBCONTINENT

The water issues in the subcontinent range from intraprovincial to transboundary levels. The purpose of this section is to look at issues related to water governance on both levels.

13.2.1 Transboundary Issues Between India and Pakistan

Water in the subcontinent has always been linked with political imperatives of colonial rule (Gilmartin, 1994). Therefore, water politics need to be studied through the lens of cultural and economic asymmetries among state elites from national to local scales (Akhter, 2014). Water issues are gaining significance in academic debates as the population increases at an alarming rate and the climate scenario worsens. Pakistan is one of those countries where impending water scarcity posits multiple challenges to the state and society. It has already crossed below the 1000cm/capita water availability threshold, which puts Pakistan in the water-scarce country category from its earlier position of a water-stressed country. Control of water in the arid, semiarid region of the Indus Basin has been closely connected with the imperatives of power in the society. The official discourse on water has been, at times, at odds with people's perception of water and its role in society. Pakistan (47%) has a shared river basin with India (37%), China (8%), and Afghanistan (6%). India, a major upper riparian with control over water sources and Pakistan as main water end user are key beneficiaries of the Indus Basin. The water disputes in the Indo-Pakistan subcontinent date back to interstate conflicts in prepartition British India. After independence, the issue of interstate waters turned into international disputes among the newly emerged countries. The problem was further aggravated because of hostile geopolitical relations between two nations. The problem of Indus water is a typical problem of upper and lower riparian water allocation.

The narratives produced in this discourse tend to calculate the available natural resources—water in this case—imagining them as fixed and "natural" and juxtapose them with population growth. Such narratives are replete with such warning words as resource threshold, carrying capacities, red-lines, and water barriers. This is the discourse that holds sway among the policy makers and officials connected with institutions for water management and development. It is argued that with the population increasing at current rates, annual per capita surface water availability has been reduced from 5260 m^3 in 1951 to the present acute shortage level of below 1000 m^3 (Seckler et al., 1998, 1999). India, like Pakistan, is a semiarid country, but has better availability of water due to its proximity and access to water resources and being upper riparian on the Indus and major basins of waters Ganges, Brahmaputra, and Meghna. Despite the apparently water-rich economy of India, it is facing a looming water crisis due to high population growth and an energy crisis. The per capita water availability was above 5000 m^3 in 1950, which declined to 1800 m^3 now and is estimated to reach water-starving status of 1000 m^3 by 2025. The increasing demand is given as a strong reason for dams to fulfill the water requirements of the agro-based economy of Pakistan.

These narratives tend to deny the fact that the human relationship with nature is not merely consumptive but also productive as human beings try to transform and produce nature through human labor to achieve certain ends. Thus the question of scarcity may not necessarily be natural but signify a lack of human ingenuity or the will to transform it to achieve desired ends or reorient its needs and ethic. The critics of the state's narrative of "ecoscarcity" say that this rhetoric shows the state's lack of capacity and the will to change its understanding about nature and environment.

Pakistan and India share not only a 1610km border, but also the waters from six watercourses: the Indus, Jhelum, Chenab, Ravi, Sutlej, and Beas Rivers, along with their numerous tributaries (Swain, 2004). The Indus main stem, Jhelum, Chenab, Ravi, and Sutlej cross the Indian territory of Kashmir before flowing into Pakistani territory. The growing populations of both countries and the resultant increased demand for water have made the sharing of transboundary water increasingly complex.

13.3 DAMS' CURSE-DAMN CURSE!

The modern Indus Basin Irrigation System (IBIS) was the technological marvel of the British Empire in the subcontinent, constructed with explicit strategic, political, and economic purposes. Technological control over water across time and space made deserts bloom and the land thus made available was distributed to create a patron-client relationship between the British Empire and the local elite (Ali, 1988). The power configuration in this "hydraulic society" was thus mainly dictated by the control of water and land. The British engineers constructed the IBIS as a natural hydrological unit; however, partition of the subcontinent resulted in political boundaries of the states running across the natural boundaries of the Indus Basin. The bilateral hostilities, in the wake of partition of the subcontinent, resulted in the formulation of mutually exclusive identities and interests. So, the struggle for security and interests between the two has turned out to be a zero sum game. The new political realities necessitated different socially constructed water discourses on both sides of the border. Although the Indus Water Treaty (IWT) provides an institutional mechanism to negotiate water-related conflicts (Zawahri, 2009), it seems to be teetering under the pressure of increased demand and supply (Sinha, 2010) and water is more frequently linked with identity and security.

Most literature on the transboundary water issues is anthropocentric and sees natural environment through a human lens, thereby ignoring nature altogether (Mirumachi and Chan, 2014). In the same context, the official discourse on water in Pakistan sees water as a "resource" that can be captured and controlled to put it to productive use. This view entails an ontological distinction between nature and society, where nature is characterized as wasteful and its energies need to be controlled to put them to productive use for mankind. The history of this discourse in the Indus Basin dates back to the colonial period when modern technologies were put to use to entrap the natural potentials of Indus rivers for greater productivity by the colonial regime. This technical and scientific discourse employed the language of science and mathematics to render it unquestionable and thus privileged the engineers' conception of water over the organic conception of water held by the general public, encouraging an attitude of disresponsibilization in which people tend to leave the job of maintaining the relationship with water to experts. This positivist conception of water perceives water as passive and calculable, unchanging and devoid of history (Selby, 2003). This discourse must be understood within the broader Malthusian tradition of the relationship between population and resources. Whereas nature is portrayed as a finite repository of resources, human population is portrayed as growing infinitely, thus foreseeing a situation where human population outstrips natural resources. It is in this context that the official discourse on water in Pakistan must be assessed (Table 13.1).

TABLE 13.1 Water Availability Status in the Indus Basin

		Per Capita Water Availability (m^3)			
Indus Basin	**Total Renewable Water Resources (km^3)**	**1990**	**2000**	**2025**	**2050**
Indus-India	78.6	2487	2109	1590	1132
Indus-Pakistan	154	1713	1332	761	545

Adapted from: International Union for Conservation of Nature (IUCN), 2011. Indus Water Treaty and Managing Apportioned Rivers for the Benefit of Basin States—Policy Issues and Options. IUCN, Pakistan, Karachi.

The eastern rivers of Pakistan were expected to compensate for waters of the eastern rivers given to India. The World Bank mobilized support in the form of grants and loans from the governments of Australia, Canada, Germany, New Zealand, the United Kingdom, and the United States. The Indus Basin Development Fund (IBDF) agreement was established to allocate funding for the Indus Basin Project (IBP), one of the world's largest irrigation projects comprising two dams for storage (Mangla and Tarbela at Jhelum and the Indus rivers respectively), six barrages, about 400 km long, link canals and irrigation works to divert 14 million acre feet of water from the western to eastern tributaries (Michel, 1967).

Under the auspices of IWT, the Indus basin witnessed massive construction of dams and reservoirs in the last half of the 20th century. The IWT, in essence, was based on the division of rivers and compensation of lost rivers in the form of sponsored building of dams for storage (Table 13.2).

TABLE 13.2 Large Dams in the Indus Basin

Country	**Name**	**River**	**Year of Construction**	**Height (m)**	**Capacity (million m^3)**	**Purpose**
India	Bhakra	Sutlej	1963	226	9620	Irrigation, hydropower
	Nangal	Sutlej	1954	29	20	Irrigation, hydropower
	Pandoh	Beas	1977	76	41	Irrigation, hydropower
	Pong	Beas	1974	133	8570	Irrigation, hydropower
	Salal	Chenab	1986	113	285	Hydropower
	Baghliar	Chenab	2008	-	33	Hydropower
Pakistan	Mangla	Jhelum	1968	116	10,150	Irrigation, hydropower
	Tarbela	Indua	1976	137	11,960	Irrigation, hydropower
	Chashma Barrage	Indus	1971	–	870	Irrigation

Data from: Food and Agricultural Organization (FAO), 2012. Indus river basin. In: Irrigation in Southern and Eastern Asia in Figures: AQUASTAT Survey 2011.

Pakistan and India are currently harnessing 11.1% and 33.65% of hydropower potential, respectively, and 0.03% and 14.22% of capacity, respectively, at the Indus Basin is under construction (GoI, 2012; WAPDA, 2011). The dams have created a huge controversy over the already threatened water of the Indus.

The IWT seems to have become inadequate in its ability to address climate change, environmental flows, and groundwater quality and quantity due to the transboundary affects of abstraction. The unpredictable variations of river inflow necessitate the storage of water and at the same time increase the need for improved transboundary water cooperation and the exploration of shared opportunities for joint climate change adaptation within the basin to avoid shared climatic threats, such as glacial lake outburst floods (Bates et al., 2008). It is estimated that by 2020 about 1.5 billion people will suffer due to water-stressed river basins, which may increase to 3 billion by 2080 (IPCC, 2008). This will lead to increased risks of conflicts over shared transboundary river basins, particularly in Asia (Cooley et al., 2009; Fraser, 2009). The transboundary issues all over the world, in general, and on the Indian subcontinent, in particular, need special attention in an era of changing demographic, economic, and climatic scenarios. This study will now discuss international guidelines for transboundary issues and water distribution in the Indus Basin.

The IWT, a trilateral agreement between Pakistan, India, and the World Bank, has the unique characteristic of dividing the rivers of the Indus Basin. The main points of the treaty include

- Pakistan has unrestricted use of three western rivers, the Jhelum, Chanab, and Indus, with the minor exceptions of mainly nonconsumptive hydropower use for India from these rivers.
- India has the right to absolute consumption of three eastern rivers: Ravi, Sutlej, and Beas rivers.
- As a compensation for the waters of three eastern rivers given to India, a fund was created to build three dams, eight link canals, three barrages, and 2500 tube wells in Pakistan.

Pakistan's concerns over the waters of the Indus Basin are about a lack of fairness in the IWT, Indian manipulation of the clauses of the existing treaty in its favor, and the future demographic, environmental, and technological concerns of Pakistan vis-à-vis the limited scope of the treaty to tackle emerging threats.

13.3.1 Groundwater

Over time, groundwater has emerged as an exceedingly important water resource and its increasing demand for agriculture, domestic, and industrial uses in the Indus Basin ranks it as a resource of strategic importance (Qureshi et al., 2010). It is one of the most important components of water balance in arid and semiarid environments as it provides an assured supply of water at the time of need. However, its management is always a great challenge for a shared aquifer as is the case of the Indus Basin. Its hidden nature, lack of a monitoring mechanism. and data are the major causes. Aquifers that date back thousands of years are under threat due to agricultural uses for growing populations the world over (Foster and Chilton, 2003; Shah et al., 2000). The long term, continuous abstractions are adversely affecting the overall water balance when the average value consistently exceeds the recharge over a long period. It may result in potential changes of the quantity and quality of natural aquifers and can pose severe aquifer sustainability threats and conflicts in the absence of groundwater rights at local levels and lack of instruments in water-sharing treaties in the case of transboundary aquifers. The process of groundwater abstraction and lowering of water tables is accelerating in Balochistan Province of Pakistan and the Indian states of Punjab, Haryana, and Rajasthan (Shah et al., 2000). Soil salinity associated with the use of poor-quality groundwater for irrigation has further aggravated the problem (Qureshi et al., 2010).

There are a number of transboundary aquifers between Pakistan and India. It is in the interest of both countries to use these in a sustainable manner. Overdraft in northern India, sharing a border with Pakistan, has taken on alarming proportions mainly due to subsidized energy for groundwater pumping. A National Aeronautics and Space Administration (NASA) study describes the significant depletion of water in northern India during the period 2000–8 (Rodell et al., 2009). This could have serious impacts on the depletion of aquifers on the Pakistan side. Pakistan is already experiencing rapid depletion of groundwater on the fringe of the border with India. However, there is not even a single study providing information regarding the absolute amount of water in the aquifer, how much has been pumped out, and its transborder effects. This creates the need to cooperate for joint studies in this direction. Global climate change (increasing temperatures, changing rainfalls patterns, or droughts) is expected to further accentuate the groundwater depletion issues due to decreasing and erratic supply of surface water and increasing water demands for different agricultural and domestic uses. Besides groundwater depletion, water quality is another important dimension. Entrance of effluents, due to excessive use of agrochemicals and heavy industrialization near the border, is another issue that needs addressing. Drainage water coming from India is heavy loaded with chemicals causing harmful effects on livelihoods, livestock, and groundwater quality. Up to now there has been no scientific study about the extent of damage and negative impacts on groundwater quality at the India-Pakistan border. Besides quantification of extraction, the health and quality of groundwater is another area that needs to be addressed without compromising relations between the neighboring countries.

13.3.2 Water Quality

There is no globally binding environmental agreement for transboundary water quality (Shmueli, 1999), and the only nonbinding instrument of UN Agenda 21 emphasizes integrated approaches to ensure the adequate supplies of good quality water. At regional levels, there are number of agreements to address the issue of water quality and 47% of these agreements have water quality control and monitoring as one of the key themes. Article IV (9) of IWT deals with pollution issues and advice to take measures to keep the Indus waters clean from sewage and pollution. Article IV (10) of the treaty does refer to the intent of each riparian to conserve the quality of waters of the Indus Basin, but does not provide for an appropriate monitoring system. This is one of the overlooked areas in the IWT, unlike some other treaties (like the Mekong River Commission) where water quality monitoring is undertaken throughout the basin under an environmental program.

Pakistan debates water quality from its position that its emerging water needs are mainly for drinking, domestic, industrial, agricultural, and ecosystem services. The quality of both surface and groundwaters in Pakistan is degrading mainly due to increased urbanization (resulting in increased sewage), excessive and future insensitive groundwater mining, alteration in the agriculture landscape, uneven use of agrochemicals, and industrialization on eastern rivers and deforestation in the watersheds on the western rivers in India. River Ravi has become a wastewater carrier from Hadiara Nullah and poor-quality flows from Baglihar can lead to heavy silt loads in Jhelum and Chenab ecologies.

13.4 INTERPROVINCIAL WATER GOVERNANCE ISSUES

Control of water in the arid and semiarid region of the Indus Basin has been closely connected with the imperatives of power in the society. Despite the existence of well-elaborated legislation and organizations since British rule, water management in Pakistan has remained an irresolute issue at all levels (from the canal to basin level) and faces a myriad of institutional problems. The IBIS is currently operating at much higher cropping intensities than its design, and the benefits of fresh flows go to head-end users, while the tail-end farmers bear the high energy costs to abstract poor-quality groundwater, which results in lower profitability and uncultivatable saline lands.

The provinces, despite the presence of a reasonably well elaborated and agreed upon Water Apportionment Accord (WAA) in 1991, remain in constant conflict over water share. The accord gives autonomy to the provinces to use their share of water in the provinces on 10 daily basis criteria. However, the looming conflicts between Sindh and Punjab over the interpretation of the accord where Sindh Province objects to Punjab's assertion to use its share of Indus water for two off-take canals for a tributary zone regardless of water shortages in Sindh. At the national level, there is little coordination between federal and provincial agencies and there is duplicity of functions performed by institutions without complementation. The Water and Power Development Authority (WAPDA), Indus River System Authority (IRSA), and Federal Flood Commission perform functions related to the Ministry of Water and Power at the federal level. There is a mismatch and lack of interorganizational coordination, which is further aggravated by the lack of financial and technical expertise. The provincial agriculture, public health, and irrigation departments lack coordination with their allied ministries and departments at the federal level (Ministries of Climate Change; Science and Technology; Food Security and Research) and also lack coordination with the district level departments of On-Farm Water Management (OFWM), Water and Sanitation Agency (WASA), public health, and Provincial Irrigation and Drainage Authorities (PIDAs). The divergent objectives of key organizations like WAPDA, IRSA, PIDAs, OFWM, WASA, and public health departments, need to be aligned with stated organizational objectives and improve linkages with allied departments. The provincial and federal agriculture- and water-related departments, despite high claims, have not been able to improve water productivity and quality. The institutional and legislative failure to keep the water charges too low to even recover O&M cost further denigrates the efficiency of existing mechanisms.

13.5 SUSTAINABILITY OF THE INDUS BASIN WATER TREATY VIS-À-VIS INTERNATIONAL GUIDELINES AND OSTROM'S DESIGN PRINCIPLES

13.5.1 International Guidelines for Transboundary Water Management and Status of the IWT

More than 3600 transboundary water treaties and agreements have been signed in the world (Kliot et al., 2001) with varying degrees of success due to the variation in geopolitical conditions (Song and Whittington, 2004). The major common principles for successful water management in these agreements and treaties include reasonable and equitable utilization, significant harm, notification and consultation, and peaceful dispute resolution (Kliot et al., 2001; Caponera, 1995; Housen-Couriel, 1994). Of the five basic tenets of sustainable transboundary water management, four principles have been discussed in the IWT (Table 13.3).

It is worth repeating that the IWT cannot be considered to be a water-sharing treaty, which could facilitate "equitable and reasonable utilization"; rather, it is a mechanism of dividing rivers among the two countries. The principle seems to be violated if India's actual use of 3 million acre-feet (MAF) at the time of the treaty is considered with its ability to gain additional and 30 MAF for future development on eastern rivers. On the contrary, Pakistan lost water rights over the eastern rivers as well as having to give up part of the western rivers and having to build storage for its actual needs.

Some critics view the IWT as static and incapable of resolving the emerging issues of climate change, water quality, shared groundwater aquifers, and so on. However, Article VII of the IWT maintains the possibility of future cooperation for optimum harvesting of potential basin flows. Despite the possibility of future cooperation in Article VII, the security concerns promote conflict rather than cooperation. Similarly, Article VI of the treaty allows exchange of hydrological data between the two parties, which has remained lacking due to issues of mistrust.

TABLE 13.3 Principles of Transboundary Management in IWT and International Conventions

Principles	Helsinki Rules (1966)	UN Water Convention (1997)	Indus Water Treaty
Reasonable and equitable water use	Articles IV, V, VII, X, XXIX	Articles 5–7, 15–17, 19	Not applicable[a]
Significant harm principle	Articles V, X, XI, XXIX	Articles 7, 10, 12, 15, 16, 17, 19– 22, 26–28	Article IV (9)
Cooperation and information sharing	Articles XXIX, XXIX, XXXI	Articles 5(2), 8, 9, 11, 12, 24, 25, 27, 28, 30	Articles VI–VIII
Notification, consultation and negotiation	Articles XXIX, XXIX, XXXI	Articles 3, 6, 11–19, 30	Articles VII [2], VIII
Peaceful conflicts resolution	Articles XXVI–XXXVII	Article 33	Article IX, Annexure F and G

[a] *The IWT divided rivers to India and Pakistan and is not a water-sharing treaty in the true sense.*

The IWT provides a detailed mechanism for conflict resolution. Under Article VIII of the IWT, the office of the Permanent Indus Commission (PIC), comprised of two commissioners appointed by the respective governments of India and Pakistan, work as state representatives to channel treaty implementation and peaceful conflict resolution. The two commissioners meet annually and submit an annual report on the measures taken to resolve conflicts, build cooperation, and develop Indus waters. The treaty has a very comprehensive mechanism for conflict resolution and provides both parties different options to resolve conflicts. Questions related to the treaty are dealt with by the PIC; the differences (ie, unsettled questions) are dealt with by a neutral expert; and the disputes (unsettled differences) are dealt with by the International Court of Arbitration.

The water issues between India and Pakistan post-IWT primarily emerged from a trust deficit and are triggered by India's overambitious campaign to build hydropower, technological change to support such projects, emerging concerns over environmental flows, and climate change. With improved technology and reduced costs, particularly in building hydropower projects, there are increasing threats of flow diversion because of increased capability of upper riparian to build dams to an extent to significantly affect lower riparian flow. Similarly, technological advancement for water abstraction has posed serious challenges to shared groundwater aquifers. Keeping in view the security relations between India and Pakistan, these concerns can't be ruled out. The office of the Indus commissioner exists on both sides of the border to resolve conflicting issues. However, due to limited human, financial, and technical resources the role of Indus commissioner remains minimal. Trust deficit is mainly triggered by the lack of coordination between water-related organizations, reluctance of data and information sharing, lack of competency for hydrodiplomacy, sour diplomatic relations due to security and terrorism, and other issues.

13.6 APPLICATION OF OSTROM'S DESIGN PRINCIPLES FOR TRANSBOUNDARY WATER COMMONS

Ostrom (1990) proposed "design principles" to provide the characteristics found in long-enduring resource systems. The design principles have demonstrated their validity as evaluative criteria by their application to successful, long-enduring resource systems during the last two decades (Gautam and Shivakoti, 2005; Cox et al., 2010; Kamran and Shivakoti, 2013). Successful mapping of self-governing small-scale common-pool resources (CPRs) underscores the importance of information about resource boundaries, collective formulation of rules based on resource information, opportunities to violate rules, and consequences for violation, and all have the potential to govern transboundary CPRs like the river systems in this study.

The CPR issues are identical at both local and global scales as states can't force local authoritative hierarchies to implement agreements, while at the local scale individuals face a lack of cooperation from distant public authority to monitor and enforce rules (Keohane and Ostrom, 1995). The design principles have been applied to gauge and compare performance of small-scale systems (Cox et al., 2010; Kamran and Shivakoti, 2014). There have only been a few efforts regarding application of design principles on a basin-level water management scale (Rowland, 2005; Dietz et al., 2003). This section looked at design principles, their interpretation and modification to meet transboundary commons' scope, and to critically evaluate the IWT and give suggestions for further improvement of effective water governance.

13.6.1 Clearly Defined Boundaries

A well-defined boundary of the resource is important and the principle requires that the individuals or households with rights to withdraw resource units from the CPR and the boundaries of the CPR itself must be clearly defined (Ostrom, 1990). The principle can also be described from information and knowledge of the resource point of view, and the boundaries may also change with better technology (eg, groundwater monitoring). In large-scale commons, the countries or provinces might have a clear idea about the political boundaries, but in such instances knowledge about the flow of water commons, their dependence on the upper riparian source (glaciers in the case of the Indus or transboundary groundwater aquifers between India and Pakistan) becomes essential. To make the principal fit for a large system, resource information needs to be more explicit to understand resource boundaries. A closer analysis of issues between India and Pakistan shows that lack of information about availability and consumption of water resources (both surface water and groundwater) is the reason for conflicts. There is a need to shift focus from political boundaries to ecosystems boundaries for effective water management.

13.6.2 Congruence

The principle states that "The distribution of benefits from appropriation rules is roughly proportionate to the cost imposed by provision rules. Appropriation rules restricting time, place, technology, and quantity of resource units are related to local conditions" (Ostrom, 1990). In a majority of the existing transboundary water agreements the basis for resource distribution are physical quantities of water rather than returns from the resource, ignoring the basin conditions and ecosystem requirements. This principle is not followed in the case of the IWT and is suggested to be part of the treaties so that the tragedy of water commons in the form of salinity, waterlogging, sea intrusion, and so on, in deltas can be prevented through congruence of basin-level distributions and policies to match governance considerations.

13.6.3 Collective Choice Arrangements

The principal states that "Most individuals affected by operational rules can participate in modifying operational rules" (Ostrom, 1990). However, transboundary governance considers the voice of national governments rather than direct users or the smaller units, provinces, head and tail users, and others, and therefore the principal needs to be modified to consider that the voice and rights of small groups are considered in international agreements. The modified principal may state that "the voice of all sections of resource-dependent sections will be addressed by national governments under the treaty." In most cases, agreements are done by consulting people at the top (as in the case of the IWT), and soon smaller sections of society raise their voice for water rights, which in turn puts pressure to gain more water from transboundary rivers and revise the treaty or break the existing one.

13.6.4 Monitoring

This principle states that the monitors, who actively audit CPR conditions and appropriators' behavior, are accountable to the appropriators or are the appropriators themselves (Ostrom, 1990). In this case the respective countries' governments monitor fairness of resource distribution. Therefore, the principle may be modified as "joint monitoring to ensure fair distribution of flows and basin health parameters."

13.6.5 Graduated Sanctions

Appropriators who violate operational rules are likely to receive graduated sanctions (depending on the seriousness of the context of the offense) from other appropriators, from official(s) accountable to these appropriators, or from both. This principle for transboundary water commons can be interpreted to mean that the sanctions must be in accordance with the level of infractions. This principle is not considered in the IWT and can benefit the governance if the body to impose sanctions is empowered and has technical expertise to make decisions for graduated sanctions.

13.6.6 Conflict Resolution Mechanism

Appropriators and their officials have rapid access to low-cost, local arenas to resolve conflicts among appropriators or between appropriators and officials. This principle desires low-cost conflict resolution and is also followed in the Indus Basin and other major treaties on transboundary commons. Despite the mechanism to resolve conflicts through neutral experts of the World Bank and the International Court of Justice, there are increasing disagreements over new initiatives and external decisions. Broad-based engagements of stakeholders to resolve conflicts locally and empowerment of the institution of Indus commissioners for local-level conflicts resolution are needed.

13.6.7 Minimal Recognition of Rights to Organize

The design principle states that the rights of appropriators to devise their own institutions are not challenged by external government authorities. The description of the principle for transboundary water commons is that the treaties are respected by regional and international agencies and other countries. The IWT has withstood wartime and has been recognized and supported by other countries.

13.6.8 Nested Enterprises

Appropriation, provision, monitoring, enforcement, conflict resolution, and governance activities are organized in multiple layers of nested enterprises. In transboundary water agreements the water-related activities are organized in multiple layers of nested enterprises across different scales and different sectors. The IWT, however, focuses only on water issues at the levels of Indus commissioners, leaving aside the linked areas of food security, environment, groundwater quality, sea intrusion, and so forth. Water conflicts in the Indus basin, by inclusion of this principle, need to be linked with bilateral trade, benefit sharing, and collective basin development initiatives to create the possibility for win-win positions (Table 13.4).

TABLE 13.4 Ostrom's Design Principles for Application to Transboundary Water Commons

Ostrom's Design Principles	Interpretation and Suggested Revisions of Design Principles	Application to Indus Basin Water Treaty	Current Issues	Possible Solutions Using Institutional Approach	Suggested Action
Clearly defined boundaries	User countries/provinces have commonly agreed knowledge about resource and resource dynamics	IWT divided the basin to follow political boundaries. The problem related to data remained dubious	Conflicts over flow, data availability, validity and lack of transboundary aquifer data	Telemetry system, remote sensing	Jointly managed water data bank
Congruence of rules to local conditions	Basin-level accords match subbasin conditions for best ecological and economic benefits	IWT considered geopolitical needs rather than ecosystem requirements and focused on distribution of water rather than its maximum economic and ecological benefits	Salinity and dead rivers at lower riparian	Decision support system and holistic modeling to understand effect of different flow regimes on user countries and ecosystem needs	Paradigm shift from water sharing to benefit sharing as basic criteria
Collective choice arrangements by resource users	Collective decision making by nations considering long-term relationships and resources sustainability. A variant in this principle is that the rules must consider choice of subgroups and represent their voice	The voice of provinces and subgroups was not considered in the treaty	Intraprovincial conflicts and head-tail issues at canal levels	Revised treaty should include wider dialogue at national level to accommodate needs of all groups	Putting humans and ecology ahead of political interests
Monitoring	Joint monitoring to ensure fair distribution of flows and basin health parameters	Permanent Indus Commission (PIC) comprising one commissioner each from India and Pakistan for monitoring	Growing concerns over receding glaciers and changing flow patterns in changing climate scenarios. Limited authority of PIC	Joint monitoring of glaciers and strategic environmental assessment of new basin projects	Strengthening technical expertise of PIC and provision of financial autonomy

TABLE 13.4 Ostrom's Design Principles for Application to Transboundary Water Commons—cont'd

Ostrom's Design Principles	Interpretation and Suggested Revisions of Design Principles	Application to Indus Basin Water Treaty	Current Issues	Possible Solutions Using Institutional Approach	Suggested Action
Graduated sanctions	Cooperative agreement on sanctions and mutual implementation agreement or agreement by a third party	Not applicable	Not applicable	Not applicable	Not applicable
Conflict resolution mechanisms	Low-cost local-level conflict resolution to all countries	There are mechanisms to resolve the conflicts through Indus commissioners and negotiations and the unresolved conflicts go for external arbitration	Increasing disagreements over new initiatives and external decisions	Broad-based engagements of stakeholders to resolve conflicts locally	Empowerment of local arbitrators and their financial autonomy
Minimal recognition of rights to organize	Treaties/accords are respected by regional and international agencies and other countries	World Bank and ICA enforce rules and other countries accept the rights under treaty	There are no issues related to external acceptance of the rules	Not applicable	Not applicable
Nested enterprises	Water management activities are organized in multiple layers of nested enterprises across different scales and different sectors	IWT only focuses on water issues at the levels of Indus commissioners leaving aside linked food security, environment, groundwater quality, sea intrusion, etc.	Issues related to water quality, groundwater, salinity, sea intrusion, climate change are on the rise	Inclusion of a broad range of sectors and shift of focus from only interstate relations to multistakeholders involvement	Water needs to be linked with bilateral trade, benefit sharing, and collective basin development initiatives to create possibility for win-win position

13.7 CONCLUSIONS AND THE WAY FORWARD

The mismatch between the scales at which a problem is experienced (ie, the ecosystem scale), and the scale where decisions are made (ie, the political and management scale), is a major source of concern in transboundary resources management. The scale problem becomes very relevant as water use crosses ecosystem, political, and social scales and is debated at local, national, and international arenas. Local-level and cross-boundary institutions are human artifacts and effective management; therefore, they demand redefining the roles of existing institutions and building new effective institutions. Revitalization of existing institutional arrangements or creating new institutional mechanisms for effective transboundary water management demands a critical review of existing institutions and drawing lessons for a new institutional contract.

Water conflicts at all levels (from local to transboundary), based on the stated facts, stem from knowledge gaps due to a lack of credible and reliable data aggravated by an unwillingness to share available data among the provincial- and national-level agencies, transboundary organizations, and between the science and policy arenas. A national "Water Policy" to address surface and groundwater resources should be built involving a wider range of stakeholders from scientific organizations, users, civil society, ministries, media, and others. There is a need to build a very strong knowledge hub to bridge the information and knowledge gap causing conflicts at different boundary levels. Interprovincial and bilateral exchange of scholars and the launch of projects of mutual interest to establish scientific authenticity on conflicting issues can be some of the initial steps to be taken by the relevant governments. The role of boundary organizations becomes very crucial to bridge the gaps between science, society, and policy. There is an urgent need to build cooperative mechanisms and connect missing links across science, society, planning, and policy through social and knowledge networks. The major focus for transboundary cooperation has been on formal institutions, while the role of informal institutions has been neglected. Vibrant civil society and media communication can serve as formal institutions to achieve cooperation. The academic institutions can bridge both the knowledge gaps, bring scholars from provinces and countries together, and provide a framework for dialogue and intellectual discussion. Informed policy based on research and knowledge created by academic institutions is necessary to achieve cooperation.

The IWT is regulated by the PIC, with a commissioner each for India and Pakistan. Despite the relative success of the treaty, unilateral development of water infrastructure by both India and Pakistan, along with changing environmental and climatic conditions and technological improvements, have led to a stalemate in water diplomacy. At present, variable river flows exacerbate tension. Enhancing the IWT's ability to cope with climate change would involve a more holistic approach—beyond the current technical focus—which would take account of social and environmental issues and involve dialogues among other, nontechnical, stakeholders such as river communities. Furthermore, the treaty could encompass provisions regarding environmental flows while provisions regarding information exchange and water measurements could be strengthened.

To make water governance more effective, there is a need to redesign the administrative setup along the lines of other successful coriparian institutions the world over. There is a lot more to learn from existing agreements such as the Mekong River Commission and Israel-Jordan water agreements. These agreements highlight the need to develop a joint committee across the two countries to foster meaningful diplomatic relations, develop a benefit-sharing mechanism, and encourage track-II diplomacy to build consensus. Broad-based institutions with the scope of climate, water, peace, trade, and human development can be helpful for broad-based collaborations. A reliable and real-time flow of data sharing through a telemetry system; joint management of upper riparian hydropower projects and benefit sharing; targeted trade based on the comparative water efficiency principal; and compensation for virtual water in bilateral trade can be potential solution to transboundary water issues (Alam et al., 2009). There is a need for regional networks, say, South Asian Waters on the pattern of the International Center for Integrated Mountain Development (ICIMOD) and the South Asian Association for Regional Cooperation (SAARC) to resolve water issues through diplomatic relations. The knowledge networks like the South Asian Network for Development and Environmental Economics (SANDEE), the Global Change Impact Study Center (GCISC), and Leadership for Environment and Development (LEAD) can be used to build scholarship, scientific solutions, the science-based approach to build trust and eradicate stakeholders' misperceptions over key transboundary ecosystem issues in general and water issues in particular. National organizational like IRSA, Pakistan Agricultural Research Council (PARC), Pakistan Council of Research in Water Resources (PCRWR), and GCISC can help to build science-based policy and information sharing to resolve interprovincial conflicts from water distributional to environmental flows. Civil society engagement and enabling policies for shared management of water in a sustainable manner at the national and regional levels can help resolve the water problem and turn conflicts into peace. The adaptive organizations and institutions must be promoted that can withstand changing biophysical, social, and knowledge systems and can easily adapt to a new set of challenges (Dietz et al., 2003). Water diplomacy is a key strategy both at the regional (among Indus Basin stakeholders) and national (among provinces) levels to ensure that water quality plays its due role in poverty alleviation and human well-being. It is not a one-time exercise and its dynamic nature requires that this should be institutionalized and supplemented by future-oriented policies. It also requires a commitment from provincial and district governments and other environmental laws enforcement and research agencies. This will increase the access to safe water, promote equity, and reduce the risk of future hazards for the citizens of Pakistan.

Keeping in view the conflicts between Indian and Pakistan, despite the fact that the IWT is meeting major principles set by international guidelines (Table 13.3) there is a need to reconsider the international guidelines and the IWT. The analysis of the IWT using Ostrom's design principles shows that the major conflicts between India and Pakistan can be resolved if lessons are drawn from the design principles put forth by Elinor Ostrom. By using IWT Article XII related to the revision of the IWT with the mutual consensus of India and Pakistan, a new "Indus Water Treaty-Revised" may be devised for successful and sustainable governance of the Indus Basin and other transboundary water commons. The discussion in this chapter about issues of the Indus Basin can be relevant to transboundary water commons issues faced by other countries.

Acknowledgments

The authors are grateful for the support of LEAD Pakistan in providing fellowship to the lead author of the manuscript under its "Future Water Leaders" program of Cohort 18. Thanks are due to the Ford Foundation for generous support in the form of a travel grant and AIT Thailand and Hanoi University of Agriculture for holding a book workshop and presentation in Hanoi. The authors are also thankful to Dr. Mariam Rehman and Ms. Zareen Khan for constructive comments and improvement in write-up, and an anonymous reviewer for valuable suggestions to improve the quality of the manuscript.

References

Akhter, M., 2014. The hydropolitical Cold War: The Indus Waters Treaty and state formation in Pakistan. Polit. Geogr. 46, 65–75.

Alam, U., Dione, O., Jeffery, P., 2009. The benefit-sharing principle: implementing sovereignty bargains on water. Polit. Geogr. 28 (2), 90–100.

Ali, I., 1988. The Punjab under Imperialism. Princeton University Press, Princeton.

Bates, B.C., Kundzewicz, Z.W., Wu, S., Palutikof, J.P., 2008. Climate change and water, IPCC Technical Paper VI.

Bulloch, J., Darwish, A., 1993. Water Wars: Coming Conflicts in the Middle East. Gollancz, London.

Caponera, D., 1995. Shared waters and international law. In: Blake, G., Hildesley, W., Pratt, M., et al. (Eds.), The Peaceful Management of Transboundary Resources. Graham and Trotman, London/Dordrecht, pp. 121–126.

Cooley, H., Christain-Smith, J., Gleick, P.H., Allen, L., Cohen, M., 2009. Understanding and Reducing the Risks of Climate Change for Transboundary Waters. Pacific Institute, California.

Cox, M., Arnold, G., Tomas, S.V., 2010. A review of design principles for community-based natural resource management. Ecol. Soc. 15 (4), 38.

Dietz, T., Ostrom, E., Stern, P.C., 2003. The struggle to govern the commons. Science 302, 1907–1912.

Dinar, S., 2002. Water, security, conflict, and cooperation. SAIS Rev. 22 (2), 229–253.

Dolatyar, M., 2002. Hydropolitics: challenging the water-war thesis. Confl. Secur. Dev. 2 (2), 115–124.

Foster, F.F.D., Chilton, P.J., 2003. Groundwater: the processes and global significance of aquifer degradation. Philos. Trans. R. Soc. B 358, 1957–1972.

Fraser, B., 2009. Water wars come to the Andes' Scientific American. Viewed on 12-03-2015, http://www.scientificamerican.com/article/water-wars-in-the-andes.

Furlong, K., 2006. Hidden theories, troubled waters: international relations, the 'territorial trap', and the Southern African Development Community's transboundary waters. Polit. Geogr. 25 (4), 438–458.

Gautam, A.P., Shivakoti, G.P., 2005. Conditions for successful local collective action in forestry: some evidence from the hills of Nepal. Soc. Nat. Resour. 18 (2), 153–171.

Gilmartin, D., 1994. Scientific empire and imperial science: colonialism and irrigation technology in the Indus Basin. J. Asian Stud. 53 (4), 1127–1149.

Government of India, 2012. Status of H. E. Potential Development – Basin Wise, Viewed 12-12-2014, http://www.cea.nic.in/reports/hydro/he_potentialstatus_basin.pdf.

Housen-Couriel, D., 1994. Some Examples of Cooperation in the Management and Use of International Water Resources. The Harry S. Truman Research Institute for the Advancement of Peace, The Hebrew University of Jerusalem.

Intergovernmental Panel on Climate Change (IPCC), 2008. Climate Change and Water. IPCC, Geneva.

Kamran, M.A., Shivakoti, G.P., 2013. Design principles in tribal and settled areas spate irrigation management institutions in Punjab, Pakistan. Asia Pacific Viewpoint 54 (2), 206–217.

Kamran, M.A., Shivakoti, G.P., 2014. Institutional response to external disturbances in spate irrigation systems of Punjab, Pakistan. Int. J. Agric. Resour. Gov. Ecol. 10 (1), 15–33.

Katz, D., 2011. Hydro-political hyperbole: examining incentives for overemphasizing the risks of water wars. Global Environ. Polit. 11, 12–35.

Keohane, R., Ostrom, E., 1995. Local Commons and Global Interdependence. Sage Publications, London.

Kliot, N., Shmueli, D., Shamir, U., 2001. Institutions for management of transboundary water resources: their nature, characteristics and shortcomings. Water Pol. 3, 229–255.

Michel, A., 1967. The Indus Rivers: A Study of the Effects of Partition. Yale University, New Haven, CT, USA.

Mirumachi, N., Chan, K., 2014. Anthropocentric hydro politics? Key developments in the analysis of international transboundary water politics and some suggestions for moving forward. Aquatic Procedia 2, 9–15.

Ostrom, E., 1990. Governing the Commons: The Evolution of Institutions for Collective Action. Cambridge University Press, New York.

Qureshi, A.S., McCornick, P.G., Sarwar, A., Sharma, B.R., 2010. Challenges and prospects of sustainable groundwater management in the Indus Basin, Pakistan. Water Resour. Manag. 24 (8), 1551–1569.

Rodell, M., Velinconga, I., Famiglietti, J.S., 2009. Satellite-based estimates of groundwater depletion in India. Nature 460 (173), 999–1002.

Rowland, M., 2005. A framework for resolving the transboundary water allocation conflict conundrum. Ground Water 43 (5), 700–705.

Seckler, D., Amarasingha, K., Molden, D., Desilva, R., Barker, R., 1998. World water demand and supply: scenarios and issues. Research report 19. International Water Management Institute, Colombo, Sri Lanka.
Seckler, D., Barker, R., Amarasinghe, U., 1999. Water scarcity in the twenty-first century. Int. J. Water Resour. Dev. 15, 29–42.
Selby, J., 2003. Dressing up domination as 'cooperation': the case of Israeli-Palestinian water relations. Rev. Int. Stud. 29 (1), 121–138.
Shah, T., Sakthivadivel, D.M.R., Seckler, D., 2000. The Global Groundwater Situation: Overview of Opportunities and Challenges. IWMI, Colombo, Sri Lanka.
Shmueli, D.F., 1999. Water quality in international river basins. Polit. Geogr. 18 (4), 437–476.
Sinha, U.K., 2010. 50 Years of the Indus Water Treaty: an evaluation. Strateg. Anal. 34 (5), 667–670.
Song, J., Whittington, D., 2004. Why have some countries on international rivers been successful negotiating treaties? A global perspective. Water Resour. Res. 40.
Swain, A., 2004. Managing Water Conflict: Asia, Africa and the Middle East. Routledge, London.
Ward, D.R., 2002. Water Wars: Drought, Flood, Folly, and the Politics of Thirst. Riverhead Books, New York.
Water and Power Development Authority (WAPDA), 2011. Hydro Potential in Pakistan, Viewed 15-01-2015, http://www.wapda.gov.pk/pdf/brohydpwrpotialapril2011.pdf.
Wolf, A., 2007. Shared waters: conflict and cooperation. Annu. Rev. Environ. Resour. 32, 241–269.
Wolf, A., Kramer, A., Carius, A., Dabelko, G.D., 2006. Water Can Be a Pathway to Peace Not War. Woodrow Wilson International Center for Scholar, Washington, D.C.
Zawahri, N.A., 2009. Third party mediation of international river disputes: lessons from the Indus River. Int. Negot. 14 (2), 281–310.

SECTION VII

NEW WAY OF THINKING TO MANAGING COMPLEX NATURAL RESOURCE SYSTEM

CHAPTER

14

A System Dynamics Approach for Integrated Natural Resources Management

M.V. Thanh

International Centre for Applied Climate Sciences, University of Southern Queensland, Toowoomba, QLD, Australia

14.1 INTRODUCTION

Natural resources are the backbone of every economy and play a central role in promoting sustainable development. Management of natural resources for sustainability is, therefore, a globally endorsed principle. Nevertheless, the making of decisions in efforts to achieve this goal is challenged by a number of factors. First, most natural resources systems are embedded in socioecological systems processing all the characteristics of complex systems (Edward et al., 2013), while knowledge about these systems is often fragmented and owned by a multitude of experts (Giupponi and Sgobbi, 2013). Policy makers and planners are often faced with the prospect of policy resistance—a tendency for policy interventions to be delayed, diluted, or defeated by responses from the system (Sterman, 2000). In other words, policy makers and planners often overestimate or underestimate the effectiveness of policies and plans, and fail to anticipate the side effects and unintended consequences of their actions. Second, a lack of common understanding and shared vision among stakeholders about how the complex issues of natural resources management (NRM) are addressed. Natural resources systems contain a diverse array of stakeholders (decision makers, scientists, nongovernmental organizations (NGOs), and various other stakeholders), each of whom has different management objectives, agendas, and interests (conservation versus economic development) (Mai and Smith, 2015). These different expectations potentially lead to unforeseen conflicts among stakeholders that necessitate compromises, collaboration, and shared visions based on a better understanding of each other's needs and requirements. Finally, necessary information for NRM is uncertain and tends to reflect reduced scenarios rather than whole system situations (Bosch et al., 2007). It is, therefore, difficult for policy makers and planners to fully utilize available data and information. It is often the case that decisions are made on an unsound basis.

Redefining Diversity and Dynamics of Natural Resources Management in Asia Volume 2
http://dx.doi.org/10.1016/B978-0-12-805453-6.00014-0

Given the complex, multidimensional and multistakeholder problems that are the nature of NRM, past approaches to decision making and problem solving in managing natural resources are reductionist, fragmented, and linear (Bosch et al., 2007). These approaches tend to oversimplify situations, focusing only on a few aspects and disregarding the existing relationships between the components of natural resources systems. They are simply "quick fixes" or "treating the problem symptoms"; thus, they don't deliver desirable outcomes. To address the dynamic complex issues of NRM, it has become clear that a new paradigm is required. The purpose of this chapter is to demonstrate the utility of system dynamics to NRM through the lens of a case study in natural resources-based tourism in Cat Ba Island of Vietnam.

14.2 SYSTEM DYNAMICS

System dynamics (SD) is an approach to understanding the structure, feedback mechanisms, and behavior of complex systems (Kelly et al., 2013). It is a radical approach to enhance cross-sectoral and organizational communication and collaboration to ensure that solutions are found at the level of the root causes. Generally, there are five main steps within the SD approach. These include problem articulation, formulation of a dynamic hypothesis, formulation of a simulation model, model testing, and policy design and evaluation (Sterman, 2000).

14.2.1 Problem Articulation

Problem articulation is the first and most important step in SD. In this step, the situation or issue at hand is defined, and the scope and boundaries of the study are identified. Problem articulation often consists of identification of problems or policy issues to management. There are a diversity of methods that can articulate the problem. The methods that we used in the following study including an in-depth interview, group discussion, and workshop.

14.2.2 Formulation of a Dynamic Hypothesis

In this step, a conceptual model of the problems, known as a dynamic hypothesis, is formulated. A causal loop diagram (CLD) is most commonly used to develop the dynamic hypothesis because it allows for the simple visualization of feedback mechanisms within a system that control system behavior, and is therefore relatively easy for stakeholders to review.

A CLD is a network of variables that connects with arrows, which show causal relationships between the variables (Sterman, 2000). Each arrow in a CLD is labeled with a "+" or a "−", where a "+" means that the two variables move in the same direction, while a "−" means they move in opposite directions. In a CLD, groups of variables linked together form different loops. The two types of loops are the reinforcing (R) or positive feedback loop, and the balancing (B) or negative feedback loop. Reinforcing loops are positive feedback systems, which represent growing and declining action, while balancing loops keep everything in equilibrium. A CLD may include many combinations of R and B loops as well as time delays (represented by //). The interplay between R and B over time, time delays, and shifts in loop dominance over time have generated a complex system of behavior. It is useful to identify the main feedback loops when interpreting a complex system as a way of telling the story of what might unfold within the system in the future and to name each loop with a mnemonic to represent the underlying feedback process.

Obviously, a CLD can be quite complex as it may include many combinations of R and B loops as well as time delays (represented by //). The interplay between R and B over time, time delays, and shifts in loop dominance over time have generated a complex system of behavior. It is, therefore, necessary to identify a core system structure that can explain general system behavior. The core system structure is referred to as a "system archetype." A system archetype is a generic system model that present a wide range of situations; it also provides good indications of the potential leverage points within a complex system, "where a small shift in one thing can produce big changes in everything" (Meadows, 1999). The identification of leverage points can therefore be facilitated in conjunction with the identification of system archetypes. In spite of their common existence, it is not easy to identify the leverage points as they are not intuitive. However, once a CLD is developed, system archetypes become more apparent.

14.2.3 Formulation of a Simulation Model

Although CLDs are very useful as they assist in understanding the structure of complex systems and feedback mechanisms with the systems, there are a number of advantages to be gained from developing a computer simulation model to investigate more deeply the dynamic issues that are a concern to management. A stock and flow model (SFM) builds a simulation model to quantitatively model system behavior over time. As the name suggests, a SFM consists of stocks (also known as accumulators or levels) that represent the system state variables while flows (also known as rates) are the processes that influence change in the stock levels (Sterman, 2000). A simulation engine runs the numerical model, and simulates the change in the values of stocks and flows over time.

14.2.4 Model Testing

The aim of model testing is to improve users' confidence in SFMs. The testing process generally consists of (a) structural tests that test whether the structure of the model is an adequate representation of the real structure, and (b) behavior tests that test whether the model is capable of producing acceptable output behavior (Barlas, 1989). In the following case study, we used both structural and behavior tests to validate the SFM.

14.2.5 Policy Design and Evaluation

After testing, a SFM formulates and evaluates alternative development scenarios. In other words, formulating and testing scenarios that would allow identifying a scenario that is a trade-off between economic development and sustainable use of natural resources, while maintaining environmental resilience.

14.3 APPLICATIONS OF THE SYSTEM DYNAMICS APPROACH FOR MANAGING NATURAL RESOURCES-BASED TOURISM ON CAT BA ISLAND, VIETNAM

14.3.1 The Study Area

Cat Ba Island is located in the northeast of Vietnam belongs to Cat Hai District of Hai Phong City. The island has many unique ecosystems and remarkable historical and archaeological values. In 2004, the United Nations Educational, Scientific, and Cultural Organization (UNESCO) designated Cat Ba Island as one of the world's biosphere reserves. The biosphere is considered as one of Vietnam's most favored and beautiful places, has not only become a preferred destination for tourists from all parts of Vietnam but also attracts tourists from around the world. The number of tourists visiting Cat Ba Island has dramatically increased over the last decade (Fig. 14.1) resulting in the tourism sector making a significant contribution to the local economy (Cat Hai People's Committee, 2005). However, there are a number of challenges that pose a significant threat to the development of sustainable tourism in the biosphere reserve.

14.3.2 Problem Articulation

A series of focus group discussions, in-depth interviews, and workshops were conducted on Cat Ba Island with the local communities, business operators, and local authorities. Discussion and consultation with stakeholders revealed a number of issues that are potential threats to sustainable tourism development on the island, such as a shortage of natural resources, chronic poverty, and environmental degradation. Fig. 14.2 depicts a general picture of tourism development on Cat Ba Island. These graphs show that the island is experiencing strong growth in tourism, while environmental degradation is declining and poverty levels persist. This indicates that achieving a sustainable tourism system on the island is still far from becoming a reality. The next section sheds further light on this by analyzing the conceptual model of the tourism system of the island.

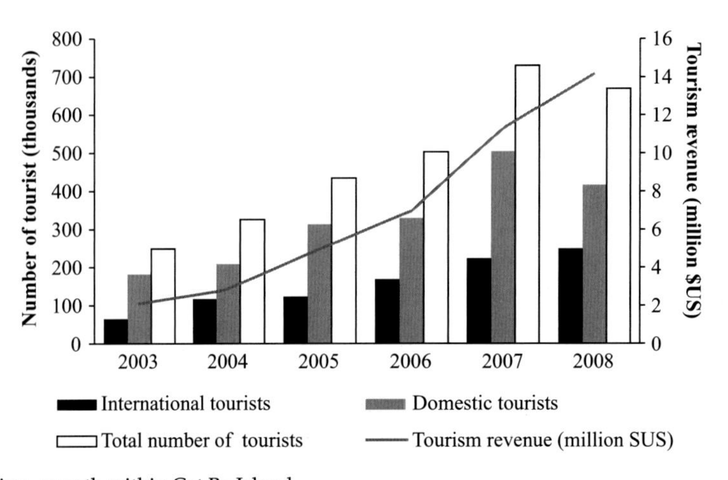

FIG. 14.1 Tourism growth within Cat Ba Island.

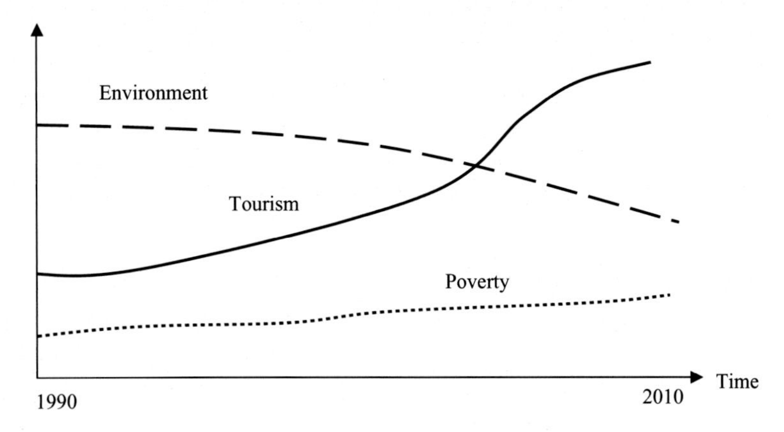

FIG. 14.2 Behavior over time of key variables of the tourism system on Cat Ba Island.

14.3.3 Formulating Dynamic Hypothesis

In this study, the development of a CLD went through three steps. First, a preliminary CLD was developed by the research team based on prior knowledge of tourism development on Cat Ba Island. Second, a working CLD was then constructed based on the preliminary CLD and consulting with individual stakeholders. Finally, we organized a joint stakeholder workshop to review the working CLD to produce a final CLD (Fig. 14.3).

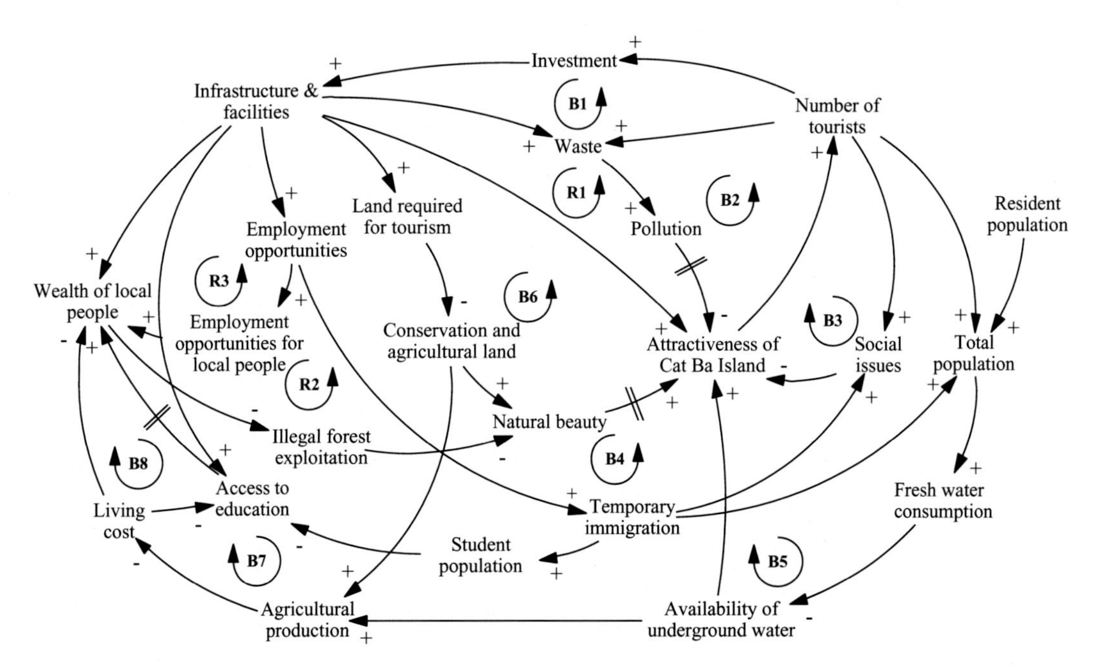

FIG. 14.3 Final CLD for tourism development within Cat Ba Island.

There are eleven feedback loops in the final CLD, including three reinforcing loops (R1 to R3) and eight balancing loops (B1 to B8). The reinforcing loop R1 represents the process of tourism growth. This process is fueled by loops R2 and R3. The balancing loops work against tourism growth through different limiting factors, including pollution (loops B1 and B2), immigration and employment (loops B3 and B4), limitations of natural resources (loops B5 and B6), and population and poverty (loops B7 and B8). Details of these loops are described below.

14.3.3.1 Tourism Growth Loop (R1)

Loop R1 links the variables: number of tourists, investment, infrastructure and facilities, and attractiveness of Cat Ba Island (Fig. 14.4). This simple reinforcing loop portrays the drivers behind the exponential tourism growth on the island during the last few decades.

Loop R1 shows how the provision of tourism-related facilities and infrastructures affect the attractiveness of the island. Rapid growth in tourism during the last decade has resulted in a large amount of tourism revenue. This served as encouragement for the government and private sectors to invest in developing tourism-related facilities and infrastructures on the island to accommodate the growing demand. Part of this is the road network, including the main road through the island that connects different communes and tourist attractions. Many tourist resorts and recreational facilities were constructed and transport facilities improved in both quantity and quality. The improvement of facilities and infrastructure has created a more comfortable environment that has led to the attraction of more tourists to the island.

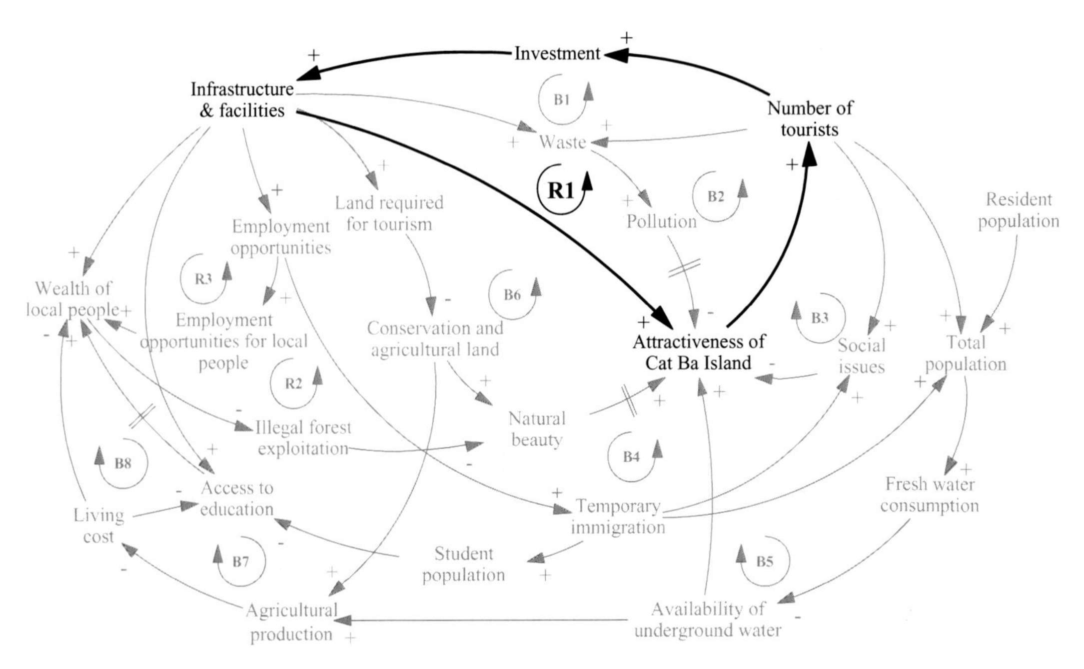

FIG. 14.4 Tourism growth loop (R1).

14.3.3.2 Poverty Reduction Loops (R2 and R3)

Loops R2 and R3 (Fig. 14.5) describe the effects that tourist numbers are having on the increasing need for infrastructure and facilities and subsequently the increase in employment opportunities for the local people (ie, construction of hotels, various tourism-related activities, small businesses, and home stays). This could clearly contribute to poverty reduction for the local people.

The improvement of infrastructure and tourism facilities, such as road systems and public facilities, leads to an increase in the value of local products, as they enable the local people to sell their products. This could contribute to poverty alleviation. In addition, the improvement of infrastructure and tourism facilities stimulates other social, cultural, and educational activities, which in turn will enrich the lives of the local people. The alleviation of poverty could subsequently lead to a reduction in illegal forest exploitation activities and thus to maintaining or restoring the natural beauty of the island.

Loops R2 and R3 also indicate that tourism development and poverty reduction provide counteracting dynamics. To ensure a sustainable tourism system on the island, local poverty needs to be reduced. Therefore, in loops R2 and R3, the poverty variable (wealth of local people) plays a critical role for the sustainable development of tourism on the island.

14.3.3.3 Pollution Loops (B1 and B2)

There are two important balancing loops highlighted in Fig. 14.6 that explain the long-term effects of tourism development on the environment and attractiveness of the island as

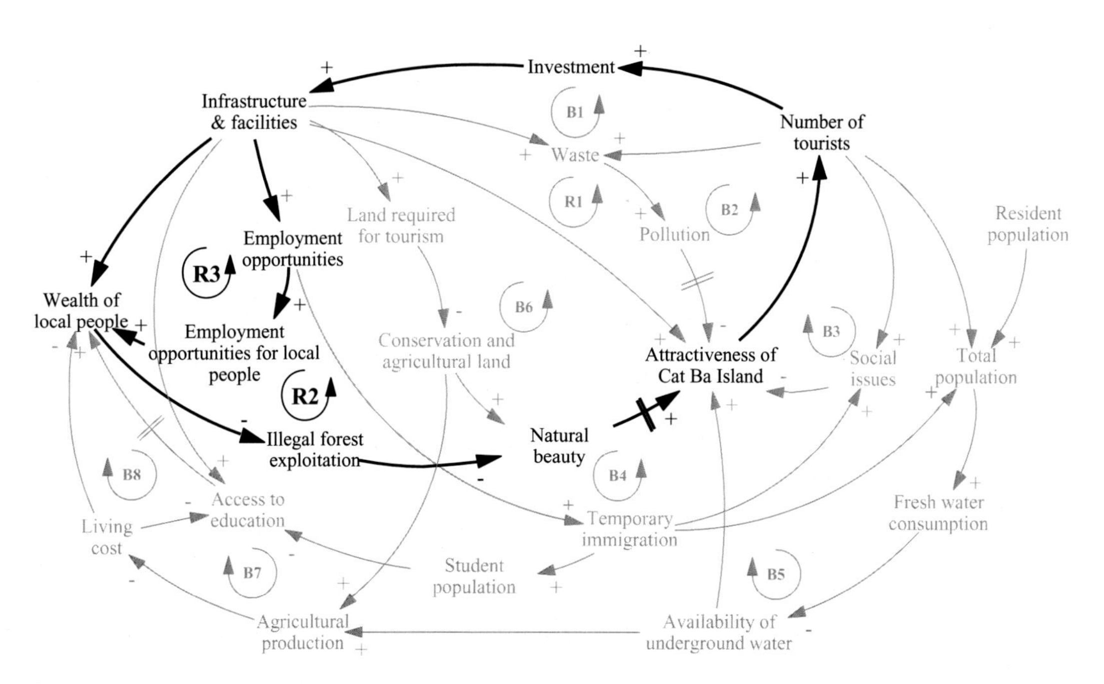

FIG. 14.5 Employment and poverty loops (R2 and R3).

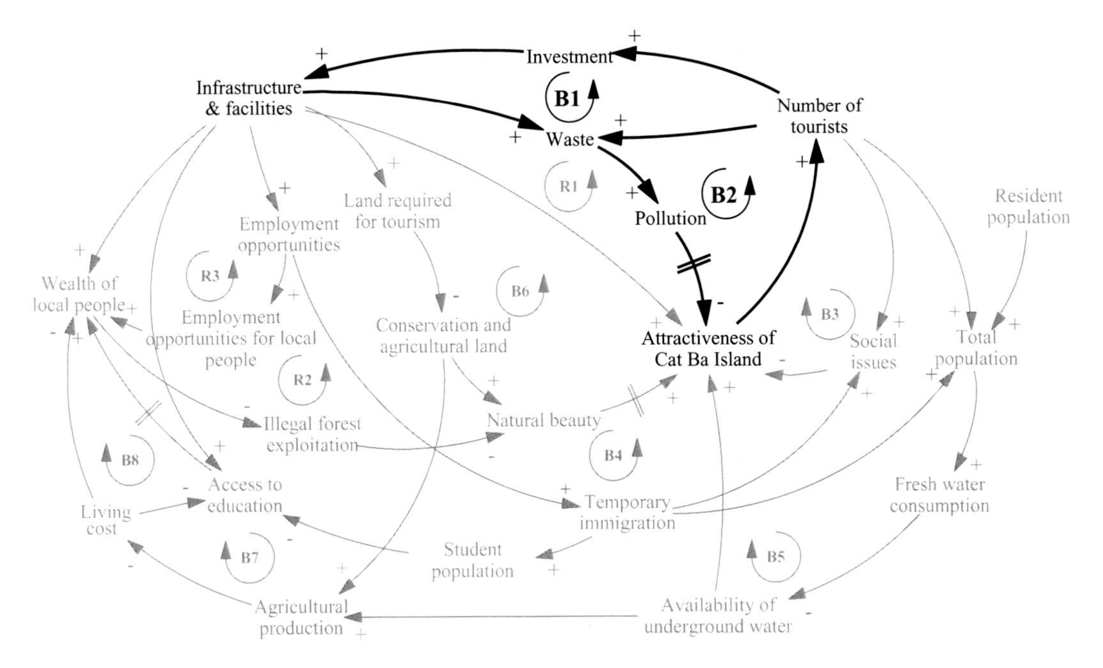

FIG. 14.6 Pollution loops (B1 and B2).

a tourist destination. These loops include B1 (number of tourists, investment, infrastructure and facilities, waste, population, and attractiveness of Cat Ba Island); and B2 (number of tourists, waste, pollution, and attractiveness of Cat Ba Island).

It is clear that the increased number of facilities lead to increasing amounts of waste (solid and liquid) generated by hotels and restaurants. These are discharged into the environment. In addition, the increased number of tourists also generates more litter from their recreational activities. Another indirect source of pollution caused by the concentration of tourist numbers in the peak season is the overexploitation of the limited underground water resources and the many hotels that have to drill their own bore wells for fresh water supply. This leads to saltwater intrusions along the coast of Cat Ba Bay. Other pollution factors include traffic congestion and sanitation problems. The increasing number of vehicles, both in marine and terrestrial areas, not only generate more waste but also increases oil leaks and CO_2 emissions on the island.

Inappropriate landfill and the nonexistence of a sewage treatment plant on the island to accommodate the large amount of solid and liquid waste that is generated daily by hotels, restaurants, and other tourism facilities are critical issues for the environment. It expects that more pollutants will discharge onto the island while environmental degradation will continue to increase due to the lack of appropriate waste treatment plants and disposal facilities. The accumulation of waste will reduce the attractiveness of the island, which subsequently will lead to a reduction in the number of tourists to the island. Therefore, the B1 and B2 balancing loops are critical feedback loops that currently counteract the growth of tourism on the island.

14.3.3.4 Immigration and Employment Loops (B3 and B4)

Loops B3 and B4 are highlighted in Fig. 14.7. Loop B3 indicates the number of tourists, social issues, and attractiveness of Cat Ba Island, while B4 shows that an increase in the number of tourists and infrastructure and facilities influences employment opportunities, which in turn will increase temporary immigration that could lead to negative social issues. This could be detrimental to the attractiveness of Cat Ba Island. Similar to the first two balancing loops, loops B3 and B4 also play an important role in controlling tourism growth on the island.

The rapid tourism development has brought more development to Cat Ba Town, as well as increased the disparity between the town itself and other areas on the island. A wide range of employment opportunities are generated through the various tourism-related services, which attract a large number of people from rural areas who are seeking a better livelihood to migrate temporarily to Cat Ba Town.

There is no doubt that temporary migration has improved the living standards of many poor families on the island and adjacent areas. In the case of unskilled laborers, migration has helped to improve their food security, access to finance for agricultural activities, and accumulation of minimal assets. These have all helped them to offset the high costs of their children's education and living expenses. On the other hand, the existence of migrants and tourists has created many social issues on the island such as drugs, crime, and prostitution. These factors could negatively affect the image and reputation of Cat Ba Island and ultimately reduce the number of tourists to the island.

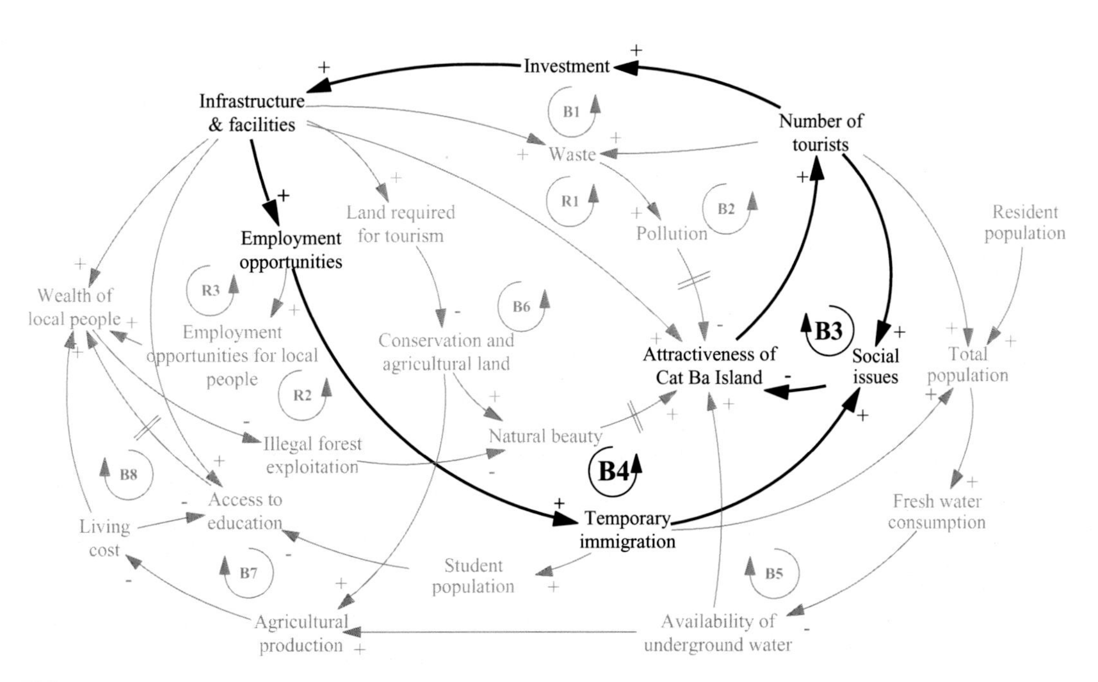

FIG. 14.7 Immigration and employment loops (B3 and B4).

14.3.3.5 Limitation of Natural Resources Loops (B5 and B6)

Loop B5 depicts how the number of tourists, the island's population, and immigration to the island are interlinked with fresh water consumption and availability of underground water (Fig. 14.8). Loop B6 describes how the number of tourists affects infrastructure and facilities, which in turn puts a demand on available land for tourism to the detriment of the natural beauty of the island and its attractiveness to tourists. These loops describe the effects of scarce resources (particularly land and fresh water) on tourism development.

The rapid development of tourism on the island over the last few decades has required more land for expansion of tourism-related facilities and infrastructures. The physical development of these facilities has created negative impacts on the environment. The most prominent of these are the large areas of agricultural and conservation land that have been cleared to build the main road and the construction of recreation facilities such as tourist resorts and resettling areas for local people.

It is expected that a large area of land will be required in the near future to build golf courses and recreational resorts, which can only be constructed on current agricultural and conservation lands. These developments will result in further reduction of the natural beauty of the island. In addition, it will have long-term effects on the availability of underground water, because of the change in land cover. This all makes the availability of land a critical factor that could limit the growth of tourism on Cat Ba Island.

A shortage of freshwater has been identified as an important factor that will limit the growth of tourism. Cat Ba Island relies almost entirely on underground water (about 95%). Water consumption varies and depends on the number of tourists that visit the island. The

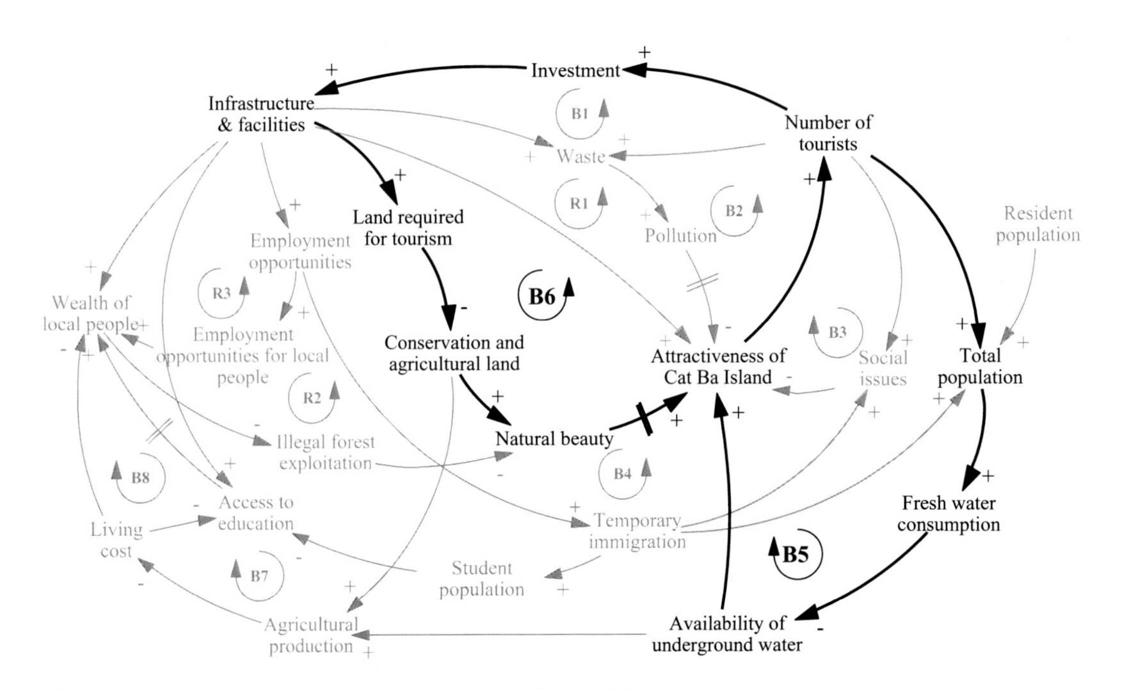

FIG. 14.8 Freshwater and land use balancing loops (B5 and B6).

water source only meets about 35% to 40% of the needs of the tourism industry during the peak season. The constraint of freshwater has already been demonstrated by several incidences where tours to the island had to be canceled due to the lack of freshwater.

It could be expected that the effects of the balancing loops demonstrated in Fig. 14.6 will finally reduce the attractiveness of Cat Ba Island, which could lead to a decline in the number of tourists to the island. In other words, loops B5 and B6 also control tourism growth.

14.3.3.6 Population and Poverty Loops (B7 and B8)

Loops B7 and B8 (Fig. 14.9) describe the effects of tourism development on the social demography and environmental attractiveness of the island. As the number of tourists primarily affects the availability of underground water, increasing tourist numbers would lead to less water being available for agriculture.

Reducing agricultural land is another factor that contributes to the reduction of agricultural production and therefore poses a negative effect on the economic well-being of the local people. As a result, it pushes local residents to exploit forests illegally to make a living. These activities include illegal hunting (including threatened species), the cutting of timber, collecting nontimber forest products, and illegal fishing inside and around the Cat Ba National Park. It also leads to an increase in the number of deliberate fires (used for the collection of natural honey) and the extent of deforestation. It is expected that when the livelihood of local people improves, there will be a reduction in these illegal activities. This will undoubtedly contribute to healthier and conserved ecosystems and a more attractive island.

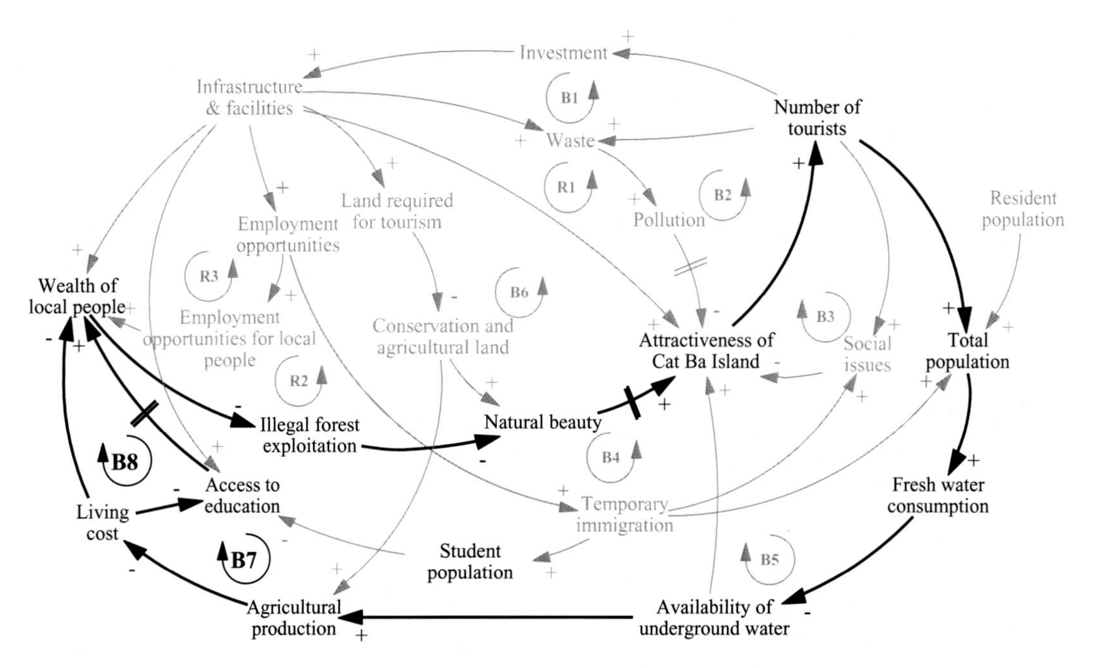

FIG. 14.9 Population and poverty loops (B7 and B8).

The quality of and access to education are also affected by temporary immigration, living costs, and infrastructures with long-term effects on local poverty levels. This loop is thus a balancing loop, which shows the strong interrelationships between the local population, poverty levels, and tourism development.

In summary, the conceptual model or CLD of the tourism system on Cat Ba Island (Fig. 14.3) provides a fair idea of the integration of the mental models that are shared and discussed by the relevant stakeholders. The model was able to present the systems in their entirety, visualizing and capturing the most important factors (drivers and inhibitors) and feedback mechanisms (reinforcing and balancing loops) that constitute, and are embedded in, the system. The process of developing the CLD brings together a wide range of concerns relating to the past and expectations for the future of the system. It can significantly assist stakeholders, who share responsibility for the management of the systems, to develop a better understanding of the dynamics and complex relationships between the systems' components that underlie the systems' behavior. The CLD can therefore serve as a common language for diverse stakeholders to engage in deep dialog and consensus building.

14.3.4 Leverage Points and Systemic Intervention Strategies

The most obvious archetype seen in the tourism conceptual model is limited to growth (Fig. 14.10). This archetype represents situations where improvements in performance or growth are limited and cannot go on forever. The lesson learned from this archetype is that some element always pushes the system back; thus, if we do not plan for limits, we are planning for failure (Senge, 2006). To anticipate future problems and eliminate them before they become threats, the growth engines and potential limiting or constraining forces need to be identified. This implies that in the "limits to growth" archetype, the advantage point is not only placed in the reinforcing loop, but also in the balancing loop. This reminds managers that they should not only focus on the growth engine factors but also take the time to examine what might be limits or constraints that could push back against their efforts.

In the case study on Cat Ba Island, most of the tourism policies and planning programs have to date been focused on the growth engine to attract more tourists to the island, rather than the constraints or negative impacts of tourism described in previous sections. In other words, the reinforcing process of growth through massive building programs of tourism

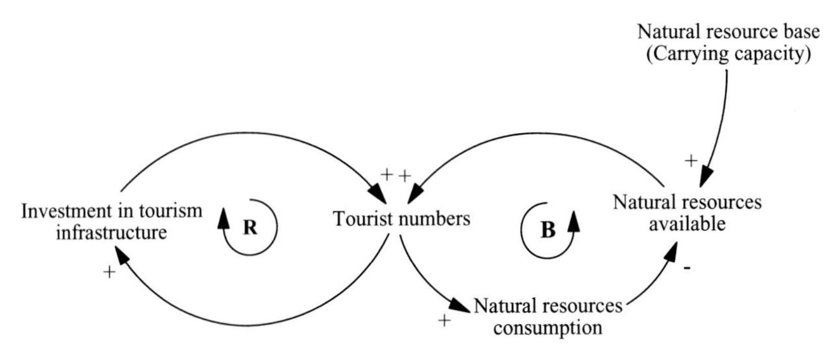

FIG. 14.10 Limits to growth archetype for tourism development within Cat Ba Island.

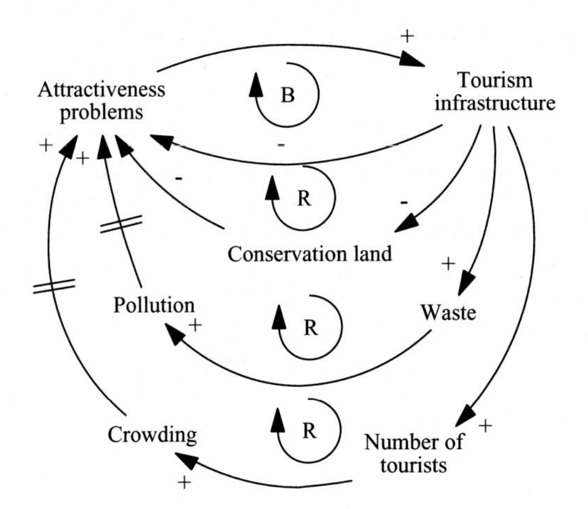

FIG. 14.11 Fixes that fail archetypes for tourism development within Cat Ba Island.

infrastructure and facilities has received a lot of attention, without recognizing the constraints, such as pollution, environmental degradation, and poor service quality. These constraints are a significant threat to the sustainability of tourism on the island. In other words, tourism development must be controlled and sensitive to the limits of Cat Ba Island's natural resources, particularly freshwater and land.

Another archetype that is within the conceptual model is "fixes that fail" (Fig. 14.11). This archetype represents situations where the managerial response to a problem is a quick fix. This fix works in the short term (balancing effect); however, it has unintended and often harmful consequences that exacerbate the original problem (reinforcing effect) and the system is reverted to its original or a worse condition after a delay (Senge, 2006). This archetype is a good reflection of reductionist thinking that leads to a continually worsening scenario. The initial problem is exposed as a symptom, which, through reductionist thinking, is perceived as only requiring a "quick fix" or short-term solution. Results will be achieved faster and initially appear to cost less. As a result, the problem symptom is temporarily diminished or removed but the short-term fix solution can have unintended consequences that in the long term may return to the previous level or even exacerbate the problems.

In the case study on Cat Ba Island, the increasing tourist numbers have created a high pressure on tourism services. Much of the tourism infrastructure and facilities, such as hotels, restaurants, and means of transportation, were established as a response to the pressure. However, these quick-response solutions have created unintended consequences that are leading to environmental degradation, fewer resources being available for development, and resulting in an acceleration of the pressure on the demands of the fast growth in tourism on the island. This approach will ultimately create negative effects on the tourism industry such as pollution and social issues, which will make Cat Ba Island less attractive and will result in a reduced number of tourists in the long run. Obviously, such decisions carry long- and short-term consequences, and the two are often diametrically opposed, because they focus on identifying and removing the fundamental cause of the problem symptom. If a temporary, short-term solution is needed, developing a two-tier approach simultaneously would be useful. That is, while a short-term fix is applied, planning needs continued to find a fundamental long-term solution.

14.3.5 Formulating Simulation Model

The SFM of tourism system on Cat Ba Island developed based on its conceptual model. Note that not all feedback loops in the conceptual model simulated because we did not have enough information on all feedback loops. Fig. 14.12 shows the systems map of main sectors and loops included in the SFM. The arrows indicate connections (material and information linkages) between the sectors. Details of SFM for each sector is seen in Mai (2012).

The intention of the tourism economic sector is to simulate the demand for accommodations and subsequent employment opportunities caused by tourism growth while the population sector estimates population growth in Cat Ba Town through births, deaths, and net immigration. The focus of the natural resources sector models land and water availability, while the environmental sector models the production of domestic and hotel solid waste and wastewater.

14.3.6 Model Testing

The SFM has undergone rigorous validity procedural tests. These include structural verification, fitting historical data, extremely condition test, and dimensional consistency test (see details in Mai, 2012). The results of the tests consistently showed that the SFM is sound and sufficiently robust used as a tool for policy analysis and decision making.

14.3.7 Policy Design and Evaluation

We ran a workshop with high-level executive managers to develop possible tourism development scenarios on Cat Ba Island. In total, five scenarios were proposed (Table 14.1).

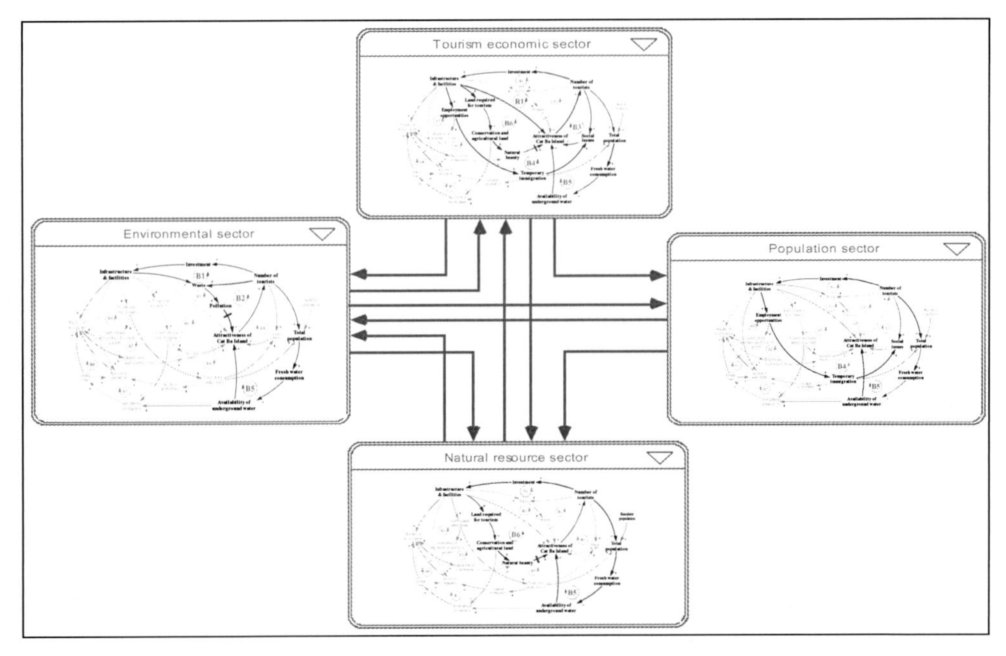

FIG. 14.12 Systems map of main sectors in the tourism system of Cat Ba Island.

TABLE 14.1 Development Scenarios for Tourism on Cat Ba Island

Policy Parameters	Scenario 1 (Base Case)	Scenario 2 (Best Case)	Scenario 3	Scenario 4	Scenario 5 (Worse Case)
Allowable land for tourism resorts	15 ha	20 ha	25 ha	30 ha	35 ha
Allowable groundwater extraction	10%	10%	10%	10%	10%
Fraction of leaking water	40%	10%	20%	30%	50%
Water use per family per year	175 m^3	Decrease 15%	Decrease 10%	Decrease 5%	Increase 5%
Water use per room by star hotel per year	160 m^3	Decrease 15%	Decrease 10%	Decrease 5%	Increase 5%
Water use per room by nonstar hotel per year	210 m^3	Decrease 15%	Decrease 10%	Decrease 5%	Increase 5%
Proportion of treated wastewater	0%	75%	75%	75%	0%
Proportion of treated solid waste	0%	75%	75%	75%	0%
Reservoir capacity	0 m^3	1,000,000 m^3 (two dams)	500,000 m^3 (one dam)	500,000 m^3 (one dam)	0 m^3

Scenario 1 (a base-case scenario) that represented the current situation, Scenario 2 (a best-case scenario) that promoted sustainable tourism development, Scenario 5 (a worst-case scenario) that did not promote sustainable tourism development, and Scenarios 3 and 4 (intermediate scenarios) that represented alternatives between the best and worst cases.

We used our SFM to evaluate the five abovementioned scenarios by using multiple criteria analysis based on scenario performance indicators. The indicators included the number tourism jobs; water supply; land area consumed by resorts; pollution index; and scenario robustness. The results for each development scenario are summarized in Table 14.2. Scenario 1 (base case) produced relatively low jobs growth and the largest amount of pollution. Scenario 2 (best case) produced the largest number of jobs and the largest clean water supply. Scenario 4 produced moderate jobs growth, the least amount of pollution, and was the most robust.

The relative performance of the five development scenarios (Fig. 14.13), along with the radar chart polygon area and symmetry (Table 14.3) indicate that scenarios 1 (base case) and 5 (worst case) were the worst performing overall. Scenario 3, while not the most symmetrical

TABLE 14.2 Development Scenario Evaluation Results (Calculated at the Year 2030)

Development Scenarios	Tourism Jobs	Water Supply (Million m^3)	Land Area Consumed by Resorts (ha)	Pollution Index	Robustness
Scenario 1	20,056	1.191	15	6.68	0.31
Scenario 2	40,333	1.778	20	2.43	0.38
Scenario 3	31,773	1.558	25	2.21	0.33
Scenario 4	24,516	1.399	30	2.01	0.29
Scenario 5	13,673	0.996	35	5.54	0.00

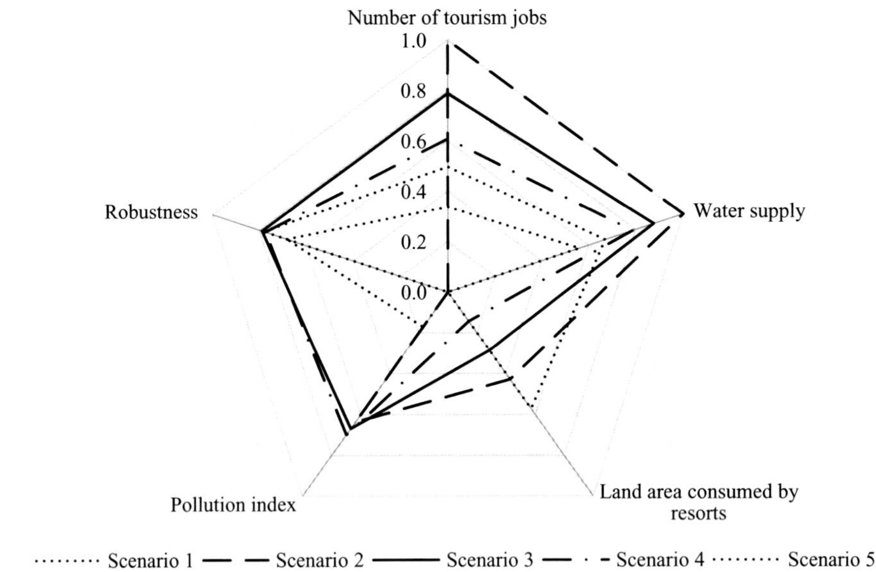

FIG. 14.13 Radar chart showing the relative performance of each development scenario.

TABLE 14.3 Polygon Area and Symmetry for Each Development Scenario Shown in Fig. 14.14

Development Scenario	Radar Chart Polygon Area Relative to Maximum Area (%)	Radar Chart Polygon Symmetry
1	19.1	0.26
2	32.5	0.27
3	45.1	0.25
4	31.7	0.27
5	9.2	0.24

(balanced across all performance indicators), has the largest polygon area, indicating the best performance overall. Scenarios 2 and 4 have similar polygon areas and symmetries; however, scenario 2 has a slightly larger polygon area, indicating better overall performance compared to scenario 4. Surprising, scenario 2 (best case) was not the best performing scenario overall.

To assist policy makers and tourism managers see the consequence of their decisions and actions, Fig. 14.14 provides a managerial-friendly interface with the computer model or control panel that allows them to conduct policy experiments. One of the features of the control panel is "sliders," such as boxes identified as "cap on Cat Ba town population"; "cap on tourism numbers"; "solid waste treated"; and "wastewater treated." These sliders present decision variables of the model. The main graph pad in this panel will show the behavior over time of output variables or key performance indicators, such as "unemployment," "tourism jobs," "pollution index," and "Cat Ba town population." The control panel can therefore be

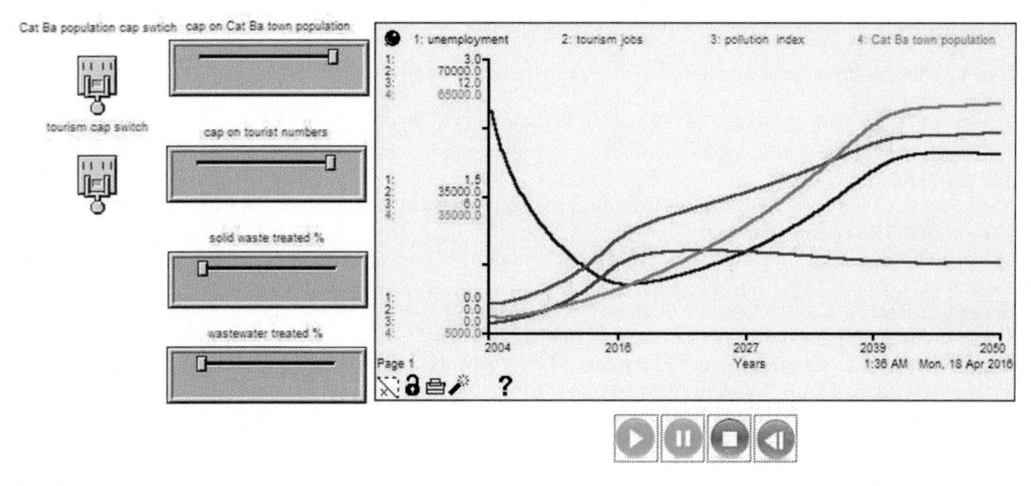

FIG. 14.14 Control panel of tourism system on Cat Ba Island.

used to enhance learning and to provide deeper understanding and insights into why the system behaves the way it does. In addition, it can test theories and metal models of stakeholders to discover inconsistencies in policies and strategies

14.4 CONCLUSIONS

A CLD developed by using the system dynamics approach could help provide a comprehensive understanding of the dynamic complexity of the natural resources-based tourism system through visualizing the underlying system structure and feedback mechanisms. This diagram potentially serves as a useful platform for dialog, collaboration, and decision making among different social actors involved in the management of a natural resources system. The simulation model developed by using the same approach potentially provides a practical tool to enhance collective learning and the design of more effective generic plans for the management of natural resources. Specifically, the model allows users to investigate more deeply the dynamic issues that are a concern to management. Through the simulation model, causal relationships and assumptions can formulate clearly and unambiguously, and experiments can perform readily with different policy structures. The simulation model can test alternative development options to identify the best plausible scenario that balances economic development, local livelihoods, and conservation goals. The simulation model provides a laboratory tool for learning about the behavior of the real world, and a potentially practical and powerful tool for systematically based management and design of strategic plans.

To this end, the system dynamics approach could help to develop a toolkit to manage dynamic complexes of natural resources systems through providing a comprehensive understanding of dynamic feedbacks between the system's components. It allows decision makers to anticipate the long-term consequences of their decisions and actions, and the unintended consequences of polices and strategies. It also provides a common language for diverse stakeholders, enabling them to engage in a deep dialog and consensus building.

References

Barlas, Y., 1989. Multiple tests for validation of system dynamics type of simulation models. Eur. J. Oper. Res. 42 (1), 59–87.

Bosch, O.J.H., King, C.A., Herbohn, J.L., Russell, I.W., Smith, C.S., 2007. Getting the big picture in natural resource management - systems thinking as 'method' for scientists policy makers and other stakeholders. Syst. Res. Behav. Sci. 24, 217–232.

Cat Hai People Committee, 2005. Master Plan for Social Economic Development of Cat Hai District (2006-2010) and Vision 2020. Hai Phong, Vietnam.

Edward, T.G., Erik, M., Douglas, S., Eve, M.M., 2013. Conservation in a Wicked Complex World; Challenges and Solutions. Conserv. Lett. 7 (3), 271–277.

Giupponi, C., Sgobbi, A., 2013. Decision Support Systems for Water Resources Management in Developing Countries: Learning from Experiences in Africa. Water 5, 798–818.

Kelly, R.A., Jakeman, A.J., Barreteau, O., Borsuk, M.E., ElSawah, S., Hamilton, S.H., Henriksen, H.J., Kuikka, S., Maier, H.R., Rizzoli, A.E., Delden, H., Voinov, A., 2013. Selecting among five common modelling approaches for integrated environmental assessment and management. Environ. Model. Software 47, 159–181.

Mai, V.T., 2012. Sustainable Tourism - Systems Thinking and System Dynamics Approaches: A Case Study in Cat Ba Biosphere Reserve of Vietnam. The University of Queensland, Brisbane.

Mai, T., Smith, C., 2015. Tourism in Cat Ba Biosphere Reserve Vietnam: addressing the threats to sustainability using systems thinking. J. Sustain. Tour. http://dx.doi.org/10.1080/09669582.2015.1045514.

Meadows, D., 1999. Leverage Points: Places to Intervene in a System. The Sustainability Institute, Hartland, VT, USA.

Senge, P.M., 2006. The Fifth Discipline: The Art and Practice of the Learning Organization. Random House, Inc., New York, USA.

Sterman, J.D., 2000. Business Dynamics: Systems Thinking and Modelling for a Complex *World*. Irwin/McGraw-Hill, Boston.

CHAPTER

15

Navigating Complexities and Management Prospects of Natural Resources in Northern Vietnam

S. Sharma*, G. Shivakoti[†,‡], M.V. Thanh[§,¶], S.J. Leisz[||]

*WWF-Nepal, Kathmandu, Nepal [†]The University of Tokyo, Tokyo, Japan [‡]Asian Institute of Technology, Bangkok, Thailand [§]International Centre for Applied Climate Sciences, University of Southern Queensland, Toowoomba, QLD, Australia [¶]Center for Agricultural Research and Ecological Studies, Hanoi, Vietnam [||]Colorado State University, Fort Collins, CO, United States

15.1 INTRODUCTION

With its diverse landscape, Vietnam is among the richest countries in biodiversity and natural resources with a large population directly dependent on agriculture, land, and forestry. Vietnam has high hills and mountains in the north and a long range of forest from north to south. The mountains in northern Vietnam are mainly settled by ethnic minority groups who recently find their livelihoods threatened by a number of vulnerabilities. The cases studied in this book focus on these vulnerabilities and how they are linked to the natural resources system, the institutional responses, and new challenges in northern Vietnam. These cases as well focus on different management strategies put forward to address the challenges and that have led to changes in the performance of the resources system, changes in the institutional arrangements and in the livelihoods of the people. The aim of this chapter is to review and study prevailing issues in the northern part of Vietnam and to understand the sustainability of the efforts that have been put in place. The first section of this chapter highlights existing resources systems in northern Vietnam, linked with economic dynamics and livelihood dependency. The second section focuses on the major threats and disturbances to the systems, while the third section illustrates the coping strategies. Recommendations built on these conclusions are discussed at the end of the chapter.

http://dx.doi.org/10.1016/B978-0-12-805453-6.00015-2

15.2 MAJOR NATURAL RESOURCES IN NORTHERN VIETNAM

Highlands with dense forest cover make up a large share of northern Vietnam, an area that also is one of the least inhabited parts of the country. The major natural resources existing in this portion of Vietnam are forest and a variety of land resources. These resources play a significant role in the livelihoods of people living in northern Vietnam.

15.2.1 Forests and Biodiversity

The forestland covers about 12.6 million hectares of total land area in Vietnam. Due to its varied climate, environments, and complex landscape, Vietnam has a variety of plants and soil types. During 1943–90 a lot of Vietnamese forests were degraded; nevertheless, forest area increased from 2000 to 2005. Some scholars point out that Vietnam's forest cover is recovering, while others suggest that this is a result of plantation forests without due consideration given to biodiversity. The Vietnamese constitution says that the forest belongs to the people with the government playing a representative role.

15.2.2 Land Resources

Land provides the basis for all human livelihood systems through its ability to produce goods and services. Vietnamese people depend on forest and agriculture systems that exist on the land, and understand the significance of their land resources. The government regulates the land tenure system and the use of resources, respectively. Approximately 75% of the labor age population works in the field (General Statistics Office, 2015) as Vietnam is one of the biggest rice exporters in the world. Presently, the trend toward plantation crops has converted natural forests into rubber and other tree plantations.

15.2.3 Home Gardens

Home gardens are grown on land areas adjacent to houses and consist of combinations of perennial and annual plants, small livestock, and fish. These are raised and managed by households for their own or commercial purposes. They are sources for food, medicine, and aesthetic plants, and essential in terms of food security and biodiversity conservation at the household level. Approximately 60 species of plants have been found in home gardens of the Tay Da Bac ethnic minority in the northern mountains of Vietnam. Home gardens along with composite swiddening of Tay Da Bac communal land has undergone changes as a result of long-term adaption of plant domestication and local ecological conditions.

15.3 RESOURCE SYSTEM DYNAMICS IN A CHANGING ECONOMY

There are clear variations since the commencement of *doi moi* in the landscape-scale developments in all parts of northern Vietnam. The variation in the patterns of change are found in the uplands, midlands, and lowlands. Specifically in the lowlands, even though private property was recognized during the post-*doi moi* era, the government continues to regulate the way irrigated paddy fields and terraces are managed. Additionally, while periurban attributes are seen in the lowlands, agriculture land use changing to urbanization has not taken place as speculated by a number of scholars.

The government of Vietnam considers adequate production of rice a major trait for defining food security and restricts land-use changes beyond 2ha of land from rice growing to nonagricultural purposes. Thus, even after the commencement of *doi moi*, the lowlands of northern Vietnam continue to be dedicated to paddy cultivation with no significant land-use changes but the nature and quality of land cover has been considerably transformed. Furthermore, additional detected changes in land cover and land use include mangrove to commercial shrimp aquaculture, which is at the same time deteriorating the resiliency of coastlines and in conjunction with extreme natural calamities has long-term ecological impacts.

Before *doi moi*, extreme logging carried out by logging companies was a primary factor for deforestation and forest degradation. This situation also provided the opportunity for small-scale farmers to practice household-based illegal logging. While the logging companies controlled trees, the villagers in the uplands practiced swiddening and increased the area of their swidden fields, which led to further deterioration of the forest conditions.

After commencement of *doi moi*, a number of subsequent policies were formulated. Forest was classified as special-use forest, protection forest, and production forest. Enforcement of these classifications has led to conserving forest area and limiting haphazard harvesting. Although in production forests, which are intended for timber production, much of these forests were destroyed. Simultaneously, nontimber forest products (NTFPs) harvesting has become a substitute within livelihood systems and has led to a decrease in the reliance on timber harvesting from natural forests.

Despite the enforcing of forest policies, swiddening is still practiced as a major form of cultivation. However, in some places, swidden-fallow land-use systems have changed to tree crops and sometimes to permanent plantations, reflecting a mosaic of grasses, rice fields, forests, and fallow areas. The land cover that has replaced the swidden fields is tea, rubber, timber, coffee, or protected forests often with less diversity in terms of age and species. Trees planted within swidden are fruit trees and expanded to get better market opportunities. Permanent croplands dominated by legumes and maize are also found, while fallow lands are growing in size to provide fodder for cattle rearing in some places.

The land-cover change from forest to agriculture and agriculture to periurban mosaic has implications for biodiversity and carbon mitigation. Research conducted earlier (Bruun et al., 2009) found swidden-fallow systems forms a better means for undertaking carbon mitigation projects at the landscape level. The same study revealed "The mosaic of land-cover types associated with swidden-fallow systems are thought to interact with the soil of an area in such a way that the carbon remains sequestered in it, compared to how soil is managed in the other land-uses" (Bruun et al., 2009). These land-use conversions in the midlands and uplands also have implications regarding the decline of biodiversity, except for protected natural forest areas. Generally, the work suggests that these changes have been prominent since the inception of *doi moi*, leading to a condition that may result in lower carbon sequestration and a decline of biodiversity in the uplands and midlands. Additionally, the transition that took place in lower areas is not as expected at the level of global growth that is taking place.

15.4 MAJOR DISTURBANCES TO RESOURCE SYSTEMS

Disturbances include biophysical disasters such as forest fires, droughts, and floods while policy changes, population pressure, market forces, and migration are socioeconomic changes having an impact on forestry, land, and agriculture. In this segment, major disturbances affecting the resource system are discussed based on the case studies in this book.

15.4.1 Climate Change

Vietnam is one of the countries thought to be most affected by climate change. Especially in northern Vietnam climate change is projected to affect agricultural production and also increase the frequency and intensity of natural disasters. The Cuu Long River Basin, which produces the highest amount of rice, is projected to lose 7.6 million tons of its total rice yield due to climate change. Studies also project that salt intrusion will salinize more than 2.4 million ha of rice-growing area, while mountainous areas and the Central Highlands are likely to suffer from drought and water shortages that will threaten agriculture production and food security. The cases studied in this book have also highlighted that there will be a decrease in mangrove area with sea level rise. Mostly, climate change studies focus on the projected impacts, while vulnerability assessments are not considered. Vulnerability assessments provide feedback that can help with mitigation and adaptation policy formulation. Vulnerability assessments in Vietnam are based on risk assessments or refer to studies conducted in other countries, and hence no uniform methodologies for carrying them out in Vietnam can be found and effective policies have not been formulated.

15.4.2 The Scales at Which Problems Occur and Decisions Are Made Are Mismatched

There is a mismatch between the scale at which problems occur and management decisions are made. For instance, issues related to water conservation occur at the ecosystem scale, while they are addressed at political and management scales of city, commune, district, and province. Local institutions are human-defined and do not often coincide with the natural divisions that are relevant to questions of ecosystem services and natural resources management. A restructuring of the institutional roles and regulations is needed to reconcile the scales at which these management decisions are made. Recognizing this and concurrently creating new institutional arrangements requires a critical review of the existing institutions, something that is lacking in the northern provinces of Vietnam.

15.4.3 Ineffective Land-Use Planning

Land use is an interaction between biophysical realities and socioeconomic forces where different people use land diversely to meet both present and future needs. Conflicting land uses have induced a number of conflicts due to the uncertainty of land tenure, poor land-market development, deforestation, farming systems, and migration of people and livestock. Effective land-use planning, recognizing the local society's needs and cultural preferences, is a potential solution for managing land sustainably, but land-use planning in northern Vietnam is not in line with environmental, cultural, and economic transformations. A study conducted by the joint Swedish-Vietnamese "Strengthening Environmental Management and Land Administration" Project (SEMLA, 2009) shows Vietnam's land-use planning to have poorly integrated diverse sectors in the planning process. There is no proper baseline database unto which the land-use planning process and results can be mapped. Land-use planning instructions provided to lower-level administrative bodies are complex and the institutions have poor planning and implementation expertise. The inflexible plans that result have no scope for incorporating conflicting interests and environmental considerations. However, in spite of these weaknesses a study carried out to analyze correlations between land-use plans and actual land-use modification in Vietnam shows that land-use planning has contributed to socioeconomic development. Especially in underdeveloped areas with less financial assistance available, the effective use of land resources is an essential key to altering the economic framework toward off-farm activities.

15.4.4 Complexity of Resource Systems

Management of resources involves diverse people with different views. These views can sometimes lead to conflicts through mutually exclusive interests related to resource utilization. Decision making together with incorporating changing environment needs have become complex and uncertain. This has led to unintended consequences and communities actively resisting government policies. Earlier management approaches and problem-solving mechanisms were reductionist, linear, and fragmented, usually targeted to instantaneous problems without considering feedback mechanisms that could create potential problems in the future. Vietnam's policy makers have tended to focus on "quick fixes" and have been "symptom solving" rather than considering the underlying causes of the problems related to effective and long-term resource management.

15.4.5 Disregarding Traditional Food Systems in the National Database

Home gardens are sources of food, medicines, and animal fodder and are essential in terms of fulfilling household food security needs and promoting agrobiodiversity conservation at the household level. These systems are often ignored as are other traditional upland farming systems by scientists, extension workers, and policy makers because of their relatively small size, compared to lowland farming systems, and their low market returns. Home gardens are often managed by women, who work to increase the gardens' productivity, but at the same time they are ignored by the national extension mechanisms as are other traditional upland systems. Though home gardens are not labor intensive, they are not included in irrigation schemes, nor are they included in extension efforts to promote fertilizer application.

15.4.6 Food Shortages

The cases from the book reveal that there are still instances of food shortages in the uplands of northern Vietnam. Specifically, communities of Thai and Kho Mu in Ky Son District of Nghe An Province face food shortages as challenging issues. Approximately 92% of households had food sufficiency for three months in a particular village located within a 2 km vicinity of the district center. The major reason for the food shortages are labor shortages, unfertile land, and the frequency of natural disasters. Food shortages are often linked with labor quality due to inadequate education, poor health, and limited access to goods and services. Often, ethnic minorities residing in these areas are resource poor and they lack the capability to expand cropping areas and to diversify household income, significantly affecting the household food status. With no options left, these communities exploit available natural resources through illegally harvesting them, abandon their fields, and migrate into cities. Population increments and market fluctuations further aggravate the situation.

15.4.7 Fragile "Policy-Practice" Linkage

Although the government has taken steps to encourage environmental sustainability and to involve local beneficiaries in natural resources management, not much attention is given to the design and implementation of the process. Forestland allocation (FLA) is implemented throughout the country following the same module and does not consider the diversity among populations, farming communities, and ecological conditions. Even though the policy has the broader objective of community benefits, it is more suited to communities with market orientations rather than to the traditional hamlets of northern Vietnam. The FLA process followed is one of top-down development with little opportunity for local participation. It is coupled with poor management. inactive planning, and high dependency on the state government resulting in very restricted success. This is one of the reasons for the decline in the quality of forest cover despite two decades of FLA polices being implemented.

15.4.8 Forest and Biodiversity Loss

Biodiversity loss is often linked with deforestation and land-use changes from natural forests to plantation estates. The government of Vietnam has formulated several policies and laws to regulate this loss, but the attempts have been intermittent and the results incomplete. Specifically the Biodiversity Law of 2008 has several deficiencies that have led to multiple interpretations and created difficult situations for local institutions to implement. A proper database of local biodiversity and cultivars is missing and this has triggered additional burdens for local institutions. As a result, incidences of overexploitation and illegal hunting are common and the frequency of the overuse of pesticide and fertilizers in the field are recurrent. The overuse and misuse of pesticides and fertilizers have simultaneously polluted underground water resources and led to the loss of a number of water species. The quality of the water and environmental condition of natural lakes in northern Vietnam are declining in the face of the demand for infrastructure and industrial development, which has led to the threat of large-scale aquatic ecosystem extinctions. Simultaneously, dangerous exotic species are also increasing in the area through poorly regulated introductions.

15.5 COPING STRATEGIES

In this section we focus on how northern Vietnam is coping with these issues or whether they fail to do so. Some of the policies are believed to improve the performances of resources systems while others further increase the systems' vulnerability and negatively impact the ability of the systems to maintain their basic characteristics and ecosystem and other services.

15.5.1 Decentralization and Community-Based Forest Management

Decentralization is the process of transferring management responsibilities from the central government to local institutions. One of the major objectives of decentralization is to promote local participation and reduce transitional costs. Equity and effectiveness are the primary desired outcomes of the decentralization process. In natural resources management, decentralization is complex and difficult to implement as power relations among sociopolitical and socioeconomic interests revolve around benefits from these resources. In Vietnam, particularly in the northern provinces, rapid deforestation and forest degradation, and degradation of other natural resources, which were previously under government authority, posed high protection costs; devolution of the management of these resources was then envisioned as a way to reduce the cost of protection of these resources while simultaneously benefiting the forest-dependent communities.

During the 1950s through to the 1980s, state forest enterprises (SFEs) were established by the government with the view of exploiting and protecting the forest. Theoretically, SFEs protected forests through enforcing a timber harvest quota; but this excluded local people from the management of the forest resources. By the end of this period, SFEs had arguably led to a forestry disaster, with no good timber to harvest and a decrease at an alarming rate of the forest cover throughout the country. To address this, the government implemented FLA provisions, which transfer forest management rights to local communities and local people. Under this law, certain parts of a forest area is allocated to local authorities for a fixed period. Supposedly, FLA could recognize customary laws and rules. This was to encourage local people and communities to revitalize customary laws in natural resources management in such a way that they were aligned to a newly formulated state legalistic framework. Decentralization led to forest and biodiversity conservation and incidences of food shortages decreased. FLA can be linked with community-based forest management (CBFM) as measures for socializing conservation and for the development of forest, rather than considered merely as forest conservation tools. Community participation is considered the key to the success of forest management.

However, there are a number of problems associated with the decentralization process through Vietnam's FLA. Case studies from Nghe An and Son La confirm that power in decision making devolved to the local government is restricted to the implementation of decision making of what the state government decided. The power of decision making still rests with the state government. Government allocated certain funds to cover implementation costs, but this created financial dependency of the local governments to the central government, creating a situation where the central government is top-down subsidizing its approach to forest management. The hamlets practicing FLA have no financial resources and are least able to exercise decision-making power and they have to rely on very small budgets contributed by villagers. Benefit sharing is not clearly defined in FLA legal provisions. Small payments and loans received for the initial forest plantations do not provide adequate incentives for the reforestation program. Implementation of FLA depends on the accountability of people with power. Poor planning and insufficient monitoring were commonly found in the case studies and may be considered as the main cause for the poor policy implementation. Many areas where forest allocation was carried out have no clearly defined borders. This has caused disputes among land users and at times severe disputes have been reported that are still not solved. Through the case studies in this volume it is clear that only upward accountability exists in FLA; there is no mechanism for downward accountability.

Regardless of the theory of the FLA implemented regarding local participation, the cases revealed that local leaders and ethnic communities had no voice in the planning process, while the decisions are still made at the national level. There is a huge communication gap between the central government and local communities and this has led to a misunderstanding of how subsidies for seedlings and technologies work. This has in turn resulted in the waste of seedlings and the destruction of recently established plantations. Similarly, there has been even less documented participation of female-headed and resource-poor households. As such, the needs and aspirations of the poorest and most vulnerable people are not included in the planning process.

15.5.2 Policy Framework for Climate Change Responses

Vietnam has a comprehensive policy system for climate change adaptation and mitigation whereby agriculture and rural development are considered as vulnerable sectors. The Ministry of Agriculture and Rural Development (MARD) has focused on developing sustainable agriculture corresponding to a national strategy and achieving the commitments made to international communities on reducing greenhouse gases (GHGs). Reviews of the outcomes to date of the objectives of the climate change action plan (2011–15) show limited outcomes. The action plan has provisions for stakeholder involvement with most projects focused on rehabilitating and developing infrastructures that required huge investments. However, funding for the action plan was short due to the lack of capacity of agencies in financial mobilization.

The authors in this book confirm the policy system's inability to address local communities' involvement in the process with no mechanism to encourage participation. There exist no guidelines to incorporate climate change into socioeconomic development plans and it is not foreseen that future financial investment will be adequate. However, there are a number of programs targeted toward mitigation and adaptation in agriculture by a diverse set of donors; unfortunately, there are no similar simultaneous government-funded projects. The projects implemented are attentive at evaluating climate change influences and suggesting adaptation measures. In some parts of northern Vietnam the government has also piloted initiatives like mangrove expansion and a system of rice intensification (SRI) as a means of addressing both mitigation and adaption issues. But farmers still practice traditional farming systems as needed technology transfer is limited. Agriculture production practices that are associated with development programs lack long-term vision and do not have the needed benefit-sharing provisions among stakeholders, explicitly farmers.

15.5.3 Internalization of Externalities

Northern Vietnam is a biodiversity hotspot that offers watershed functions beyond boundaries. Ever since implementation of the *doi moi* economic policy reforms, land allocation and population changes have led to changes in agricultural areas and cropping patterns. This has led to pressures on environmental services. Payment for Environment Services (PES) is a new method introduced to influence how natural resources including forests, within watershed areas, are managed and protected, to encourage environmental services such as watershed protection, carbon sequestration, and landscape beauty. "PES converts conservation principles through expensive requests executed on land owners to possible bases of revenue conditioned to conservation targets" (Kinzing et al., 2011). The underlying goal is to financially motivate groups and communities to internalize negative externalities through agreements on resources management.

In the process, communities are asked to set aside sloping areas for reforestation and maintain these areas as forest, and payments are then made in terms of cash and irrigation facilities. Through the cases analyzed, PES programs focused on developing ways to increase food production (explicitly irrigation in a case study) in the hope of increasing local participation in PES programs. It is believed that these provisions will create win-win situations and reduce inequalities in benefits gained from protecting forests in rural communities of mountainous Vietnam. One feedback from this process is with improvement in forest conditions, these resource-poor forest-dependent households will be able to extract NTFPs that are not considered in PES and further reduce the incentives for land users to pursue other land-use changes. Farms with direct owners will have less possibility to undergo land-use changes to rubber and palm plantations. PES programs also do not affect off-farm work undertaken by households. However, smaller resource-poor households may be affected due to the reduction of off-farm work, such as timber harvesting, and forests are protected.

15.6 SUSTAINABILITY OF THESE COPING STRATEGIES

Humans are the key forces in the changing resource dynamics at diverse spatial and temporal scales. In the recent past northern Vietnam has seen a change in resources management policies from SFEs to the implementation of FLA policies, or in other words, from centralized management to local community management. The dynamism between forest, livelihood, and governance is a crucial element binding the integrity of ecosystem services. Changing this paradigm has triggered drivers across scales; the perceived local knowledge gained through local management is given up to comply with national rules and laws. New challenges have occurred in balancing livelihood systems and fulfilling economic needs. Adaptive livelihood strategies focused on agriculture diversification and livelihood system changes are seen, while uncertainties in environmental stressors impact on this process.

The main issue pertaining to northern Vietnam is how this paradigm shift and transformation of coping strategies in resources management deals with changes in the Social-Ecological System (SES). It is now crucial to observe if the coping strategies trigger sustainable SESs. Research has shown some government polices quicken resource ruin while some hasten managers to capitalize on possibilities to accomplish sustainability (Ostrom, 2009).

SESs are the complex relationships among resource systems, resource units, governance systems, and user dynamics in a given economic and political setting. Analyzing this complex system is difficult. Ostrom (2009) presented a version of a nested framework for analyzing the sustainability of SESs. This framework identifies variables for analyzing SESs, without which it becomes difficult to organize the study of and empirical research from diverse cases. Ten second-level variables are considered to positively or negatively affect the likelihood of self-organization in efforts to achieve a sustainable SES. Table 15.1 shows the

TABLE 15.1 Application of Ostrom's Sustainability Framework in Analyzing SES Under FLA and PES

Ostrom's Variables	Interpretations	Application to Resource Systems Under FLA	Application to Resource Systems Under PES
Size of the resource system	Very large territories are unlikely to self-organize	Medium-sized territories allocated	Medium-sized territories
Productivity of resource system	Has curvilinear effect on self-organization	There is discrepancy among resource users. Some users were allocated productive forest while others not. This is a major issue prevailing in FLA	Mostly assumed to be productive. Additionally, with formation of terraces, the productivity enhanced
Predictability of system dynamics	System dynamics requires predictability, whereby users can estimate consequences if harvesting rules and harvest territories established	Forests under FLA are predictable	Predictable
Resource unit mobility	Self-organization less likely with mobile resources	Stationary resources	Stationary resources
Number of users	Self-organization is negative with very small and very large resources; because of higher monitoring and management costs	NA because a portion of land allocated to single household. However, with larger users monitoring is likely to be costly	Self-organization would be easy
Leadership	Local leaders with entrepreneurial skills will have positive attribute in promoting self-organization	Not available in the case study	Not available in the case study
Norms/social capital	Reciprocity, trust among common moral and ethical standards lowers monitoring and transaction costs	Often, inquiry of equitable benefit sharing has created conflicts among users. As mostly, reciprocity and trust is missing in the community	PES is based on performance outcomes; this does not affect reciprocity; some level of positive interactions observed
Knowledge of the SES	Common knowledge of SES promotes self-organization	There exists common knowledge on resource condition and resource unit	There exists common knowledge on resource condition and resource unit
Importance of resource to users	Cost of self-organization likely to be lower in forest-dependent communities	Forest important for livelihood generation	Forest important for livelihood generation
Collective choice rules	Autonomy to craft and enforce own rules	This is relatively lower as they cannot craft constitutional rules as most regulations are crafted centrally. They are, however, allowed to plant trees and harvest according to their needs	Regardless of general legal provision; local participating households are allowed to plant trees as per own need; because it is performance-based, one's action affects one's income

overall interpretation of the coping strategies in terms of Ostrom's framework. Because the cases reviewed in this book do not consider climate change policy application consequences, this section does not consider it under this analysis.

The information gathered above is a proxy measure for analyzing whether self-organization is possible for sustainable resources management, but the real data analysis of these interactions is difficult. These variables are highly dependent upon each other in a nonlinear fashion. With FLA, rules establishment is dependent on government policies, which may not be sufficient in the long run. Owing to the fact that some rules established under FLA are not appropriate to local contexts, long-term sustainability would be difficult to achieve. However, rules used in PES mechanisms are flexible and the mechanisms are voluntary and specially equipped with incentives to continue the benefits attained earlier; long-term sustainability may be achieved because rules are then set in terms of resource systems, units, and users. Moreover, because PES is based on individual performances, monitoring is likely to be strong.

FLA has the same blueprint approach throughout Vietnam, and does not consider diversity among populations, farming communities, and ecological conditions. Even though the policy has broader objectives of community benefits, it is more suited to communities with market orientations rather than to the traditional hamlets of northern Vietnam's uplands. FLA is a top-down development strategy with little to no opportunity for local participation. It is coupled with poor management, inactive planning, and high dependency on state government. The results of FLA to date is one of the reasons for the declining forest-cover quality in northern Vietnam over the past two decades

15.7 RECOMMENDATIONS

Based on the findings from the case studies, we suggest the following recommendations:

1. A unified method for assessing vulnerability to climate change is a prerequisite for formulating any policy provisions. As such, inputs from communities must be solicited based on participatory rapid assessments. Policy development regarding climate change should include the participation of the private sector to create incentives to participate in climate change adaptation and mitigation. Including appropriate legal provisions into development plans needs to be institutionalized.
2. To make natural resources governance effective, administrative redesign to take into account coriparian institutions is vital. PES programs could be a sustainable approach to overcome diverse biophysical impacts on resource systems as well as help fulfill local aspirations and needs.
3. Regional and knowledge networks need to build science-based approaches to reduce misperceptions over problems occurring at an ecosystem scale.
4. Land-use planning (LUP) significantly contributes to socioeconomic development and environmental protections. To anticipate efficiently in LUP it is crucial to analyze diverse development scenarios within different settings, explicitly physical environment and socioeconomic criterions. Particularly, environmental factors that concomitantly affect development and livelihood of people needs to be addressed.

5. Decision making should consider long-term consequences and unintended consequences through consideration of systematic thinking and dynamic feedbacks between system's components.
6. Home gardens as a means of conserving agrobiodiversity and fulfilling household nutritional requirements for ethnic minorities needs to be supported and expanded. Scientific research on improving these systems should be initiated. Food security must be a part of strategic goals, whereby household climate change adaptation can be linked through species diversification, fostering a green revolution and broadening and diversifying income opportunities.
7. Strong foundations for decentralization and CBFM have been created and future efforts require expanding this to wider beneficiaries, strengthening tenure rights, and facilitating regeneration of tangible forest benefits.
8. Mechanisms for decentralizing and implementing power should be provisioned with access to information, participation, and formulating a legal framework for acknowledging assigned tasks. Benefit sharing needs to be clearly defined; simplification of who power and decision making is delegated to needs to be carried out; and the local communities' privileges and accountabilities should be clarified.

References

Bruun, T.B., de Neergaard, A., Lawrence, D., Ziegler, A., 2009. Environmental consequences of the demise in swidden agriculture in Southeast Asia: carbon storage and soil quality. Hum. Ecol. 37, 375–388.

General Statistics Office, 2015. Agriculture, Forestry and Fishing Statistics. General Statistics Office of Vietnam. Available online on http://www.gso.gov.vn/default_en.aspx?tabid=774.

Kinzing, A.P., Perrings, C., Chapin III, F.S., et al., 2011. Paying for ecosystem services-promise and peril. Science 334, 603–604. Available online on http://perrings.faculty.asu.edu/pdf_papers Perrings/Kinzig_et_al_Science_(2011).pdf.

Ostrom, E., 2009. A general framework for analyzing sustainability of social-ecological systems. Science 325 (5939), 419–422.

SEMLA, 2009. Integrated Land Use Planning: Results and Lessons Learnt. Strengthening Environmental Management and Land Administration, Vietnam – Sweden Cooperation Programme, Hanoi, Vietnam.

Index

Note: Page numbers followed by *f* indicate figures and *t* indicate tables.

K

L

M

N

W

X

Printed in the United States
By Bookmasters